Springer Series in Wood Science

Editor: T. E. Timell

M. H. Zimmermann
Xylem Structure and the Ascent of Sap (1983)

J. F. Siau
Transport Processes in Wood (1984)

R. R. Archer
Growth Stresses and Strains in Trees (1986)

W. E. Hillis
Heartwood and Tree Exudates (1987)

S. Carlquist
Comparative Wood Anatomy (1988)

L. W. Roberts/P. B. Gahan/R. Aloni
Vascular Differentiation and Plant Growth Regulators (1988)

C. Skaar
Wood-Water Relations (1988)

J. M. Harris
Spiral Grain and Wave Phenomena in Wood Formation (1989)

Springer Series in Wood Science

Editor: T. E. Timell

M. H. Zimmermann
Xylem Structure and the Ascent of Sap (1983)

J. F. Siau
Transport Processes in Wood (1984)

R. R. Archer
Growth Stresses and Strains in Trees (1986)

W. E. Hillis
Heartwood and Tree Exudates (1987)

S. Carlquist
Comparative Wood Anatomy (1988)

L. W. Roberts / P. B. Gahan / R. Aloni
Vascular Differentiation and Plant Growth Regulators (1988)

C. Skaar
Wood-Water Relations (1988)

J. M. Harris
Spiral Grain and Wave Phenomena in Wood Formation (1988)

Christen Skaar

Wood-Water Relations

With 198 Figures

Springer-Verlag
Berlin Heidelberg New York
London Paris Tokyo

CHRISTEN SKAAR
Department of Forest Products
Virginia Polytechnic Institute and
State University
Blacksburg, VA 24061, USA

Series Editor:

T. E. TIMELL
State University of New York
College of Environmental
Science and Forestry
Syracuse, NY 13210, USA

Cover: Transverse section of *Pinus lambertiana* wood. Courtesy of Dr. Carl de Zeeuw, SUNY College of Environmental Science and Forestry, Syracuse, New York

ISBN-13:978-3-642-73685-8 e-ISBN-13:978-3-642-73683-4
DOI: 10.1007/978-3-642-73683-4

Library of Congress Cataloging-in-Publication Data. Skaar, Christen. Wood-water relations / Christen Skaar. p. cm. – (Springer series in wood science) Bibliography: p. 263 Includes index. 1. Wood-Moisture. I. Title. II. Series. TA419.S53 1988 620.1'2–dc19 88-20012 CIP

Softcover reprint of the hardcover 1st edition 1988

Typesetting: K+V Fotosatz GmbH, Beerfelden

2131/3130-543210 – Printed on acid-free paper

Preface

Wood is formed in an essentially water-saturated environment in the living tree, and the cell wall remains in this state until the water flow from the roots is interrupted, such as by felling the tree. The wood then begins to lose most of its moisture by drying, resulting in changes in most of its physical properties. These changes, and their relationship to the environment to which the wood is subsequently exposed, are the subject of this book.

The text consists of six chapters. The first chapter discusses certain empirical relationships between wood and water, methods of measuring wood moisture content, factors which affect its equilibrium moisture content, and the effect of moisture content on wood strength. The second chapter treats the thermodynamics of moisture sorption by wood, including enthalpy, entropy, and free energy changes. The third chapter discusses some of the theories which have been proposed to explain the sorption isotherms for hygroscopic materials such as wood. Chapter 4 considers hygroexpansion or the shrinking and swelling of wood associated with moisture change. Chapter 5 is concerned with how moisture moves through the cell wall of wood in response to both moisture and temperature gradients. The sixth and final chapter discusses the theoretical and practical aspects of the electrical resistance and dielectric properties of wood, including the principles involved in their application in electrical moisture meters.

There is no detailed discussion of wood structure, nor of the mechanism of moisture movement through the gross wood structure, such as migration between cells, since these are covered in an earlier book in this series, *Transport Processes in Wood*, by J.F. Siau (1984). The subject of stabilizing wood against hygroexpansion is covered only superficially, since other books, such as *Wood and Cellulose Science* by A. J. Stamm (1964) give thorough treatment of this subject.

I am indebted to the Virginia Polytechnic Institute and State University for providing the facilities required to complete this book, particularly to my colleagues and students in the Department of Wood Science and Forest Products, chaired by G. Ifju. Also my long association with the Department of Wood Products Engineering of SUNY College of Environmental Science and Forestry at Syracuse provided the opportunity to develop much of the required background material. The editor of this series, T. E. Timell, was particularly helpful in expediting the completion of the book, especially in checking the references.

Thanks are expressed to Syracuse University Press for releasing permission to use much of the material from my earlier book *Water in Wood* (1972). Thanks are also due to McGraw-Hill Book Company for permission to use Figures 4.2 on page 125 and 5.2 on page 180 from Panshin and de Zeeuw (1980) *Textbook of Wood Technology*, fourth edition. Figures from other sources have been redrawn in all cases, with credit given in the individual legends or in the text.

Particular appreciation is due to Mrs. J.F. Buckner, M.N. Yopp, D.E. Garnand and W.W. van Winkle for typing the manuscript. Finally, I wish to especially thank my wife Dorothy, and my children, Eric, Steven, Dee and Janet for their inspiration and help. This book is dedicated to them.

September 1988 Christen Skaar

Contents

Chapter 1

Wood Moisture and the Environment

In order to understand the interaction of wood with the water in the atmosphere, it is necessary to consider first some of the basic properties of water itself. In this chapter some of the important characteristics of water such as its phases, vapor pressure, relative vapor pressure or activity, and the effect of temperature on these parameters, will be treated. These are the particular properties of water which are important to describe qualitatively the general interaction of wood and water. In later chapters, other characteristics of water appropriate to the topics covered in the particular chapter will be discussed.

Siau (1984), in another book in this series, gives an extensive treatment of relative humidity and the vapor pressure of water, including methods of measuring these parameters. This section will complement his treatment of the subject, emphasizing those aspects which are most useful in developing the various topics covered in the present work.

1.1 Humidity and Vapor Pressure of Water

Water can exist in three general states or phases – solid (ice), liquid, or vapor,– depending upon the temperature and pressure to which it is exposed. In Fig. 1.1 the phase diagram for water (not drawn to scale) is shown. It can exist in each phase only when the temperature and pressure are within the limits shown. At equilibrium two phases can coexist only at the conditions of temperature and pressure represented by the line between the phases. At the triple point, all three

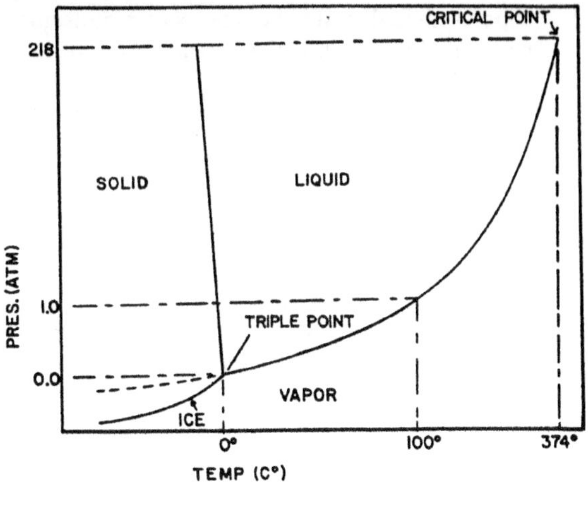

Fig. 1.1. Phase diagram for water (not to scale)

phases may exist simultaneously. At the critical temperature (374 °C) and pressure (218 atm), there is no longer any distinction between liquid and vapor.

The extension of the liquid vapor line below the freezing point actually represents the vapor pressure of supercooled liquid water. Notice that it is higher than the solid line representing the vapor pressure of ice. The regions of primary interest are those areas denoting the vapor state and the lines separating the vapor from the liquid and, to a lesser extent, the solid state.

The primary difference between the liquid and vapor state is the spacing between the water molecules. In the liquid phase the molecules are sufficiently close to each other that appreciable forces of attraction and repulsion exist among them. The individual molecules are constantly vibrating about a region of equilibrium where the force between adjacent molecules is zero.

Most molecules in the liquid state have insufficient vibrational energy to overcome the attractive forces of the closely spaced adjacent liquid molecules. However, because of the wide distribution of energies among molecules, an individual molecule may momentarily have sufficient kinetic energy to escape from the force field of its near neighbors, and become a vapor molecule. Only a small fraction have sufficient energy to break away from their neighbor molecules in the liquid form. These vapor molecules, because of their high kinetic energy, exert a pressure against an enclosure, called the *vapor pressure* of the water. The higher the temperature of the liquid water, the greater the number of water vapor molecules with sufficient energy to escape from the liquid to the vapor state, and the greater the vapor pressure of the water. Also, the mean kinetic energy of the vapor molecules increases with temperature, thus adding to the vapor pressure.

The mean or average distance between the centers of molecules in liquid water can be calculated by use of Avogadro's number (6.02×10^{23} molecules per gram mole). The gram molecular weight of ordinary water is 18 g mol^{-1}, and its molar volume is 18 cc at room temperature. The volume (cc) occupied by one water molecule in liquid form therefore is $18/(6.02 \times 10^{23}) = 30 \times 10^{-24}$, and the average distance between molecules therefore is slightly more than 3×10^{-8} cm, or 30 nm. This distance remains nearly constant with temperature, since the density and specific volume of liquid water vary only slightly with temperature.

The average distance between water molecules in the vapor state can also be calculated. However, it varies considerably with temperature and relative humidity because they affect the density of water vapor. Table 1.1 gives the density of saturated water vapor at various temperatures, the mean distance between molecules, and the saturated vapor pressure. The spacing of water vapor molecules is from 8 to 60 times larger than that of liquid water molecules, even when the air is saturated, over the temperature range from 0° to 140 °C. When the vapor pressure is lower than p_0 at any temperature, the spacing is even greater.

To a first approximation, a linear relation exists between the logarithm of the saturated vapor pressure of water (log p_0) and the reciprocal, $1/T$, of absolute or Kelvin temperature T, where $T = 273.15 + C°$ (Fig. 1.2). From the data given in Table 1.1, the equation obtained is approximately

$$\ln p_0 = 20.59 - 5205/T \ . \qquad (1.1)$$

Table 1.1. Properties of water vapor at saturation

Temperature (°C)	Density (g/l)	Molecular spacing (nm)	Vapor pressure	
			(mmHg)	(atm)
0	0.004846	1830	4.579	0.00604
20	0.017291	1200	17.535	0.0230
40	0.051164	840	55.324	0.0728
60	0.13024	610	149.38	0.1965
80	0.29338	470	355.1	0.469
100	0.5977	370	760.0	1.000
120	1.122	299	1489.1	1.959
140	1.968	248	2710.9	3.567

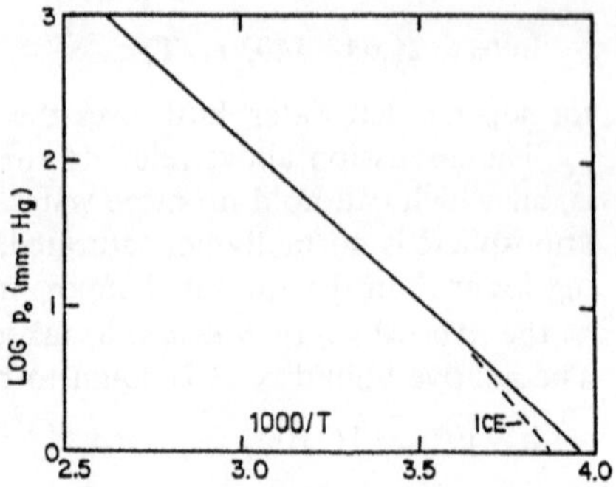

Fig. 1.2. Vapor pressure of water and ice as functions of temperature

It will be shown (Chap. 2.2) that the slope of this curve is related to the heat required to evaporate an unit mass of water.

Although Eq. (1.1) gives a reasonable approximation of the temperature dependence of the vapor-pressure p_0 of water, the addition of another temperature term in the form of the Kirchoff vapor pressure formula,

$$\ln(p_0) = 51.29 - 6651/T - 4.531\,[\ln(T)]\ , \qquad (1.2)$$

gives better agreement with experimental data and theory over the temperature range from 0° to 100°C.

Wood is often exposed to subfreezing temperatures. In this case liquid water in the wood or on its surface freezes. It was mentioned earlier that the vapor pressure of ice decreases more rapidly with decreasing temperature (Fig. 1.1) than does that of supercooled liquid water. The latter is an unstable state of liquid water, which is not normally found under natural conditions.

Under laboratory conditions, however, it has been possible to measure the vapor pressure of supercooled liquid water at temperatures as low as −50°C. Table 1.2 gives the vapor pressures of ice and of supercooled liquid water at several temperatures from −30° to 0°C. It will be shown later (Chap. 2.2) that the differences between the vapor pressures of ice and supercooled water can be used to calculate the heat of melting ice. It is also the primary reason for the phenomenon known as "coldness shrinkage" in wood.

Table 1.2. Vapor pressures of ice and supercooled water (after Troller and Christian 1978)

Temperature		Vapor pressure (mmHg)	
K	°C	Water	Ice
273.15	0	4.579	4.579
263.15	−10	2.149	1.950
253.15	−20	0.943	0.776
243.15	−30	0.383	0.286

Simple linear regressions for the data shown in Table 1.2 yield the equations

$$\ln p_s = 24.00 - 6140\,(1/T) \tag{1.3}$$

for the vapor pressure of ice, and

$$\ln p_0 = 21.64 - 5493\,(1/T) \tag{1.4}$$

for supercooled water, both over the range from −30° to 0 °C.

The discussion above refers to the vapor pressure of fully saturated air, that is, air which can hold no more water at the specified temperature. The ordinary atmosphere is normally not saturated, the existing vapor pressure p generally being lower than the saturated vapor pressure p_0, or p_s. The ratio p/p_0 is defined as the *relative vapor pressure* h, taken to be essentially identical to the *activity*. The relative humidity H is equal to h expressed in percent. Thus

$$h = p/p_0 = H/100 \tag{1.5}$$

and

$$H = 100\,h = 100\,(p/p_0)\ . \tag{1.6}$$

Any or all of these terms will be used in this book depending on convenience and the context.

Figure 1.3 shows the relationship between vapor pressure p, and temperature for various relative vapor pressures. It is clear that the activity h is dependent on the vapor pressure p and also on the temperature, since this determines the value of p_0. In a room, for example, the vapor pressure p tends to remain constant throughout the room, but the temperature fluctuates considerably from one location to another. Therefore there may be wide variations in relative humidity within the room. For example, if the temperature variation within the room is 1 °C, the relative humidity at room temperature (25 °C) may vary by 3% within the room, even if the actual vapor pressure is constant throughout the room. It is therefore important to have good temperature control in a room designed to maintain constant humidity.

Under ordinary conditions, the activity h is less than unity, that is, the existing vapor pressure p is less than p_0. If the temperature of a room or of an object in the room is lowered, p_0 is also lowered. The dew-point temperature T_d is that at which moisture begins to condense from the atmosphere onto a cold object.

For example, referring to Fig. 1.4, if the temperature in a dry kiln is 100 °C and the activity h is 0.5, the dew-point temperature is determined by the intersec-

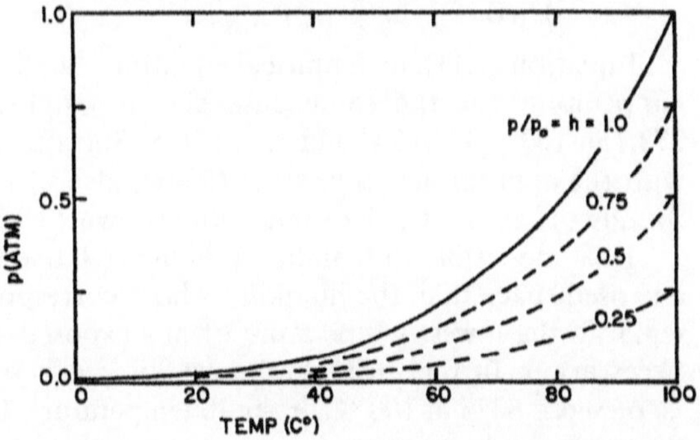

Fig. 1.3. Vapor pressure p as a function of temperature for different water vapor activities h

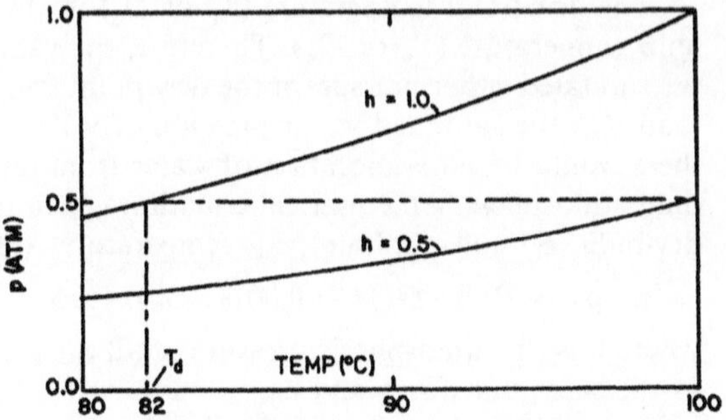

Fig. 1.4. Vapor pressure p as a function of temperature for water activities h of 0.5 and 1.0. Dew-point temperature T_d (82 °C) is shown for h of 0.5 at 100 °C

tion of the horizontal broken line with the curve for $h = 1.0$. In the figure it appears to be about 82 °C. If the walls of the dry kiln are at a temperature of 82 °C or lower there will be condensation of moisture on the walls, even when the relative humidity is as low as 50%.

Equation (1.1) can be used to calculate the dew-point temperature from the dry-bulb temperature and activity by rearranging Eq. (1.1) into the form

$$p_0 = \exp(A - B/T) \ , \qquad (1.7)$$

where $A = 20.58$ and $B = 5205$, over the temperature range from 0° to 100 °C, and p_0 is the calculated saturated vapor pressure (mmHg) at dry-bulb temperature T. Assuming that Eq. (1.7) is correct, the existing vapor pressure p is by definition equal to the saturation vapor pressure p_{0d} at the dew-point temperature T_d

$$p = p_{0d} = \exp(A - B/T_d) \ . \qquad (1.8)$$

Combining Eqs. (1.7) and (1.8)

$$p/p_0 = \exp[(B/T) - (B/T_d)] \qquad (1.9)$$

or

$$\ln(p/p_0) = (B/T) - (B/T_d) \ , \qquad (1.10)$$

which gives p/p_0, the activity h, if T and T_d are known. Equation (1.10) can be rearranged to give the dew-point temperature T_d if p/p_0 and T are known. Thus

$$T_d = B/[(B/T) - \ln(p/p_0)] \ . \tag{1.11}$$

Equation (1.11), an empirical equation, can be used to calculate the dew-point temperature for the same case shown graphically in Fig. 1.4, where T is 273.15+100 = 373.15 K and p/p_0 is 0.5. Substituting these values into Eq. (1.11), with the appropriate value of B (5204), gives $T_d = 355.48$ K, or 82.33 °C, which rounds off to 82 °C, the same value shown in Fig. 1.4.

It is impossible to maintain a higher relative humidity in a dry kiln or other enclosed space than the humidity whose corresponding dew-point temperature is equal to the lowest temperature of any exposed surface in the kiln. It would be necessary to further insulate the dry kiln walls or doors if the relative humidity is to exceed 82% at 100 °C dry-bulb temperature. Otherwise the dry-bulb temperature must be lowered.

The dew-point temperature (T_d or Θ_d) is always slightly lower than the wet-bulb temperature (T_w or Θ_w). Therefore, the vapor pressure p, equivalent to p_{0d}, the saturated vapor pressure at the dew-point temperature, is always slightly lower than p_{0w}, the saturated vapor pressure of water at the wet-bulb wick. Otherwise there would be no evaporation of water from the wet-bulb wick to the surrounding atmosphere. An empirical equation which relates the difference ($p - p_{0w}$) to dry-bulb (Θ) and wet-bulb (Θ_w) temperatures (°C) is

$$p - p_{0w} = P(\Theta - \Theta_w)(1 + 0.00115\,\Theta_w)/1515 \ , \tag{1.12}$$

where P is the atmospheric pressure of all the gases in the atmosphere, expressed in the same units as p and p_{0w}.

It is evident from Eq. (1.12) that the difference ($p - p_{0w}$) is proportional not only to the dry- and wet-bulb temperatures but also to atmospheric pressure P, as Siau (1984) points out. Equation (1.12) can be solved for the vapor pressure p by moving p_{0w} to the right side of the equation.

At normal pressures and temperatures, many gases, to a first approximation, obey the relationship called the ideal gas law. This relationship is:

$$pv = nRT = (w/M)RT \ , \tag{1.13}$$

where p is the gas pressure, v its volume, n the number of moles of gas (the mass w, divided by molecular weight M), T the Kelvin temperature, and R the universal gas constant, whose value depends upon the units in which p and v are measured.

All real gases deviate from the ideal gas law, and water vapor is no exception. However, it can be shown that over the temperature range from 0° to 100 °C the deviation is slight, in the order of 1% or less. At 0 °C, for example, saturated water vapor has a density ϱ of 0.4846×10^{-5} g cm^{-3}, or 0.4846×10^{-2} g l^{-1}, and a vapor pressure p_0 of 0.00604 atmosphere, from Table 1.1. Rearranging Eq. (1.13), and substituting in these values

$$R = (p_0 M)/(\varrho T)$$

$$R = \frac{(0.00604)(18)}{(0.004846)(273.1)} = 0.0821 \text{ liter-atm-mol}^{-1}\,\text{deg}^{-1} \tag{1.14}$$

compared with 0.08205 for an ideal gas. At 20 °C, R = 0.0819; at 40 °C, R = 0.0818; at 60 °C, R = 0.0816; at 80 °C, R = 0.0812; and at 100 °C, R = 0.0807, – all in the same units.

1.2 Measurement of Wood Moisture Content and Humidity

1.2.1 Measurement of Wood Moisture Content

The moisture content of wood is usually expressed as a fraction m or percent M of its dry weight. Thus

$$m = w_w/w_0 \tag{1.15}$$

$$M = 100\,m \ , \tag{1.16}$$

where w_w is the weight of water and w_0 the dry weight of wood. Either or both the fractional m and percent M are used in this book, depending on convenience and context. The weight of water w_w is obtained from the difference between the moist w_m and dry w_0 weights of the wood, and

$$m = (w_m - w_0)/w_0 \tag{1.17}$$

$$M = 100(w_m - w_0)/w_0 \ . \tag{1.18}$$

Newly formed wood cells in a living tree are born in a water-saturated environment, and the cell walls themselves are presumably also fully saturated with water. Those wood cells which function primarily as channels for conducting water from the roots to the branches and foliage may remain essentially water-saturated in order to provide continuous unbroken water columns from roots to leaves.

When the wood cells no longer function primarily as water transport media, the cell-cavity water may become partially or fully displaced by gases, including water vapor. The cell wall itself, however, remains fully saturated. Therefore, green wood generally contains water in three forms, liquid water partially or completely filling the cell cavities, water vapor in the empty cell cavity spaces, and water in the cell wall.

The liquid water in the cell cavities is sometimes called "free" water to distinguish it from the cell-wall water which may then be called "bound" water. This is because it is attracted or bound to the wood with stronger forces than those which hold the so-called "free" water. The terms "free" and "bound" are relative only since the water in the cell cavities is subjected to capillary forces and therefore is not in the same thermodynamic state as is ordinary liquid water in a wide container. Furthermore, cell cavity water may also contain water-soluble materials which reduce its thermodynamic state and make it less "free" than ordinary liquid water without solutes.

When green wood is dried, a given cell loses its cell cavity or "free" water first, since it is held with weaker forces than is the cell wall or bound water. The moisture content at which a given cell has lost all of its cavity water and contains only water vapor in the cavities but the cell wall is fully saturated was first designated as the fiber-saturation point by Tiemann (1906). It is represented by M_f (%) or m_f (g/g), generally ranging from 20% to 40% of the dry weight of wood. It is discussed more fully in Section 1.4.6.

There are many methods, up to 15 (Kollmann and Höckele 1962), which have been used to measure the moisture content of wood. Several of these methods are described here.

1.2.1.1 Gravimetric Method

In the basic gravimetric method of measuring wood moisture content, the moist sample is weighed and then dried until the reference weight w_0 is attained. For ordinary moisture-content determination, where high precision is not required, the dry weight w_0 is obtained by drying the sample in an oven maintained at $103 \pm 2\,^\circ C$ until constant weight is attained. This procedure reduces the sample moisture content to a low value at equilibrium with a relative vapor pressure sufficiently close to zero that the sample is assumed to have attained its dry weight w_0.

There are several objections to using the simple oven-drying method for obtaining w_0 when precise moisture contents are required. Among these are: appreciable departures from zero of the relative vapor pressure in the oven, the evaporation of volatile constituents other than water from the wood drying in the oven, and the possible effect of previous sorption history.

The activity $h\,(=p/p_0)$ in a drying oven is a function of several variables, including oven temperature, ambient room temperature and humidity, and sources of moisture within the oven. The last factor can be minimized by use of a forced-circulation oven and sufficiently long drying times so that the vapor pressure p in the oven is the same as that in the room. Under these conditions the relative vapor pressure in the oven $(p/p_0)'$ can be calculated from the equation

$$(p/p_0)' = (p/p_0)''(p_0''/p_0') \ , \tag{1.19}$$

where $(p/p_0)''$ and p_0'' are the ambient relative and saturated vapor pressures, and p_0' is the saturated vapor pressure in the oven. For example, if the room temperature and humidity are 30° and 60%, the relative vapor pressure in the oven at 103 °C can be calculated from Eq. (1.19). Expressing p_0 in mmHg, $(p/p_0)' = (0.60)\,(31.82/845.1) = 0.0226$. In the case where room temperature and humidity are 20 °C and 10%, $(p/p_0)' = (0.10)\,(17.53/845.1) = 0.00207$, less than 10% of the value obtained in the first case. The residual moisture in the dry sample in the high temperature and humidity room situation should be in the order of 0.5%, whereas at the lower room or ambient conditions it should be in the order of 0.05%.

Figure 1.5 shows curves relating the percent relative humidity in a drying oven maintained at 103 °C to the ambient humidity in the room at several room tem-

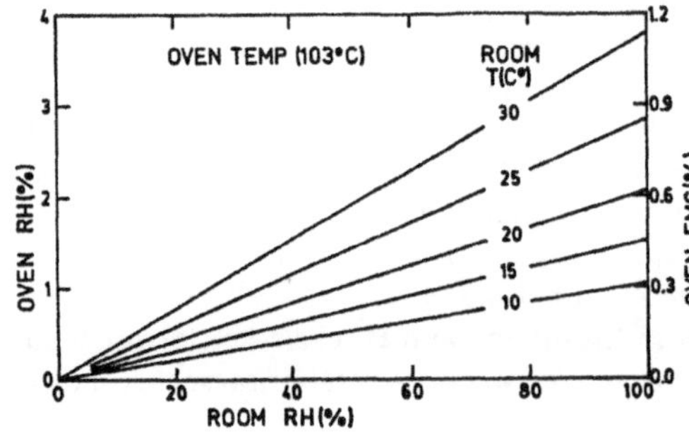

Fig. 1.5. Oven relative humidities and estimated EMC's as functions of ambient room humidity and temperature

peratures. Also shown are estimated equilibrium moisture contents of wood samples dried in the oven at the corresponding humidities in each case, assuming a linear relationship between moisture content M and relative humidity H such that $M \simeq H/5$.

Experimental verification of this effect is indicated by oven-dry weights of 38 different woods of Venezuela, which were 0.42% lower on an average when dried in an oven in Syracuse, New York, in January compared with their oven-dry weights determined in Merida, Venezuela. The difference in retained moisture at the oven-dry condition is due primarily to the higher ambient vapor pressure in Venezuela compared with Syracuse in the middle of winter. The conditions at the two locations approached those given in the numerical example above.

The effect of room conditions can be greatly reduced or eliminated by use of a vacuum oven. Lower oven temperatures may be used since the ambient relative vapor pressure $(p/p_0)''$ is essentially zero. Sometimes a strong desiccant such as phosphorus pentoxide is used instead of a vacuum because of its low vapor pressure. However, much longer drying times are required to attain equilibrium at room temperatures compared with elevated oven temperatures.

The gravimetric measurement of moisture content in wood samples in particulate form can be expedited by use of infra-red lamps combined with direct weighing. In this case a small sample of the moist material is placed on an enclosed balance, which can be heated directly by internally mounted infra-red lamps. The material dries rapidly and the balance may be adjusted to read moisture content directly. The infra-red lamps, whose radiation penetrates only a short distance into the wood surface, may be replaced by radio-frequency heating particularly in the microwave region. This has the advantage of heating the sample internally to drive out the moisture.

Both infra-red and radio-frequency electromagnetic radiation have been used to measure wood moisture content directly rather than merely to dry the wood rapidly (Sect. 1.2.1.5).

It is important in gravimetric moisture measurement that the precision of the balance be sufficient to establish the desired degree of precision ΔM in moisture measurement. As a general rule, the precision Δw of weight measurement should be equal to or better than the value given by

$$\Delta w = w_0(\Delta M/100) , \qquad (1.20)$$

where w_0 is the oven-dry weight of the sample and ΔM is the desired precision (%) of moisture content. For example, if the oven-dry weight w_0 is 10 g, and the desired precision ΔM is 0.1%, the balance should read to a precision Δw of at least (10)(0.1/100) or 0.01 g.

The effect of previous sorption history on the dry weight w_0 of small wood samples vacuum-dried at room temperatures has been reported by Hergt and Christensen (1965), working with *Araucaria klinkii* in Australia. They found, for example, that the apparent dry weight w_0 of 40-μm-thick sections was a minimum when they were dried rapidly at 27 °C under vacuum directly from the water-soaked condition. However, when the same or similar samples were first conditioned to intermediate moisture contents in the hygroscopic range before being vacuum dried, they appeared to reach equilibrium with the vacuum at somewhat higher weights than the minimum value of w_0. This indicated that a certain amount of water was retained in the samples upon vacuum drying if the samples were not completely wet just prior to rapid drying. They found that the maximum retention occurred in samples which were dried following exposure to p/p_0 from 0.4 to 0.6.

A later study by Christensen and Hergt (1969) showed that the percent of water retained upon vacuum drying, M_r, increased approximately linearly with the square root of time the samples were exposed to p/p_0 from 0.4 to 0.6. An equation which fits their data up to about 900 h is

$$M_r = 0.025\sqrt{t}\,, \tag{1.21}$$

where t is exposure time in hours. It can be seen from Eq. (1.21) that exposure of thin wood samples to p/p_0 of, say, 0.5 for 100 h results in a retained moisture content M_r after vacuum drying of 0.25% and after 900 h of 0.75%.

1.2.1.2 Distillation Method

The evaporation of volatile components other than water during drying may cause substantial errors in the gravimetric method of measuring moisture in wood. In this case it is necessary to use a modification of the gravimetric method by heating the wood in a distillation apparatus containing a water-immiscible liquid which is a solvent for the volatile extractive compounds. The water is condensed in a reflux condenser system and separated from the solvent by means of a calibrated trap (Fig. 1.6). According to Kollmann and Höckele (1962), several kinds of solvent can be used satisfactorily, including toluene, xylene, and trichloroethylene. These workers also report that the method is less exact than the

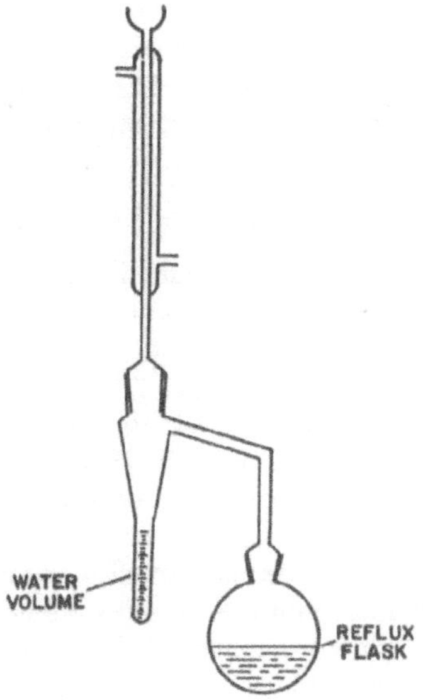

Fig. 1.6. Distillation apparatus for measuring wood moisture content

Karl Fischer method discussed below, at least partially because of the inaccuracy involved in measuring the water volumetrically.

1.2.1.3 Karl Fischer Titration Method

The Karl Fischer titration method for measuring the moisture content of wood is another technique which is particularly useful for material containing volatile extractives. Kollmann and Höckele (1962) found that this technique gave the best results among several standard methods. Resch and Ecklund (1963) found that the Karl Fischer method gave lower moisture content values than the American Society for Testing Materials (ASTM) oven-drying method for *Chamecyparis lawsoniana* and *C. nootkatensis*, presumably because of the volatile oils in these woods.

In this method the water content is measured by titration, using the Karl Fischer reagent. The classical reagent consists of a solution of pyridine (C_5H_5N), sulfur dioxide, and iodine in methanol. It reacts with water as follows:

$$3 \cdot C_5H_5N + SO_2 + I_2 + H_2O \rightleftarrows 2C_5H_5N \cdot HI + C_5H_5N \cdot SO_3 \ . \tag{1.22}$$

The end point of the titration may be determined either colorimetrically (free I_2 present) or electrically (conductivity decreases when free water is exhausted).

The Karl Fischer method is best suited for moisture measurements on liquids or on solids such as wood in particulate form, but is not practical for use in large wood samples, particularly those of high moisture content. Modern instruments are highly automated, convenient, and rapid. In some instruments coulometric titration is used, in which case the titration is accomplished electrically rather than by use of the classical volumetric buret. Furthermore, substitutions for some of the original Karl Fischer chemicals have been made in some cases (Troller and Christian 1978).

1.2.1.4 Nuclear Magnetic Resonance

Nuclear magnetic resonance (NMR) techniques have been applied to measure wood moisture contents under laboratory conditions by Nanassy (1973, 1974) and by Sharp et al. (1978). Such NMR techniques are based on the fact that the nucleus of each hydrogen atom in water is a magnetic dipole due to its characteristic spin. When subjected to an external static magnetic field of intensity H_0, the magnetic dipole precesses about the direction of the field H_0 with a frequency f which is directly proportional to H_0. For the basic hydrogen nucleus

$$f = 4.257 H_0 \ , \tag{1.23}$$

where f is in kHz when H_0 is in Gauss.

According to Sharp et al. (1978), there are two techniques for using NMR to measure moisture content, one steady state and the other unsteady state. The steady-state method was used by Nanassy (1973, 1974) with a wide line NMR spectrometer and comparison technique. In this technique an electromagnetic field at an appropriate frequency f, in the MHz range, is applied to a small wood

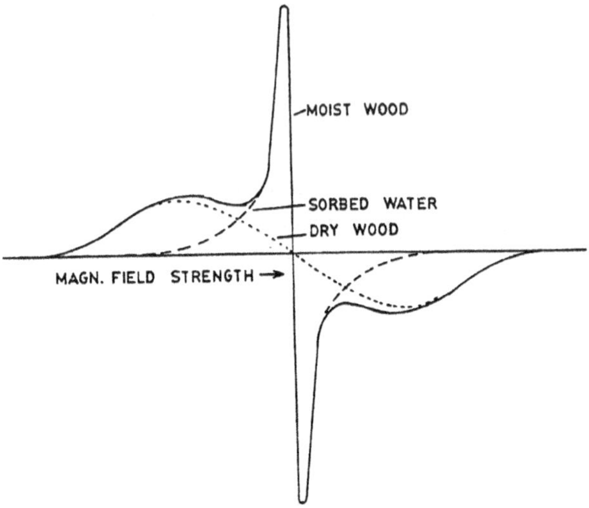

Fig. 1.7. Derivative of NMR absorption line plotted against magnetic field strength for dry and moist (5%) yellow birch wood, and for the sorbed water (after Nanassey 1973)

specimen. A magnetic field of slowly varying intensity H is also applied. At combinations of f and H close to resonance as given by Eq. (1.23), the nuclei absorb power from the system. This power absorption is plotted as a function of H at constant f. This curve and its derivative give information on the amount of water in the wood and also on the relative mobility or freedom of the water molecules to orient themselves in the direction of the applied field.

Figure 1.7, adapted from Nanassy (1973), shows the derivatives of the NMR absorption lines for yellow birch in the dry condition and also at 5% moisture content. It is clear that the moisture has a pronounced effect on the NMR spectrum. Increasing the wood moisture content resulted in increasing the magnitude and width of the narrow peak curve, and Nanassy was able to interpret this in terms of total moisture content as well as the relative mobilities of weakly and strongly bound water in the wood. Similar NMR studies have been made on cellulose (Pittman and Tripp 1972) and on foods (Troller and Christian 1978).

In the pulsed NMR method, according to Sharp et al. (1978), an initial equilibrium magnetization is first applied to the sample parallel to the external magnetic field H_0, together with the sinusoidally alternating magnetic field of appropriate frequency f, based on Eq. (1.23). Then "a short, intense burst of a magnetic field oscillating in resonance with the spin precision frequency is applied at right angles to H_0" (Sharp et al. 1978). This induces a net magnetization into a plane perpendicular to H_0 and a voltage is induced in a coil surrounding the sample. Following the short pulse, the individual magnets tend to return toward their original equilibrium magnetization configurations. The corresponding induced voltage, called the free induction decay (FID), measured 50 μs after pulse removal, increased linearly with wood moisture content for maple and spruce (Fig. 1.8). They noted that the results are independent of grain orientation and that the pulsed NMR technique is a promising method for moisture content measurement in wood and similar materials in the laboratory. They also concluded that it is probably not practical for field measurements, since it is necessary to

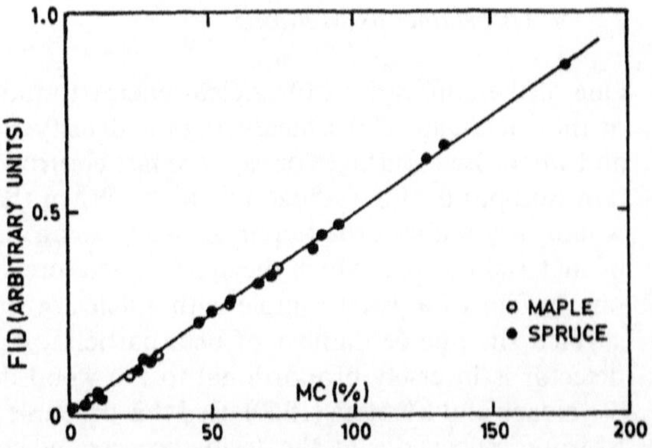

Fig. 1.8. Free induction decay (FID) as a function of wood moisture content for maple and spruce (after Sharp et al. 1978)

remove small samples from the wood in the order of 1 cm^3 or less, and the method cannot be used for in situ measurements.

Both steady-state (wide-line) and pulsed NMR have been used to investigate the state of water in frozen foods (Nagashima and Suzuki 1981). These techniques have also been used to differentiate between bound and free water at room temperature. They can give precise measurements of water content in short times (in the order of 1 min after samples are prepared). However, the costs of purchase and operation of the equipment are high.

It has been proposed by J. King and Rollwitz (1983) that a simplified NMR system might be used to measure the moisture content of flowing coal. They built a laboratory prototype of such an apparatus which combined NMR measurements with electron magnetic resonance (EMR) measurements. The latter were used primarily to determine the amount of carbon present in the coal, since carbon contains significant concentrations of unpaired electrons which are measured by EMR.

1.2.1.5 Electrical Moisture Meters

Most moisture measuring systems for wood are based on samples removed from the wood and are therefore destructive of the sample. Electric moisture meters, however, provide essentially instantaneous and nondestructive moisture measurements.

Two general types of electrical moisture meters are in use. The simplest and oldest type is the *resistance meter* which measures the DC (or low frequency) electrical resistance. The second type, designated here as the *dielectric moisture meter*, measures some function of the complex dielectric constant or electrical permittivity of the wood. It uses AC (alternating current) techniques at frequencies ranging from audio through the microwave range.

Electric moisture meters and the factors which affect the important electrical properties of wood, including moisture content, temperature, grain orientation, density, etc. are discussed in detail in Chapter 6.

1.2.1.6 Miscellaneous Methods

One of the applications of nuclear energy to industry has been the development of the "beta-gauge" for measuring the density and/or moisture content of paper and wood. Beta particles or rays are fast electrons emitted from the nuclei of certain isotopes during radioactive decay. When they pass through matter, such as wood, they are absorbed approximately as an exponential function of the mass of material through which they pass. Therefore, if a beta source is placed against one surface of a wood sample with a detector of beta radiation on the opposite surface, the rate or number of beta particles per unit of time which reaches the detector is inversely proportional to the wood density, for a given thickness.

Noack and Kleuters (1960) used this principle to measure the moisture content of wood veneers during the drying process and concluded that it might be a practical method for monitoring the moisture content of veneers or wood chips during manufacturing or drying. One basic limitation in the use of beta particles for this purpose is their high absorptivity in solid materials such as wood, thus limiting the upper range of thickness to that of veneers. It is of interest to note that this technique has been used to monitor density variations through the annual rings of wood as a tool for growth-quality studies (Phillips et al. 1962).

Another application of nuclear radiation techniques to moisture-content measurement involves the use of two different kinds of radiation simultaneously. A "fast" neutron generator is the source of high-energy neutrons which are directed into the material to be measured (Fig. 1.9). Some of these fast neutrons lose much of their energy in the material and become slow or "moderated" neutrons as they are reflected back to a slow-neutron detector. Hydrogen is a particularly effective neutron moderator, and since each molecule of water contains two atoms of hydrogen, the amount of moisture in the material has a large effect on the number of moderated neutrons received by the detector. Dry wood also contains hydrogen, of course, and it is therefore necessary to calibrate an instrument for this background.

It is also necessary to determine the density of the material in order to measure its moisture content on a weight basis, since the neutron radiation system measures the amount of water per unit volume of material. Density is measured by means of a gamma radiation and detection system. A beam of gamma rays is directed into the material; the intensity of radiation reflected or transmitted through the material to the gamma detector is inversely proportional to the density of the material (Fig. 1.9). The combined output data obtained from the two radiation systems can be combined to give the moisture content on a weight basis.

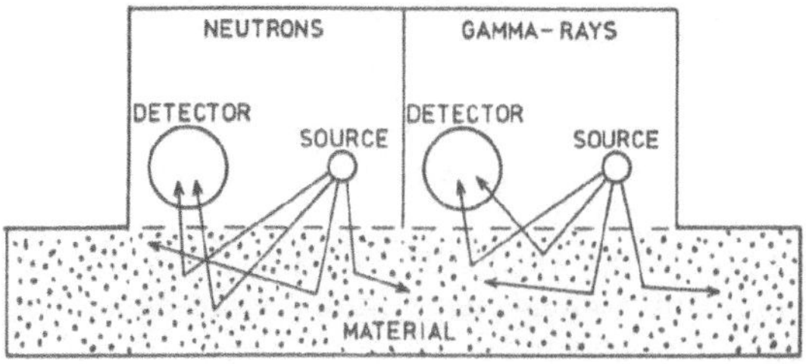

Fig. 1.9. Schematic diagram of a nuclear moisture content gage using both neutrons and gamma rays

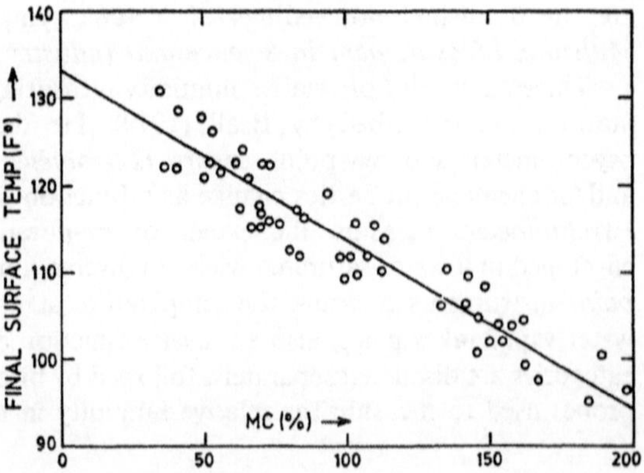

Fig. 1.10. Plotted points and regression curve of final surface temperature in relation to green wood moisture content for nominal 2- by 6-inch mixed hemlock-fir lumber (after Troughton and Clarke 1987)

This system is adaptable to measuring granular materials such as wood chips in a continuous process such as particleboard manufacture.

Troughton and Clarke (1987) proposed a wood moisture measurement method based on the cooling effect of moisture evaporating from the surface of green wood. They designed and tested a system in which the temperature rise of the wood surface, after moving past an infra-red heating source at a constant rate of speed, was monitored, using infra-red surface temperature detectors. The surface temperature rise was observed to be greater for wood of lower initial moisture content (Fig. 1.10). They concluded that their technique had "good potential for sorting unseasoned wood into moisture content ranges for more effective drying."

1.2.2 Measurement of Relative Humidity

The measurement of relative humidity H or activity h of the atmosphere is based primarily on techniques developed by meteorologists. It is often important to know the equilibrium relative humidity ERH or activity of the wood. When wood is at equilibrium with its environment, for example, its ERH is equal to that of the environment. Therefore its equilibrium moisture content EMC can be estimated from its sorption isotherm at ambient temperature if the relative humidity is known.

Several factors (Sect. 1.4) affect the sorption isotherm. Therefore, knowledge of the ambient relative humidity alone does not permit precise determination of wood moisture content. Nevertheless the ambient relative humidity is closely related to the equilibrium moisture content of the wood and is therefore an useful parameter to measure. Furthermore, it is of importance in itself (Sect. 1.1) because, together with the air temperature, it determines the ambient vapor pressure p. The magnitude of this vapor pressure compared with that of the wood surface determines whether the wood is gaining or losing moisture and therefore whether it is swelling or shrinking with time.

A primary source of information on the state of the art of humidity measurements as of 1963 is the four-volume publication edited by Wexler (1965). These are the published proceedings of a 1963 symposium entitled *Humidity and Moisture: Measurement in Science and Industry* held in Washington, DC.

Three categories of relative humidity-measuring instruments used in the wood industry are described by Beall (1979). He designates these as hygrometers, psychrometers and dew-point sensors. *Hygrometers* are instruments whose physical and/or chemical properties change as a function of the ambient relative humidity. *Psychrometers* measure the extent of evaporative cooling of a thermometer enveloped in a water-saturated wick, an inverse function of relative humidity. *Dew-point* instruments measure the temperature at or below which condensation of water vapor takes place, also an inverse function of relative humidity. These three categories are discussed separately, followed by brief descriptions of some humidity probes used to measure the relative humidity in the interior of wood.

1.2.2.1 Hygrometers

Many biological materials, including wood, are hygroscopic. *Mechanical* hygrometers operate on the principle that the material shrinks or swells in a given dimension as it loses or gains water from the atmosphere, a function of the relative humidity. Most such hygrometers use human hair as the sensing element, although animal membranes and, more recently, synthetic membranes have also been used.

There are several problems associated with mechanical hygrometers, including hysteresis effects, mechanical creep, slow response, and anomalous temperature and aging effects. Therefore they are not suitable for indicating small humidity changes, particularly those that occur rapidly. However, they are useful indicators of humidity and have been widely used in meteorological applications.

Wood elements have been used occasionally as mechanical hygrometers and also as hygrostats, that is, humidity controls, particularly in lumber dry kilns. One type of wood element hygrostat consists of a wood strip, of approximately 0.3, 2, and 20 cm in the longitudinal, radial, and tangential directions, respectively. It is mounted in an adjustable frame so that its tangential shrinking or swelling activates an electrical or pneumatic system connected to the humidifying controls. It has the advantage that it can be calibrated directly in terms of desired EMC rather than relative humidity. However, it suffers from the disadvantages inherent in mechanical hygrometers as mentioned above.

Greater sensitivity can be obtained with mechanical hygrometers by use of the principle employed in bimetallic thermometers, where thin strips of two metals of different thermal expansion properties are bonded together. A wood hygrometer consisting of two thin strips bonded together, one with the longitudinal and the other with the tangential direction oriented along the hygrometer length, can be used. The difference in hygroexpansion in the two directions bends the hygrometer accordingly.

Nonmechanical hygrometers are based on changes in physical or chemical properties other than dimensions. The properties most often affected are the electrical resistance and/or capacitance, although some sensors respond to humidity changes by changing color. The latter are more qualitative than quantitative.

A popular type of *electrical* hygrometer consists of hygroscopic salts or other moisture-sensitive films whose resistance decreases severalfold for relatively small increases in relative humidity. Increasing temperature also causes a decrease in resistance so the elements must be compensated for temperature changes. These sensors are polarizable and must therefore be used only with alternating current. They respond very quickly to atmospheric changes, with time constants of only a few seconds. However, they are sensitive to atmospheric pollutants and must not be subjected to condensation of liquid water.

Other types of surface humidity sensors, including some based on solid-state materials, have been developed in recent years. Some of these depend on resistance and others on capacitance changes, both associated with surface sorption of moisture from the atmosphere. Special instrumentation has been developed for many of the hygrometers, which makes it convenient to use them. Furthermore, the rapid response time favors their use in control applications.

Wood itself can be used as an electric hygrometer, either of the resistance or capacitance type, based on the variation of these parameters with moisture content and therefore with relative humidity (Chap. 6). The same problems with respect to hysteresis and long response time that characterize mechanical hygrometers also operate in this case. Nevertheless, such hygrometers have been used in conjunction with electrical wood moisture meters (Duff 1966).

Since hygrometers are generally indirect indicators of relative humidity, they must be calibrated both initially and after various periods of use, if their inherent accuracy is to be realized. The next two categories, dew point sensors and psychrometers, measure humidity directly, and can therefore be used as calibrating instruments, particularly dew point sensors.

1.2.2.2 Dew-Point Sensors

Knowledge of the *dew-point* temperature T_d, together with the dry-bulb or air temperature T, permits direct calculation of the activity h based on Eq. (1.9). The dry-bulb temperature can be measured with relative ease, but the dew-point temperature is difficult to measure accurately. Nevertheless, a number of instruments have been developed to measure the dew-point temperature.

The dew point is that temperature to which air must be cooled so that it is saturated with water vapor at a given vapor pressure p. Water will therefore condense on a clean nonhygroscopic surface whose temperature is at or below the dew point. If this temperature is lower than the normal freezing point of water, it is referred to as the *frost point*, since frost rather than liquid will condense on the surface.

Usually a mirror is used as the condensation surface. The mirror may be cooled by any one of a number of techniques including use of ice-water mixtures, acetone-dry ice mixtures, etc. Modern instruments use thermoelectric cooling which can be controlled with relative precision. Detection of condensation may be by visual observation or by use of a photodetector together with a collimated light source. As condensate forms on the mirror, light scattering occurs, which is detected and amplified by suitable instrumentation (Fig. 1.11). A temperature-

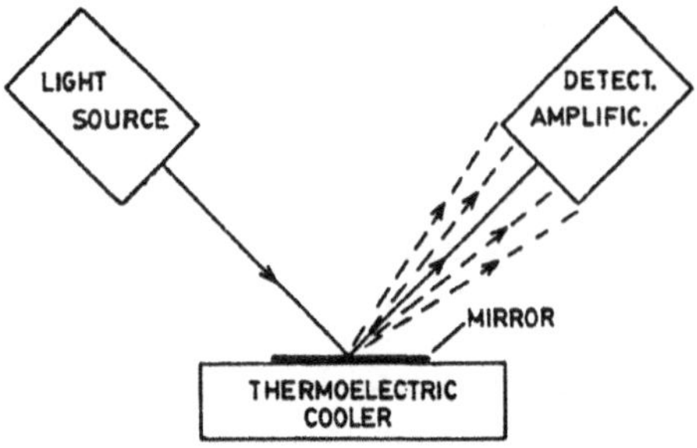

Fig. 1.11. Schematic diagram showing operation of a dew-point hygrometer

sensing device such as a thermocouple is embedded in the mirror to measure the dew point directly.

Less expansive manual techniques are also used in which condensation is observed visually on the outer mirrored surface of a "dew cup." The accuracy of such a technique is dependent on the skill and judgment of the operator, as well as on other empirical factors. It is also a discrete measurement which must be repeated each time a value is required.

1.2.2.3 Psychrometers

The wet- and dry-bulb *psychometer* is based on the principle of evaporative cooling. The wet-bulb temperature is that at which liquid water evaporates from a surface such that it saturates the ambient air adiabatically. It is obtained by enclosing a temperature-sensing element in a moistened wick and passing the air to be measured over the wick at a velocity sufficiently high (at least 5 m s^{-1}) to assure maximum rate of evaporation.

The dew-point temperature (T_d or θ_d) is always slightly lower than the wet-bulb temperature (T_w or θ_w). Therefore, the vapor pressure p, equivalent to p_{0d}, the saturated vapor pressure at the dew-point temperature, is always slightly lower than p_{0w}, the saturated vapor pressure of water at the wet-bulb wick. Otherwise there would be no evaporation of water from the wick to the surrounding atmosphere. An empirical equation which relates the difference ($p-p_{0w}$) to dry-bulb (θ) and wet-bulb (θ_w) temperature is Eq. (1.12), rewritten here,

$$p-p_{0w} = P(\theta-\theta_w)(1+0.00115\,\theta_w)/1515 \ , \qquad (1.24)$$

where P is the atmospheric pressure of all the gases in the atmosphere, expressed in the same units as p and p_{0w}.

It is evident from Eq. (1.24) that the difference ($p-p_{0w}$) is proportional not only to the dry- and wet-bulb temperatures but also to atmospheric pressure P, as Siau (1984) points out. Equation (1.24) can be solved for the vapor pressure p by moving p_{0w} to the right side of the equation. Although it is different from the equation given by Siau, it gives similar results.

Beall (1979) lists five important factors which affect the accuracy of wet- and dry-bulb thermometers. These include adequate ventilation rate (5 m s^{-1} minimum), use of clean distilled water on a clean wick, identical calibration of wet-

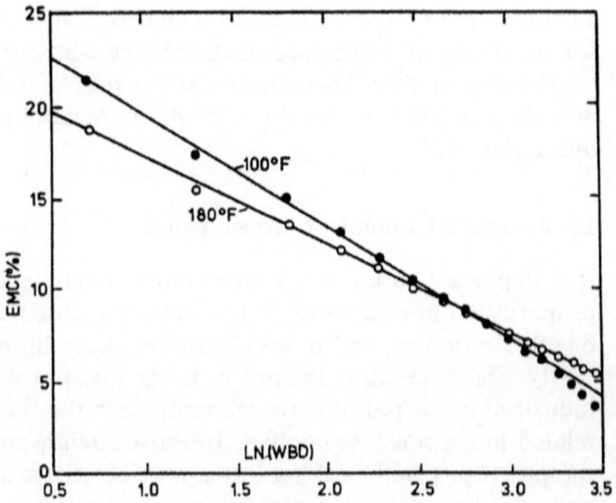

Fig. 1.12. Relationship between wood EMC and the logarithm of wet-bulb depression for temperatures of 100° and 180°F

and dry-bulb temperature sensors, and radiation shielding of the temperature sensors form each other and from extreneous sources of thermal radiation. He also points out the importance of drawing air over the sensors rather than blowing it directly from a motor-driven fan, because of the heat generated by the blower or fan motor.

With the sling psychrometer, which consists of dry- and wet-bulb thermometers attached to a hand-operated sling swung in a circle of about 1 m or more in diameter, body temperature may affect the readings. In any case, local temperature variations may cause substantial errors in humidity measurement, even in an air-conditioned room. For example, an error of 0.1 °C in wet-bulb depression may result in an error of 1% in the calculated relative humidity in the room temperature range, particularly at higher humidities.

Dry-kiln operators have been aware for many years of the general relationship between wet-bulb depression and the EMC wood, essentially independent of temperature over the normal dry-kiln range. Noack (1964) noted that the EMC of wood is a linear function of the logarithm of wet-bulb depression over most of the hygroscopic moisture range. For example, Fig. 1.12 shows such plots at two different temperatures based on sorption isotherms taken from the *USDA Forest Service Wood Handbook* (1974). Regression equations fitted over the range of EMC from about 5% to 20% or more are essentially linear at temperatures of 100° and 180°F, with only a small difference due to temperature.

1.2.2.4 Humidity Probes in Wood

Moisture probes of several kinds have been used to monitor the relative humidity in the interior of wood during use. Duff (1968), for example, has used an electric resistance wood element probe to measure the moisture distribution in wood-frame walls. Meierhofer and Sell (1979) used a humidity transducer consisting of a hygroscopic electrolyte on a semiconductor plate between two electrodes, with a temperature-correcting element included. Sell (1985) later used a solid-state sensor consisting of a hygroscopic membrane whose dielectric properties are related to relative humidity. This sensor was not subject to loss of calibration due to condensation of moisture on the element, a potential problem with the earlier transducer element.

1.3 Moisture Content of Green Wood

It is important to know the green moisture content of wood for several reasons, primarily economic. Most of the water must be removed prior to use as lumber or other products, and removal requires expenditure of energy and/or time, both costly. The renewal of interest in using wood as a fuel source for domestic and industrial use is particularly relevant, since the fuel content of wood is inversely related to its moisture content. In those circumstances where logs are floated to the processing mill, sinking losses may be excessive if moisture contents are too high. An interesting practice in this connection has been the traditional one of girdling teak trees in Thailand and Burma 2 or 3 years prior to felling, to reduce moisture content to acceptable levels for floating to the mill. Finally, the cost of transporting green logs is in direct proportion to their moisture contents.

The wood or xylem has several functions in the economy of a living tree. These include mechanical support, transport and storage of nutrients, and transport of water from the roots to the crown. The active passageways for water or sap transport are generally restricted to the sapwood, particularly that most recently formed.

It is generally accepted among tree physiologists that water in living trees is transported primarily in response to negative pressures which occur in the leaves due to transpiration of water vapor to the atmosphere. This concept was proposed initially by Dixon (1914), according to Zimmermann (1983), and is known popularly as the "cohesion theory of sap ascent." The origin of the negative pressures or tension in water in the leaves is the reduced activity as given by the osmotic pressure equation [Eq. (2.74)] in the form relating liquid water tension to the relative humidity of the ambient atmosphere.

The tension in the foliage water pulls on the liquid water in the sapwood xylem, which supports continuous columns from the roots to the foliage. Thus, water which is transpired in the foliage is replaced by water drawn from the soil through the root system. Some of the water drawn into the leaves is synthesized into carbohydrates in the vital process of photosynthesis.

When a tree is cut, the water columns are ruptured and water transport from the roots ceases. The wood moisture content at the time of felling or harvesting is called the *green* moisture content. It may change between the time of felling and processing into lumber or other products depending on exposure time, climatic conditions, kind, age, and size of tree, and on whether or not foliage is left on the tree for some time after felling.

The green moisture content of wood varies considerably among kinds of trees, between heartwood and sapwood in the same tree, and even between logs cut from different heights in the tree. In some cases there may be seasonal variations in green moisture content.

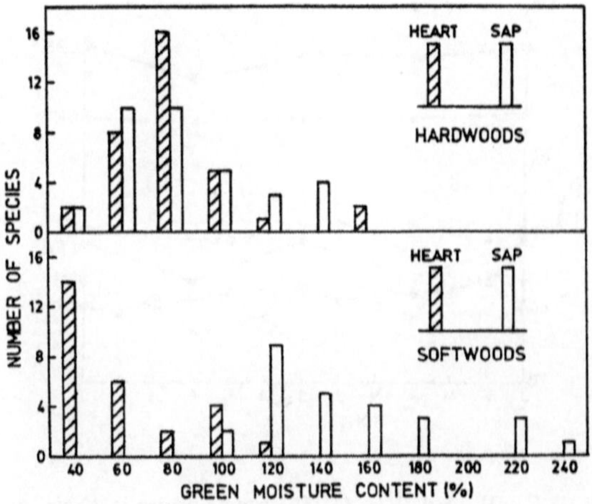

Fig. 1.13. Frequency distributions of green moisture content for sapwood and heartwood of hardwoods and softwoods of the US (based on data of Peck 1953)

Peck (1953) reported on the average green moisture content of sapwood and heartwood of 27 softwoods and 34 hardwoods native to the United States. The average for conifer sapwood was 148.9%, ranging from 98% to 249%. This was almost three times greater than the heartwood mean value of 55.4%, ranging from 30% to 121%. In contrast, the hardwoods showed no consistent difference, with a mean of 81.4%, varying between 44% and 162% for heartwood, and 82.7%, varying between 44% and 146% for sapwood (Fig. 1.13).

Manwiller (1975) tabulated the mean moisture contents of the bark and wood of the stems and branches of 22 different species (10 trees per species) of small (6-inch diameter) hardwoods growing on southern pine sites in the United States. The mean values for stem and branch *wood* were 74% (range 47 to 120) and 66% (range 46 to 109), respectively. The corresponding values for stem and branch bark were 68% (range 44 to 126) and 80% (range 56 to 134. There was no consistent pattern with respect to differences between wood and bark, except that the moisture content of branch bark was almost always greater than that of branch wood.

It is anticipated, based on physiological considerations, that the moisture content of a tree should change throughout the year at least in temperate climates, according to Zimmermann (1983). Studies by Gibbs (1958) and his co-workers have shown that the moisture contents of tree stems often decrease during the summer and early autumn months when the moisture stress is greatest. For example, J. Clark and Gibbs (1957) found that the mean moisture content of yellow birch trees grown in New Brunswick, Canada, decreased from a peak of 90% in late April to 50% in September.

L. G. Henderson and Choong (1968) reported mean moisture contents of 105%, 121%, 92%, and 98% in trees sampled in April, July, October, and January, respectively. Peck (1953) concluded that seasonal variation in tree moisture content is small and inconsistent in most species grown in the United States. Figure 1.14 shows the seasonal variation in a number of species, both European and North American, based on data of Peck.

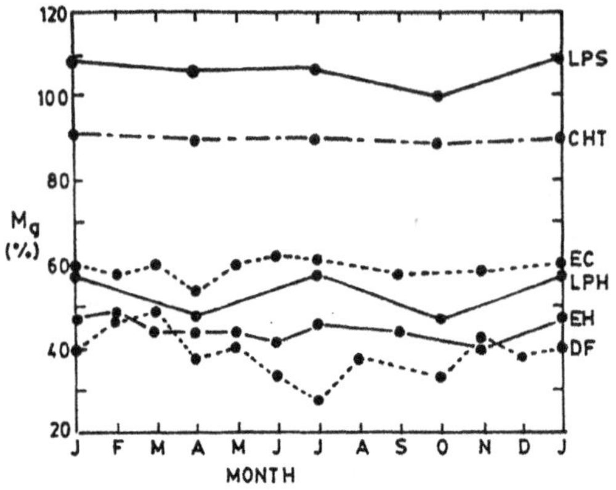

Fig. 1.14. Moisture content M_g (%) variation in green wood throughout the year from January of one year to the next, from data of Peck (1953). The woods shown are loblolly pine sapwood (LPS) and heartwood (LPH), chestnut (CHT), the means of five European conifers (EC) and of 24 European hardwoods (EH), and Douglas-fir (DF)

Chalk and Bigg (1956) have shown in their study of the comparative moisture distribution in trees of Sitka spruce and Douglas-fir grown in England that it is more instructive to express moisture content in terms of a fraction of the saturated capacity of the wood rather than in terms of dry weight. Their method eliminates the effect of differences in density of the wood from different parts of the tree and is particularly useful in studies of tree moisture-content variation from the viewpoint of the physiology of moisture movement in living trees. From the viewpoint of utilization of wood, however, the moisture content based on dry weight of the wood is more useful and conventional.

1.4 Equilibrium Moisture Content of Wood

Wood in the living tree generally has a moisture content of 30% or greater, that is, the cell wall is fully saturated (Fig. 1.15). The cell cavity generally contains some water, the amount of which varies greatly among trees and also among cells in the same tree. When the green wood is exposed to atmospheric conditions after the tree is felled, however, it loses moisture until it comes to a sufficiently low moisture content to be at equilibrium with the ambient atmosphere. This moisture content, designated as the equilibrium moisture content EMC, is approximately

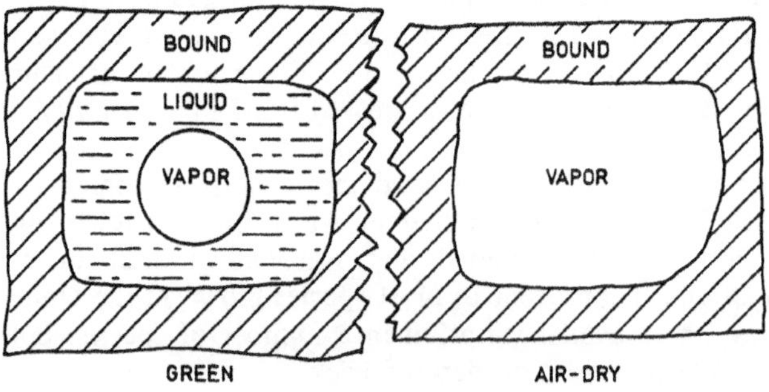

Fig. 1.15. Schematic diagram illustrating the three kinds of moisture in green wood and the two kinds in air-dry wood

proportional to the ambient relative humidity. For given exposure conditions it varies somewhat among different wood species, between heartwood and sapwood, with the kind and proportions of cell-wall constituents and extractives. It is also affected by temperature, previous exposure history and mechanical stress. Some or all of these factors interact to some extent, but it is convenient to discuss them individually, considering the relationship with other factors as they appear.

The theoretical interpretation of moisture sorption by wood is discussed in Chapters 2 and 3. The present discussion is limited to the phenomenological or descriptive characteristics of moisture sorption.

1.4.1 Effect of Relative Humidity

The most important factor affecting the EMC of wood is the relative humidity H or relative vapor pressure h (essentially equivalent to activity) to which the wood is exposed. The curve relating EMC and H or h at constant temperature is called a *sorption isotherm.*

Urquhart (1960) describes five kinds of sorption isotherms. Only three of these are discussed here since they are most applicable to the wood-water system. Figure 1.16 shows these three types, 1, 2, and 3, respectively.

Type 1 characterizes sorption of a vapor on a substrate in which only a single layer of the sorbent (vapor) is found on the sorbate (substrate). It may also be considered that the vapor forms a hydrate with the substrate. In either case the attraction between the sorbent and sorbate is greater than that between the sorbent molecules themselves in the liquid state.

A type 3 sorption isotherm is obtained when several layers of sorbent are formed on the sorbate, that is, multilayered sorption occurs. It may also be viewed as a solution of the sorbent in the hydrated sorbate. The attraction of the sorbed

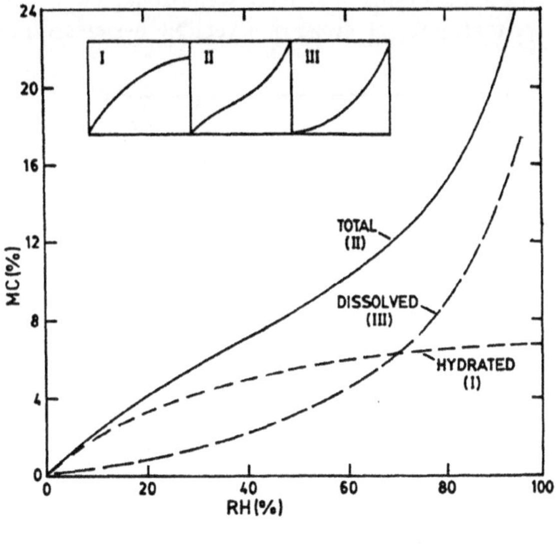

Fig. 1.16. Shapes of the types I, II and III isotherms (*inset*), showing their similarities to those of the hydrated (*I*), dissolved (*III*) and total (*II*) moisture sorption isotherms

molecules to the sorbate is considered to be minimal in this case, compared with the type 1 condition.

The type 2 isotherm is typical of the moisture sorption isotherms of wood and many other natural hygroscopic polymers such as cotton, wool, cereal grains, etc. It is typically sigmoid in shape, and appears to be a composite of types 1 and 3. For example, the lower portion of Fig. 1.16, taken from Simpson (1979) shows a typical type 2 moisture sorption isotherm for wood, together with its two components, the hydrated water (type 1) and the dissolved water (type 3) as analyzed by the Hailwood-Horrobin sorption theory (Chap. 3). Similar curves of type 1 and type 3 are obtained assuming monolayer and multilayer sorption, respectively, based on a surface sorption model such as the Dent modification of the BET sorption theory (Chap. 3).

1.4.2 Effect of History (Sorption Hysteresis)

The EMC of wood at a given relative humidity H is generally highest during the initial desorption of the green wood. Figure 1.17, taken from Spalt (1958), shows the initial desorption isotherm as well as subsequent resorption and desorption isotherms for Douglas-fir. The initial desorption isotherm was obtained by exposing 2-mm-thick sections of green wood to successively lower humidities and measuring the equilibrium weight at each humidity. All measurements were made in a vacuum system at 90 °F (32 °C).

The lowest curve in Fig. 1.17 shows the resorption isotherm obtained after the sample was vacuum dried and subsequently exposed to successively increasing humidities up to essentially 100% H. A second desorption resulted in the intermediate curve shown. It lags below the initial desorption curve down to about 50% humidity, below which it essentially coincides with it. This indicates that there is an initial irreversible loss in hygroscopicity at humidities above about 50% after the initial drying of green or water-soaked wood. In this paper, we will be referring primarily to the cyclic or reversible hysteresis loop unless stated otherwise.

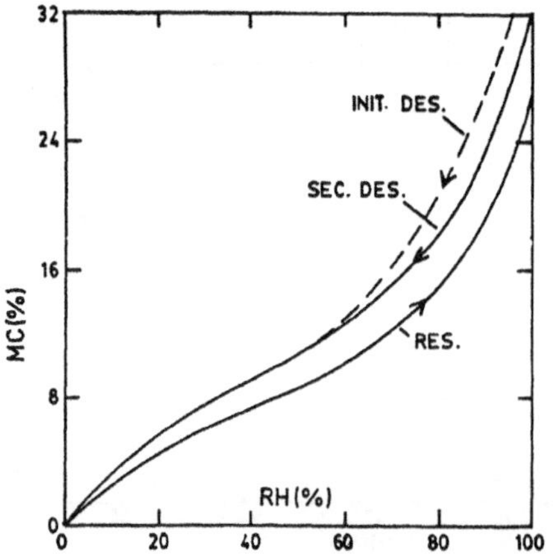

Fig. 1.17. Initial desorption, resorption and secondary desorption isotherms for Douglas-fir (Spalt 1958)

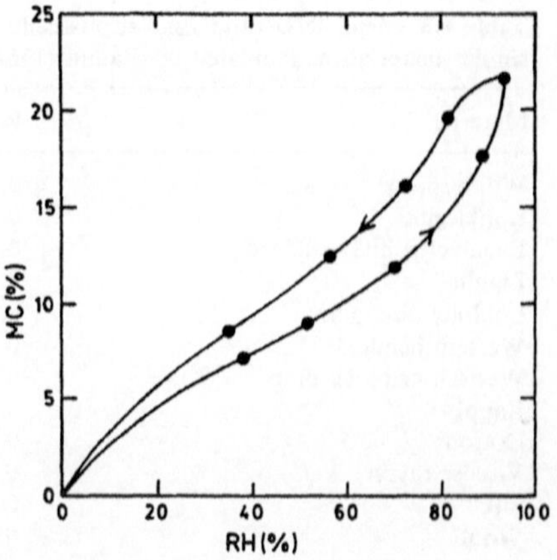

Fig. 1.18. Mean desorption and resorption isotherms for ten US hardwoods at 25 °C (Okoh 1976)

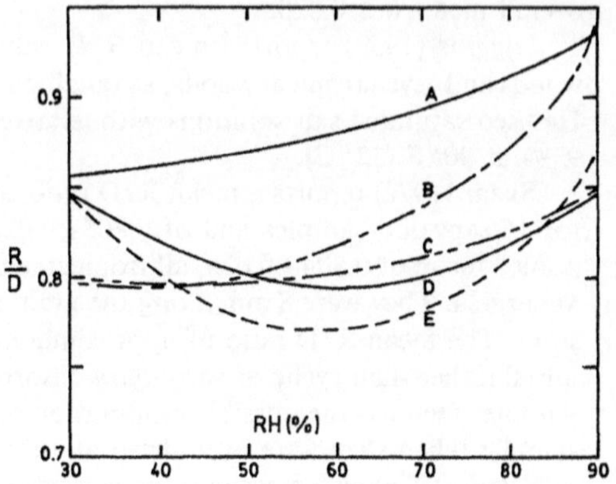

Fig. 1.19. Variation of R/D ratios with humidity for ten US barks (*A*), eight hardwoods (*B*), eight softwoods (*C*), ten US hardwoods (*D*), and ten Ghana woods (*E*). Curves *B* and *C* were measured by Spalt (1958), and *A*, *D*, and *E* by Okoh (1976)

Figure 1.18 shows the mean resorption and desorption isotherms for 10 US hardwood samples, measured by Okoh (1976) at 25 °C. The samples were 3 mm thick along the grain. Each point on the resorption isotherm was measured after vacuum drying the samples overnight. The desorption experiments, however, were carried out in successive steps following the last resorption exposure to 94% relative humidity.

Experimental evidence indicates that a closed hysteresis loop is obtained when successive resorption and desorption cycles are carried out. The magnitude of the hysteresis effect is sometimes expressed as the ratio of resorption to desorption EMC's (the R/D ratio) at a given H. The R/D ratio usually varies with H, as is shown in Fig. 1.19, taken from Okoh (1976). Table 1.3 shows some R/D ratios given by Stamm (1964).

Spalt (1958) calculated mean R/D ratios for eight hardwoods and eight softwoods over the entire hygroscopic range, from the ratio of the total area under the hygroscopic isotherm for resorption to that for desorption. The mean R/D ratios for the hardwoods ranged from 0.790 to 0.849 (mean = 0.828 ± 0.018) while

Table 1.3. Mean R/D ratio for various celluloses and similar materials as tabulated by Stamm (1964)

Material	R/D ratio
White spruce	0.83
Klinki pine	0.78
Eucalyptus holocellulose	0.86
Lignin	0.78 – 0.79
Loblolly pine pulp	0.85 – 0.88
Western hemlock	0.82 – 0.84
Western hemlock pulp	0.87
Kapok	0.81
Cotton	0.77 – 0.87
Viscose rayon	0.78
Silk	0.81 – 0.84
Wool	0.84 – 0.85

those for the softwoods varied between 0.785 to 0.844 (mean = 0.812±0.023). The overall mean was 0.820.

Higgins (1957) reported mean R/D values for 12 woods, including five US woods and seven tropical woods, as ranging from 0.80 to 0.88 with a mean of 0.84. He used saturated salt solutions with relative humidities of 21%, 43%, 81%, and 93% at 90 °F (32 °C).

Skaar (1972) reports a mean R/D ratio, at 40 °C and 76% H, of 0.803±0.032 for 15 sapwood samples and of 0.805±0.028 for 24 heartwood samples, with a pooled mean of 0.804±0.028, all tropical hardwoods from the Guayana region of Venezuela. They were 8 mm along the grain and 25×25 mm in transverse dimensions. The mean R/D ratio of approximately 0.80 was probably low because the initial rather than cyclic or secondary desorption value was measured (Fig. 1.17). Another factor is the possible evaporation of volatile constituents from the green samples when they were oven-dried at 103 °C prior to resorption.

Okoh (1976) also measured the resorption and desorption cyclic isotherms for the inner bark of ten North American hardwoods. He found that the R/D ratio was considerably higher for the bark (Fig. 1.19) than for the wood the same trees. The mean R/D value was 0.884±0.018 for the inner bark samples compared with 0.808±0.010 for the corresponding woods. Most of the difference was due to the appreciably greater desorption isotherms for wood compared with bark, since the resorption isotherms were essentially identical. Okoh also measured the sorption isotherms for ten woods from Ghana in the same study. The mean R/D ratio for these woods was 0.786±0.017.

Turc and Cutter (1984) studied sorption hysteresis on six woods from Argentina at relative humidities of 17%, 30%, 50%, 70%, 90%, and 97% at 20 °C. The mean R/D values ranged from 0.88 to 0.93, averaging 0.895. They were generally lowest at the intermediate humidities.

Several factors may enter into the variation in sorption hysteresis as measured by the R/D ratio. These factors include: incomplete attainment of equilibrium, immediate past history, such as the number of resorption steps and possibly the time at each step, temperature, physicochemical differences in the cell wall, extractive content, etc.

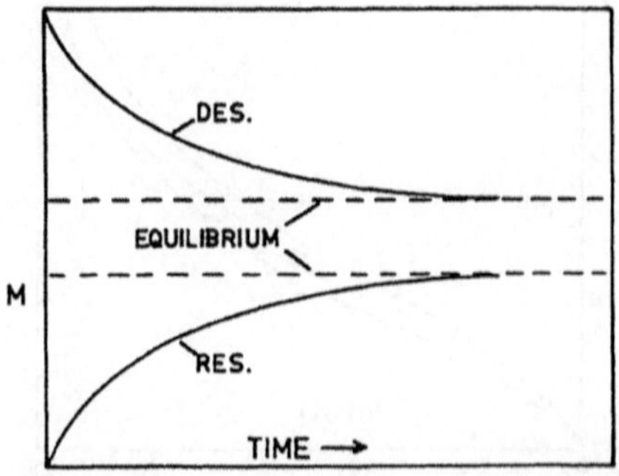

Fig. 1.20. Hypothetical curves showing approach to sorption equilibrium with increasing time (after Morton and Hearle 1962)

Table 1.4. Desorption and resorption EMC's and R/D ratios at several relative humidities for yellow birch (*Betula alleghaniensis*). Resorption EMC's were obtained in single steps or in several steps (multistep) (Prichananda 1966)

H (%)	Desorption (D) (Multistep) M (%)	Resorption (R)			
		(Multistep)		(Single-step)	
		M (%)	R/D	M (%)	R/D
40	5.47	5.16	0.734	5.47	0.778
50	7.47	7.05	0.817	7.47	0.866
60	9.11	8.60	0.819	9.11	0.868
70	10.54	10.19	0.821	10.54	0.849
80	14.19	13.69	0.846	14.19	0.877
Mean	–	–	0.807	–	0.848

Incomplete attainment of equilibrium, either in resorption or desorption, would tend to increase the R/D ratio. Figure 1.20, adapted from Morton and Hearle (1962), shows the effect of time on the attainment of equilibrium from both desorption and resorption, assuming that the process of attaining equilibrium is an exponential function of time. It may require weeks or even months to attain true equilibrium. This may be due to the fact that slow molecular rearrangements may be occurring in the wood as the structure accommodates itself to swelling forces (Skaar et al. 1970).

The effect of immediate past history, as reflected in the number of resorption steps to which a sample is exposed, on the magnitude of the resorption isotherm (and therefore on the R/D ratio) was first reported for wood by Christensen and Kelsey (1959a). Table 1.4, taken from data of Prichananda (1966), shows the direct effect of this factor on the R/D ratio. Here are shown the desorption EMC's, calculated at several values of H, together with the single-step and multistep resorption EMC's at the same humidities. The data for single-step resorption were obtained by equilibrating a sample to each humidity after first drying it in an oven at 103 °±2 °C. The data for multistep resorption were obtained by exposing each sample to increasingly higher humidities after equilibrating at the previous humidity condition.

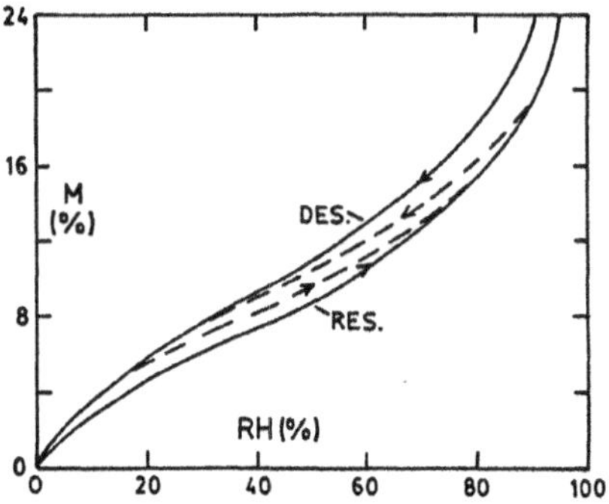

Fig. 1.21. Illustration of intermediate hysteresis loop (after Urquhart 1960)

As Table 1.4 indicates, the EMC was always higher for the single-step than for the multistep resorption. Therefore, the R/D ratio, where D is the desorption value obtained only by multistep desorption, was always higher for the one-step (mean = 0.848) than for the multistep resorption (mean = 0.807).

Sorption hysteresis in wood decreases with increasing temperature according to Weichert (1963), and disappears at temperatures between 75° and 100 °C for European spruce. Data of Kelsey (1957) show agreement with this tendency in that hysteresis in *Araucaria klinkii* of Australia was smaller at 55° than at 25 °C.

The complete resorption and desorption isotherms, that is those obtained from complete dryness and near saturation, respectively, give the greatest R/D ratio. Urquhart (1960) gives a thorough discussion of sorption hysteresis in textile materials, and points out that cyclic isotherms carried out over smaller humidity ranges tend to fall between the complete isotherms. In other words, quoting from Urquhart (1960), "the two (resorption and desorption) curves do not form an equilibrium locus but define an equilibrium area." Stamm (1964) generally concurs with this viewpoint. Figure 1.21 illustrates an intermediate hysteresis loop of the type given by Urquhart (1960).

Differences in the physicochemical constitution of the wood, including extractives, probably account for the variation in the R/D ratio among different woods. For example, the data as reported above for Spalt (1958), Okoh (1976), etc., probably reflect these differences. According to Stamm (1964) the mean R/D ratio is higher for wood pulp and holocellulose (0.86) than for gross wood (0.82). The ratio for lignin is lower, averaging about 0.78 (Table 1.3). Regenerated cellulose also gives lower values, 0.78 and 0.77 for viscose and cuprammonium rayon, respectively.

1.4.3 Effect of Temperature

Elevating the temperature of wood has two effects on the sorption isotherm. The immediate effect is to reduce the EMC at a given relative humidity, a temporary, reversible effect. The second effect is a permanent reduction in hygroscopicity after the wood returns to normal temperatures, the extent of which depends on the temperature and duration of exposure to this temperature.

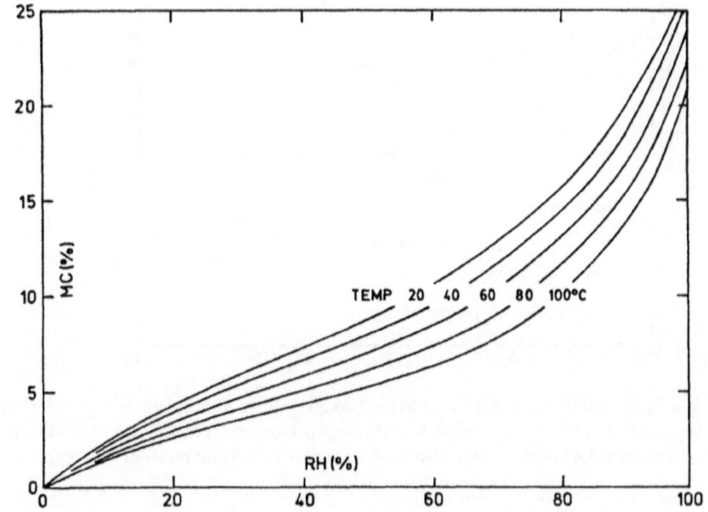

Fig. 1.22. Desorption isotherms for Sitka spruce at several temperatures (Stamm 1964)

The *immediate effect* of elevating wood temperature is to increase the activity or relative humidity of the sorbed water in wood conditioned to a given moisture content (Fig. 1.22). In other words, the hygroscopicity or EMC at a given relative humidity is reduced. This effect is anticipated based on thermodynamic considerations as discussed in Chapter 2.

If the sorption isotherm curves of Fig. 1.22 are extrapolated to unit activity or 100% relative humidity, it is evident that the fiber-saturation point M_f, thus estimated, decreases with increasing temperature. Stamm and Loughborough (1935) reported a decrease of approximately 1% per 10 °C increase in temperature between 25 ° and 100 °C for Sitka spruce. Kollmann (1959) shows curves of Lykow which indicate a 1.3% decrease per 10 °C increase between 20 ° and 80 °C. Weichert (1963) reported a 1.2% decrease per 10 °C increase between 20 ° and 100 °C. It should be noted that extrapolation of sorption data to 100% relative humidity is an approximation to M_f, at best, since data in the high humidity region are generally not very reliable. Also, the sorption isotherms tend to increase sharply in the vicinity of 100% relative humidity, theoretically approaching complete saturation of the wood, including the cell cavities (Sect. 1.4.6).

The isotherms shown in Fig. 1.22 apply only up to 100 °C. Above this temperature the same general trends apply, that is, the wood is less hygroscopic. However, it is difficult to measure sorption isotherms above 100 °C because measurements must be made at pressures above atmospheric for completes sorption isotherms. However, curves have been reported of the EMC of wood against temperature (Fig. 1.23) in pure superheated steam at atmospheric pressure. In this case the maximum attainable relative vapor pressure is equal to the ratio $1/p_0$, where p_0 is the vapor pressure of steam, in atmospheres, at the wood temperature.

It is evident from Fig. 1.23 that the EMC in superheated steam decreases rapidly with increasing temperature above 100 °C, for two reasons. The primary

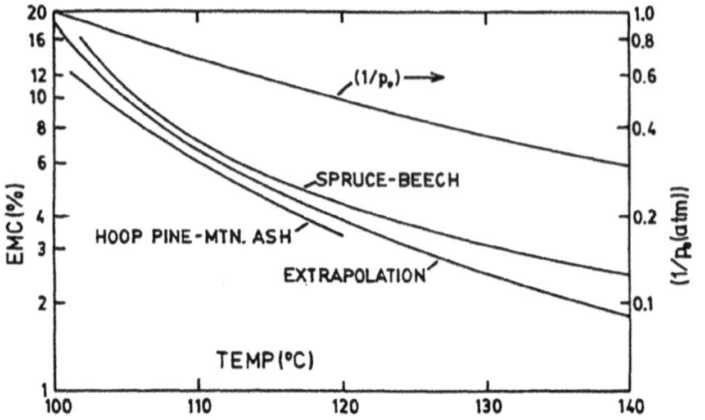

Fig. 1.23. EMC values for wood in superheated steam at atmospheric pressure. The hoop pine-mountain ash curve is from Grumach (1951), spruce-beech from Keylwerth (1949), and the extrapolation curve from Kauman (1956). Also shown is curve of $1/p_0$ in relation to temperature (after Rosen 1979)

cause is the rapid decrease in maximum relative humidity, defined as $H = 100/p_0$ if p_0 is given in atmospheres. This is because p_0 increases so rapidly with temperature. The second reason is the decrease in EMC with increasing temperature, as is shown below 100 °C in Fig. 1.22.

Barometric pressure P_b is generally lower than p_0 as elevation increases above sea level. For example, it decreases to 0.930, 0.863, and 0.801 atm at elevations of 0.6, 1.2, and 1.8 km, respectively. Therefore, the maximum relative vapor pressures of superheated steam at elevations above sea level are lower than those shown in Fig. 1.23. At 1.8 km elevation it is only 80% of the values anticipated from the curve at any temperature.

At temperatures below 0 °C, the EMC of wood is expected to decrease with decreasing temperature, opposite to the trend above freezing. This is because the vapor pressure of ice decreases more rapidly with decreasing temperature than does that of sorbed water (Fig. 1.1). This effect is responsible for the phenomenon of "coldness-shrinkage" discussed in Chapter 2.

The second effect of exposing wood to high temperatures is the *permanent decrease* in hygroscopicity associated with a partial decomposition of the wood structure. According to Stamm (1964), most of this decrease is due to the thermal decomposition of the hemicelluloses which are the most hygroscopic of the principal cell-wall constituents.

Mitchell (1981) found that reduction in the hygroscopicity of loblolly pine was usually greater than for Douglas-fir when both were heated in sealed "bombs." He applied a first-order kinetic equation, together with the Arrhenius equation (Chap. 2), to data on the reduction of hygroscopicity as a function of heating time and temperature, and obtained activation energies in the order of 23000 cal/mol. Rate constants increased from about 0.00004/h at 105 °C for dry wood to 0.04/h at 150 °C for wet wood.

It would appear from the above that heating wood to elevated temperatures is an inexpensive method for reducing its hygroscopicity and its hygroexpansion characteristics. Unfortunately, there are permanent losses in density and certain other physical and mechanical properties which limit the use of this technique. This topic is discussed in more detail in Chapter 4.

Wood which has been kiln-dried usually has a lower EMC than air-dried wood. Furthermore, the EMC is generally lower the higher the kiln temperature used. For example, Salamon (1966) found for one series of experiments that the average EMC for western hemlock at 72 °F and 60% RH was 9.65% for air-dried samples, 9.11% for samples dried at temperatures from 160° to 185 °F, and 8.39% for samples dried at 225 °F dry-bulb temperature.

Experiments designed to measure the sorption isotherm at temperatures above 100 °C must account for the permanent effect of exposure of wood to such high temperatures on the sorption isotherm, as Noack (1959) has pointed out. This factor may account for some of the discrepancies noted between sorption isotherm values measured experimentally compared with those extrapolated from measurements made at lower temperatures.

1.4.4 Miscellaneous Factors

Several other factors also affect the hygroscopicity of wood. Among the more interesting is the effect of stress, first explained by Barkas (1949). This effect predicts a lowering of the EMC of wood subjected to a compressive stress and the reverse for a tensile stress. It is most pronounced when the stress is applied in the direction in which maximum hygroscopic swelling takes place, that is, in the tangential direction. It also causes a reduced moisture sorption in wood restrained from swelling during exposure to a higher humidity. For example, Bello (1968) found that the mean EMC of five US hardwoods, restrained from transverse swelling by means of steel rings, and allowed to adsorb moisture from the dry condition to equilibrium with 87% humidity, was 16.7%, 1.44% lower than those unrestrained from swelling. This effect is discussed in more detail in Chapter 2.

Some studies have been made of the effect of nuclear radiation, particularly gamma radiation, on the hygroscopicity of wood. Figure 1.24, taken from Paton and Hearmon (1957), indicates that a dosage rate of 10^8 rad of gamma radiation consistently lowers the EMC over the entire sorption range. The reduction is less for smaller dosages of radiation.

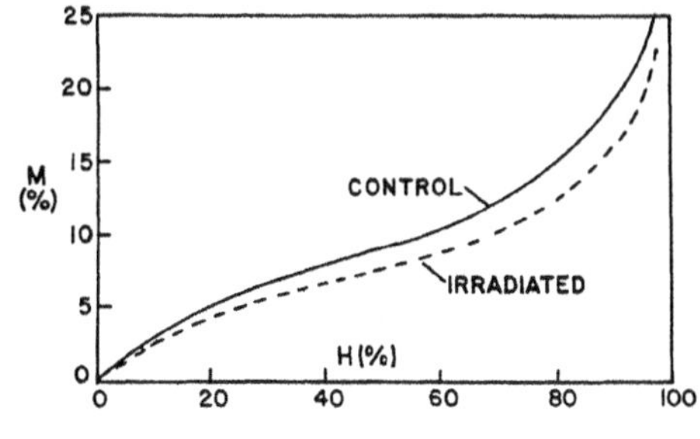

Fig. 1.24. Effect of 100 Mrad of gamma radiation on the sorption isotherm of wood (Paton and Hearmon 1957)

1.5 Moisture Content of Wood in Use

Wood used in final form as furniture, building construction, musical instruments, and other uses is generally subjected to fluctuating atmospheric humidity. The EMC is therefore constantly changing, sometimes periodically and at other times sporadically, such as when wooden decking is exposed to rain.

Wood exposed to high humidity conditions or to liquid water during use may be subjected to biological deterioration. Quoting from Kirk and Cowling (1984), "Liquid water is needed in wood cells to provide a medium for diffusion of the enzymes or other metabolites by which wood-decomposing organisms digest the wood substance. If there is no liquid water present inside the wood cells, there will be no medium for diffusion, and therefore no biological decomposition except for certain insects of relatively minor importance. Thus, as long as wood is kept below its fiber-saturation point (about 27% of its dry weight), it will never decay." Kirk and Cowling indicate that a few brown-rot fungi and also subterranean termites can transport moisture from moist soil or other sources to dry wood to maintain their activity.

It is evident, therefore, that wood subjected to moderately dry atmospheric conditions, that is, well below 100% humidity, will generally not decay. However, other deleterious effects may occur due to changes in moisture content associated with changes in atmospheric humidity and temperature. Hoadley (1980) has pointed out that moisture- (and humidity-) related changes, particularly dimensional changes, in wood during manufacturing and use of consumer products is the *single most important* problem in wood utilization.

The EMC of wood in use is affected most dramatically by the relative humidity of the atmosphere to which it is exposed (Fig. 1.22). Temperature also has a direct effect, but its most important effect is indirect in that raising the temperature of the atmosphere containing a given amount of water vapor decreases the relative humidity. This is because the saturated vapor pressure p_0 increases markedly with temperature [Eq. (1.1)] while the actual vapor pressure p remains essentially constant, the ratio p/p_0 therefore decreasing.

In temperate climates, heating buildings in winter months causes large humidity variations indoors because of outdoor temperature variations. Figure 1.25, taken from Hoadley (1967), shows the annual variation of both indoor and outdoor humidities. The indoor humidities were calculated from the monthly av-

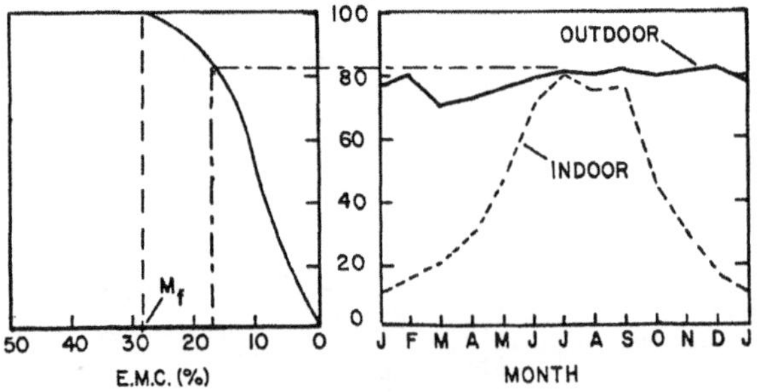

Fig. 1.25. Annual variation of outdoor and indoor humidity *H*, and of corresponding EMC of wood at Amherst, Massachusetts (Hoadley 1967)

erage outdoor values of temperature and relative humidities at Amherst, Massachusetts, assuming a constant indoor temperature of 70 °F (21 °C) and a vapor pressure indoors p′ equal to the outdoor vapor pressure p″. The indoor relative vapor pressure (p'/p_0') is therefore calculated from the outdoor relative vapor pressure (p''/p_0'') by the equation

$$p'/p_0' = p''/p_0' = (p''/p_0'')/(p_0'/p_0'') , \quad (1.25)$$

where p_0' and p_0'' are the indoor and outdoor saturated vapor pressures. Since p_0' is larger than p_0'' in the winter months, the indoor relative vapor (p'/p_0') is lower than that outdoors (p''/p_0'') even when the vapor pressures p′ and p″ are equal.

Based on Fig. 1.25, it appears that the EMC outdoors in the Massachusetts climate is essentially constant at approximately 16%, but that the EMC indoors may fluctuate from a low of 2% in midwinter to the outdoor equilibrium of 16% in midsummer. This does not necessarily mean that particular wood samples, such as the components of wooden furniture, kept indoors in Amherst will vary between the two extremes given above, for several reasons. These include the long times required for moisture to diffuse in and out of thick lumber and the retarding effect of protective coatings, as Hoadley (1967) points out.

Another factor which reduces the seasonal moisture variations is hysteresis in the sorption isotherm. This effect, discussed previously (Sect. 1.4.2), is maximum when the sorption cycle occurs over the entire range of h from 0 to 1.0. In this case the EMC during resorption is about 0.85 of the value during desorption over most of the sorption range.

When wood is exposed to smaller humidity cycles, the hysteresis effect is less pronounced, but still reduces the change in EMC associated with a given change in h. For example, in Fig. 1.21 the cyclic change in EMC between H limits of 0.80 to 0.20 is 16% to 6% rather than 19% to 6%, as it would be for desorption only, or 16% to 5%, as it would be for resorption only. This represents a cyclic change in EMC of only 10% rather than 13% for the desorption curve and 11% for the resorption curve, or an average change of 12.0% in EMC for an intermediate curve. A similar reduction in moisture change for a given humidity change is also anticipated because of the gradual nature of the cyclic changes, according to the findings of Christensen and Kelsey (1959a) (Sect. 1.4.2).

Tsoumis (1960) studied seasonal moisture content variations in previously air-dried wood test samples of four different European woods, beech, oak, chestnut, and pine, exposed to the outdoor atmosphere under shelter in Thessalonika, Greece. His results, summarized in Fig. 1.26, show a similar cyclic annual moisture content variation which is more or less opposite to the seasonal variations shown in Fig. 1.25. In other words, the EMC is maximum in winter and minimum in summer because the relative humidity in Thessalonika follows the same seasonal cycle, high in the winter and low in the summer. Figure 1.26 shows the seasonal temperature variations and also the calculated EMC, the latter presumably based on the sorption isotherm. It may be noted that observed moisture content variations are less than the calculated EMC variations, as anticipated.

It is clear that the moisture content to which wood should be dried depends on its location during use. It should be dried to equilibrium with the mean envi-

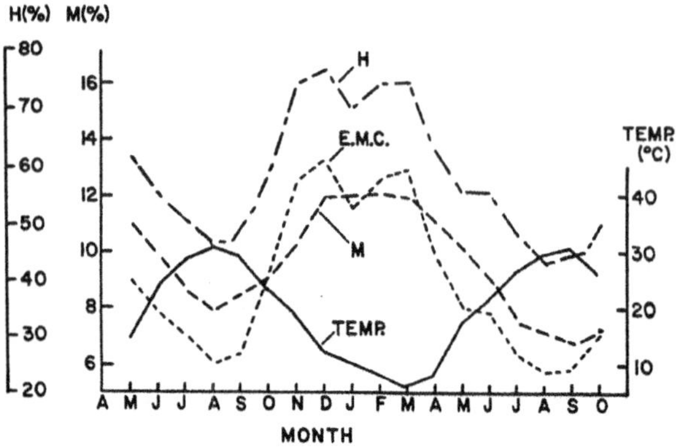

Fig. 1.26. Annual variation in humidity H, EMC and measured mean moisture content M in a sheltered outdoor location in Thessalonika, Greece (Tsoumis 1960)

Table 1.5. Estimated average moisture content of interior woodwork in 13 widely separated cities during January and July (Peck 1950)

City	Moisture content (%) interior woodwork	
	July (%)	January (%)
Atlanta, Georgia	11.5	8.5
Albuquerque, New Mexico	6.0	7.0
Boston, Massachusetts	13.0	7.0
Dallas, Texas	9.0	9.0
Duluth, Minnesota	10.5	5.0
Madison, Wisconsin	10.0	6.0
New Orleans, Louisiana	13.5	12.5
New York, New York	12.5	7.0
Portland, Oregon	9.5	9.0
Salt Lake City, Utah	4.0	7.0
San Francisco, California	10.5	10.5
Seattle, Washington	11.0	8.5
Washington, DC	11.0	8.0

Table 1.6. Recommendations for initial moisture content of lumber for dwellings (Peak 1950)

	Moisture content (%)					
	Dry south-western states		Damp southern states		Remainder of the US	
	Aver.	range	Aver.	range	Aver.	range
Interior finishing, woodwork, and flooring	6	4–9	11	8–13	8	5–10
Sheathing, framing siding, and exterior trim	9	7–12	12	9–14	12	9–14

ronmental conditions. These conditions vary geographically, even within the same country. For example, Table 1.5, taken from Peck (1950), shows the estimated average moisture content of interior woodwork in 13 widely separated cities in the United States during January and July, months which represent seasonal extremes. Table 1.6, also taken from Peck (1950), shows the recommended moisture content of lumber intended for either interior or exterior use in the three principal climatic regions of the United States.

1.6 Fiber-Saturation Point

The term *fiber-saturation point* (M_f) was originally applied to wood by Tiemann (1906). He noted in his studies on the effect of moisture content on the strength properties of wood that there appeared to be a moisture content below which the mechanical properties of wood increased with decreasing moisture content, but above which they were independent of moisture content. Tiemann considered that M_f is the moisture content at which the cell cavities contained no liquid water, but the *cell walls were fully saturated* with moisture (Fig. 1.27). He reasoned that the presence or absence of liquid water in the cell cavities had no effect on the strength properties of wood, since only the cell wall was effective for strength. Figure 1.28 illustrates the idealized behavior of several mechanical properties of wood as functions of moisture content, and the location of M_f.

Tiemann's original definition of the fiber-saturation point is elegant in its simplicity. However, it is difficult to measure M_f experimentally. One difficulty is that M_f may not be the same throughout a given wood sample. For example, Hart (1984) has shown that liquid water is retained more tenaciously in parenchyma than in prosenchyma cells, such that the latter may be below M_f while the cavities in adjoining parenchyma cells are still water-filled, particularly in parenchyma-rich hardwoods. Other difficulties in estimating M_f are discussed below.

Once the concept of a fiber-saturation point had been established, it was determined that moisture-dependent properties of wood other than strength also showed discontinuities in the vicinity of M_f. Myer and Rees (1926) explored the use of electrical resistance changes with moisture content to determine M_f, as noted in Chapter 6, finding a discontinuity in the *resistance-moisture content*

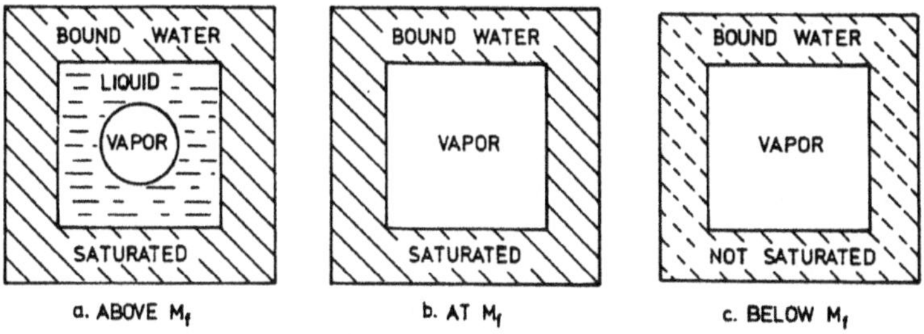

Fig. 1.27. Schematic diagram showing an idealized representation of the moisture distribution in a wood cell cross-section **a** above, **b** at, and **c** below M_f

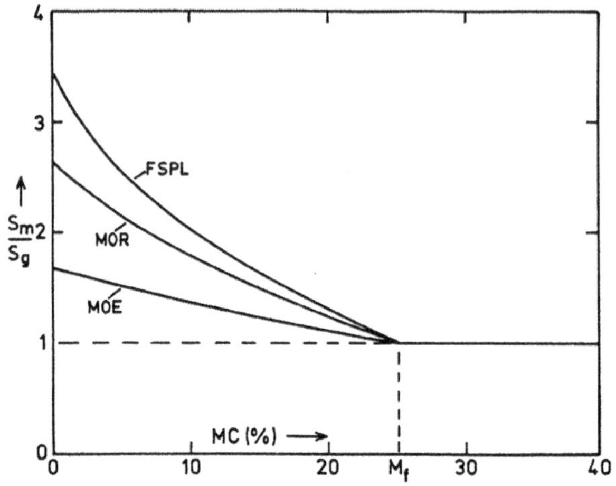

Fig. 1.28. Idealized curves showing the ratio S_m/S_g of strength at moisture content M (S_m) to its value for green wood (S_g) for defect-free wood. The properties shown are modulus of elasticity (*MOE*), modulus of rupture (*MOR*), and fiber stress at proportional limit (*FSPL*), all in flexural bending

relationship near M_f. Stamm (1929) perfected this method and also explored the *shrinkage-moisture content* technique (Chap. 4), as well as the method in which M_f is determined by extrapolating the *sorption isotherm* to unit vapor activity (Sect. 1.4.1). Hawley (1931) also discussed the fiber-saturation point and methods of estimating it.

A number of other methods, besides the four noted above, have been developed to estimate fiber saturation. Stamm (1971), for example, described nine different methods. Aroon Chomcharn (personal communication 1973) listed a total of 12 different methods, including those given by Stamm, which have been proposed for estimating M_f. Several of these are described briefly below.

The sorption-isotherm extrapolation technique is subject to error because it is based on the assumption that water will not condense in the cell cavities at vapor activities lower than unity. The Kelvin equation, however,

$$\ln(1/h) = 36\gamma/(rRT) \quad , \qquad (1.26)$$

where γ is the surface tension of the air-water interface and r is the capillary radius, predicts that condensation of water may occur in small capillaries when h is less than unity. The relationship given by Eq. (1.26) is approximated, at room temperature, by

$$H/100 = h \simeq 1.0 - 0.0011/r \quad , \qquad (1.27)$$

where r is the capillary radius in micrometers. Equation (1.27) applies for radii of 0.1 μm or *larger*. Thus it applies over the range of capillary sizes found in the microscopically visible capillaries in wood such as those of the cell cavities and pits. For example, the relative humidities H at equilibrium with capillary radii of 100, 10, 1, and 0.1 μm, are 99.999%, 99.99%, 99.9%, and 99.0%, respectively, at room temperatures.

The tapering ends of cells and the pits may have sufficiently small openings to condense water at humidities approaching 99%. Furthermore, the presence of water-soluble materials in the cavities may further reduce the humidities at which moisture may condense. This should be particularly true in parenchyma cells, whose cell contents probably include a large proportion of water solubles, thus enabling them to retain water at lower humidities than do prosenchyma cells, as Hart (1984) has reported.

Stamm (1971) has pointed out some of the problems in estimating M_f by this technique, as have Spalt (1958) and Wangaard and Granados (1967). These include capillary condensation possibilities, as noted above, but also the experimental difficulty in controlling humidity with the required degree of precision at high humidities. Barkas (1949) noted that there is no sharp discontinuity in the isotherm at high humidities, and that M_f therefore is not a point but a region of moisture contents.

The difficulty in maintaining good humidity control at high humidities has prompted use of the *porous pressure plated* method for estimating M_f. In this method the water in a thin water saturated sample is subjected to hydrostatic tension of sufficient magnitude to reduce its activity h to various values between unity and 0.99. These correspond to hydrostatic tension values from zero ($h = 1.0$) to approximately 14 atmospheres ($h = 0.99$) based on the swelling pressure equation [Eq. (2.74)]. A thin wood sample is exposed to the appropriate hydrostatic tension until equilibrium is attained, after which the sample is removed from the apparatus and weighed to determine its EMC.

Stone and Scallan (1967) applied this technique to microtomed sections of *Picea mariana*, 100 μm thick and 2 by 2 cm square, reporting the results shown in Fig. 1.29. Each of the four points shown with h of 0.99 or greater were obtained using the porous pressure plate technique. Those at lower activities were obtained by exposing samples to water vapor at the activities shown. They estimated M_f to be about 40%, corresponding to $h = 0.9975$, somewhat higher than the value obtained from the sorption isotherm extrapolation technique.

The porous pressure plate method has been used primarily to measure liquid water retention in textile and wood fiber masses in connection with water removal from such materials. Morton and Hearle (1975) described the principles and practices involved in using this technique.

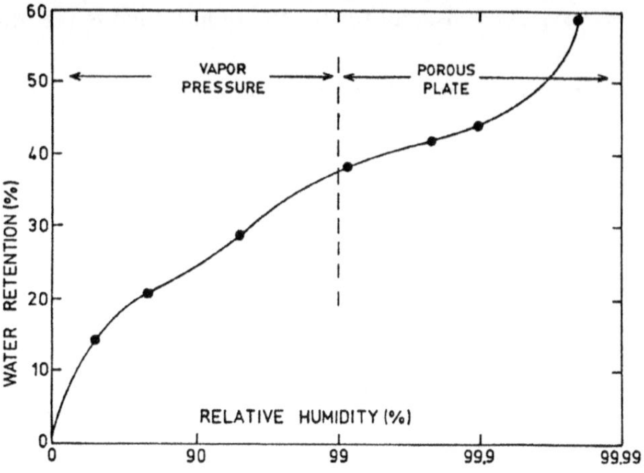

Fig. 1.29. Moisture sorption isotherm for *Picea mariana* including data obtained by standard vapor sorption techniques below, and by the porous plate technique above, 99% RH (after Stone and Scallan 1967)

Spinning water-soaked wood samples in a high speed *centrifuge* has been suggested as a means of removing capillary water from wood without affecting bound water. This is because the centrifugal forces are in the same order of magnitude as those retaining capillary water in the cell cavities, but much lower than those required to remove bound water from the cell wall. Thus, if soaked wood samples could be centrifuged until the capillary water was removed (in a vapor-saturated environment), the remaining water is all bound water at M_f, the same principle involved in the porous pressure plate method.

Perem (1954) used the centrifuge method to estimate M_f for sapwood of *Pinus resinosa* and *Picea glauca*. He measured mean residual moisture contents of 31.6%, 31.3%, and 31.1% for three green samples of *P. resinosa* centrifuged for 3 h using centrifugal forces equivalent to 11 atm for 3 h, 13 atm for 3 h, and 13 atm for 5 h, respectively, all at 20 °C ambient temperature. The residual moisture content was presumed to be equivalent to M_f. Ahlgren et al. (1972) also tested this method for estimating M_f.

The centrifuge method seems to be suitable only for relatively permeable woods. Moisture cannot be moved readily from one cell to another in impermeable woods, since it must then move through the cell walls or impermeable membranes. Similar considerations would seem to apply to the porous pressure plate method.

The *transmission of beta rays* through wood decreases as its moisture content increases, the rate of decrease being greater above than below M_f. Ecklund (1961) made such measurements on 3-mm-thick (along the grain) samples of *Quercus kelloggii* measured from the green to oven-dry condition, using beta radiation from a Strontium ninety source. Figure 1.30 shows plotted points of percent beta ray transmission against wood moisture content. The value of M_f (shown as F) was estimated from the intersection of the curve at high moisture contents with that at lower moisture contents. Values of M_f in the order of 30% were calculated based on his results, but the procedure is tedious, with several possible sources of errors.

Included in Figure 1.30 are curves of radial and tangential shrinkages and the corresponding shrinkage intersection points, R and T, obtained by Ecklund (1961). As is usually found, the radial value is the lower of the two.

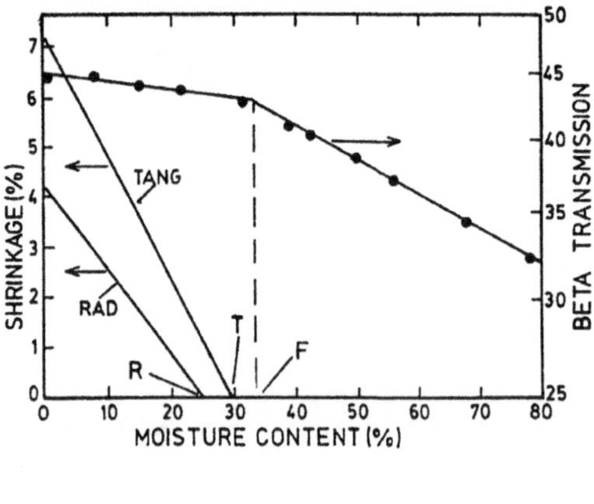

Fig. 1.30. Curves of radial and tangential shrinkages and of beta ray transmission through *Quercus kelloggii* plotted against wood moisture content (after Ecklund 1961)

The essential absence of a *heat of wetting* W or of a heat of sorption Q_s at moisture contents above M_f has been used to estimate the fiber-saturation point. This method is based on the assumption that heat is generated when water is taken up by wood from the liquid state only if the wood moisture content is below M_f (Chap. 2.3.3). Above this point there is presumed to be negligible energy of interaction between liquid water and that in the cell cavities. If either the heat of wetting W or the heat of sorption Q_s are measured as functions of wood moisture content, they approach zero at moisture contents close to M_f (Fig. 2.7). Stamm and Loughborough (1935) reported a value for M_f of 29% for *Picea sitchensis*, and Kelsey and Clarke (1956) of 31% for *Araucaria klinkii*, using this method.

A variation of the heat of wetting method is discussed in Chapter 2.3.3.2, in which the total heat of wetting W_0 appears to be related to M_f. For example, Kajita (1976) noted that W_0 was linearly related to M_f measured by the extrapolated sorption isotherm method [Eq. (2.60)] for ten Japanese woods.

Yoshimura (1961) attempted to estimate the fiber saturation points of ten tropical and two temperate zone woods based on their *rates of decay* by the brown-rot fungus, *Poria monticola.* He reported that the percent weight loss of these woods for a given exposure time to the fungus increased linearly with the initial wood moisture content at time of exposure. Extrapolation of the curve of initial moisture content against percent weight loss, to zero weight loss gave an estimation of M_f. Figure 1.31 shows such a curve obtained for *Carapa procera* (andiroba), a wood relatively nonresistant to decay. He estimated M_f to be 26.6% based on the curve, in reasonable agreement with a mean value of 27.9% based on radial and tangential shrinkage intersection points.

Three of the other woods tested also showed reasonably good agreement, giving values of 25.7% and 27.5%, 33.0% and 27.8%, and 26.6% and 26.5%, respectively, for M_f by the two methods. For the other eight woods, various factors, such as insufficient or too much weight loss during the test period, precluded meaningful estimates of M_f based on decay rate.

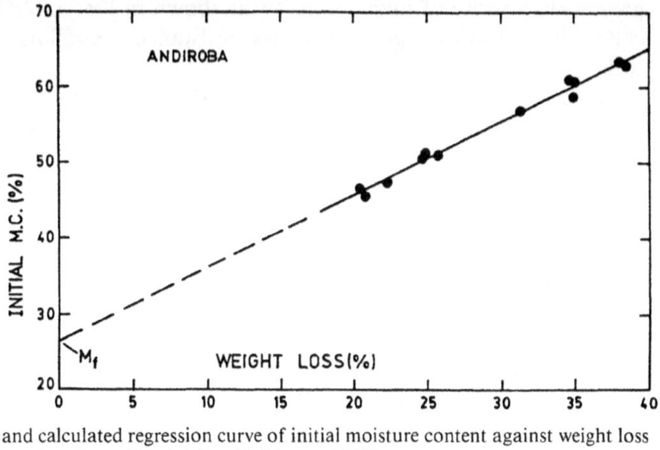

Fig. 1.31. Plotted points and calculated regression curve of initial moisture content against weight loss for andiroba wood subjected to decay fungi (after Yoshimura 1961)

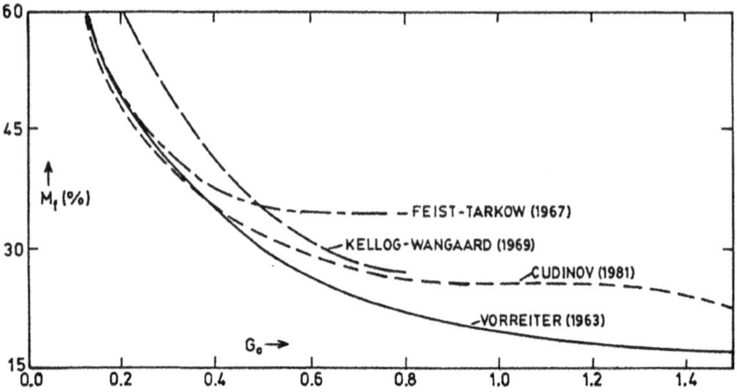

Fig. 1.32. Curves of apparent M_f against dry-volume specific gravity G_0, from various sources

Feist and Tarkow (1967) described a procedure for estimating M_f based on a *polymer exclusion principle.* In this method water-soaked wood samples are equilibrated in an aqueous solution of a large-molecular-weight polymer. They used polyethylene glycol of molecular weight 9000 (PEG-9000), the molecules of which are too large to penetrate the cell wall of wood. Thin sections, 0.5 mm along the grain, water-saturated after water extraction, were immersed individually in a large excess of a 3% solution of PEG-9000 and agitated at room temperature for 2 days. The concentration of PEG-9000 in the solution was then analyzed. From this concentration, together with the concentration in the original solution, as well as the initial wood moisture content, the amount of moisture in the cell walls at M_f was determined. It was assumed that the activity h of the PEG-9000 solution was essentially unity, and that the polymer could freely disperse in the cell cavities but could not enter the cell wall.

Feist and Tarkow found that the fiber saturation of low density woods was greater than those of higher density, as shown in Fig. 1.32. They speculated that cells with thin walls might provide less resistance to swelling than those with thick walls.

Figure 1.32 also includes curves of M_f against specific gravity obtained by Kellogg and Wangaard (1969), by Vorreiter (1963) and by Cudinov (1981). Feist and Tarkow (1967) suggest that the low values of M_f for high density woods as reported by Vorreiter (1963) may be due to the high extractive contents often characteristic of such woods (Sect. 1.4). The fourth curve shown in Fig. 1.32 is taken from Cudinov, who calculated the curve based on the amount of cell cavity surface as a function of wood density. He assumed that the total bound water consisted of two components. One of these, *adsorbed* water, he took to be constant. The secondary or *capillary condensed* water he took to be a function of the amount of cell-wall area exposed to the cell cavity, an inverse function of wood density. Therefore M_f, representing the sum of the two components, decreases with increasing wood density, as shown in Fig. 1.32. His model assumes no variation of cell-wall composition with wood density.

The reported values of M_f vary over a wide range. A few of these have been noted above under the discussion of methods of measurement, and other values as reported by various workers are summarized here. Stamm (1971) reported values for North American softwoods ranging from 21% to 32% based on the sorption isotherm extrapolation method for resorption on previously dried samples, at room temperature. For various hardwoods, most of which are tropical woods, the range was from 16% to 33%. He also noted that M_f decreases about 1% for each 10 °C increase in temperature between 0° and 100 °C. Stamm asserted that the polymer exclusion method (nonsolvent water content) used by Feist and Tarkow (1967) and also by Stone and Scallan (1967) gave the most reliable measurement of M_f. He cited values of M_f ranging from 26% to 30% based on the variation of each of 14 different kinds of mechanical properties of Sitka spruce as functions of wood moisture content, the kinds of tests from which Tiemann (1906) first defined M_f.

In the same paper, Stamm (1971) tabulated values of M_f obtained from electrical resistance measurements, for 12 North American woods. These ranged from 28% to 32%, except for one species, swamp oak. For this wood the apparent value of M_f was 67% during drying from the green condition, but after drying and resoaking M_f decreased to 32%. This discrepancy may be related to the phenomenon reported by Hart (1984) in which parenchyma cells in some hardwoods during initial drying from the green state retained water long after the prosenchyma cells were empty. In this case, if empty cells were in electrical series with full cells, the empty cells would largely determine the overall resistance and would tend to indicate that all of the wood was near M_f.

Vorreiter (1963) tabulated fiber-saturation points and dry densities of 40 different kinds of wood from a number of different sources. As is shown in Fig. 1.32, he noted that M_f decreased with increasing wood density. He also reported that the relationship was such that a plot of the logarithm of M_f against the

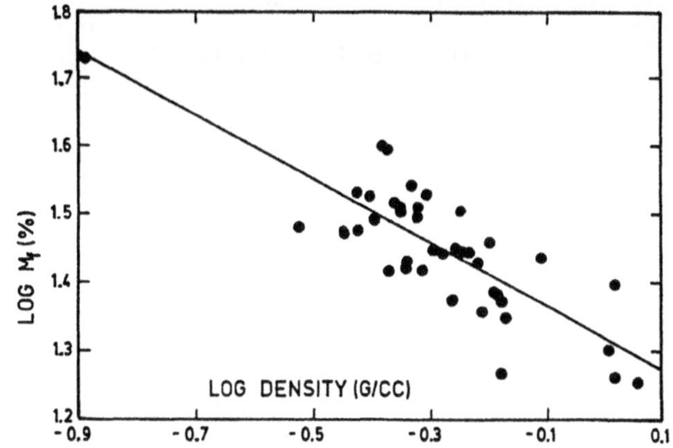

Fig. 1.33. Plotted points and calculated regression curve of log (M_f) against log (dry density) for 40 different kinds of wood, as measured by Vorreiter (1963)

logarithm of dry wood density was linear (Fig. 1.33). He proposed that the value of M_f extrapolated to the density of the cell wall is 16%.

Wangaard (1957) estimated M_f for 15 different woods, using both the strength-moisture content and radial and tangential shrinkage-intersection point methods. He modified the latter method so as to obtain values for M_f based on "stress-free" shrinkage in which the effect of *compression set* due to initial moisture gradients was minimized. The correlation between the two techniques was good when the mean value obtained from modulus of rupture and modulus of elasticity was plotted against the mean of the radial and tangential shrinkage intersection points. Wangaard also found a high degree of correlation between M_f and the ratios of air-dry to green strength and stiffness. He anticipated this based on the observation that tropical hardwoods, which have a lower M_f than those grown in temperate zones, also have a lower ratio of air-dry to green strength properties.

Kelsey (1956) measured radial and tangential shrinkage intersection points, as well as basic densities and moisture expansion coefficients, for 54 wood species from New Guinea and the Pacific Islands. She noted that because of nonlinearities in shrinkage at low and high humidities (Chap. 4.4.2) it is best to use data measured at moisture contents over the center of the hygroscopic range. She suggested that measurements be made between moisture contents at equilibrium with relative humidities of 65% and 30%, in calculating expansion coefficients and intersection points. For collapse-susceptible species, it is best to make measurements after removal of collapse. As is normally observed (Chap. 4.5.2), radial shrinkage intersection points, which ranged from 13.6 to 31.8 with a mean of 23.74% ± 3.98%, were consistently lower than values measured tangentially, which varied between 16.1 and 38.7 with a mean 27.68% ± 4.13%. In contrast to the curves shown in Fig. 1.32, Kelsey (1956) found little correlation between M_f and wood density.

1.7 Moisture Content and Strength of Wood

It has already been noted (Sect. 1.6) that the strength or mechanical properties of wood generally increase with decreasing moisture content below fiber saturation while above M_f there is no essential change with loss of moisture. In other words, loss of moisture from the cell walls results in an increase in the mechanical properties, while loss of capillary or "free" water from the cell cavities has no effect.

The mechanical properties of clear wood are usually measured in the laboratory for wood in the green and also in the air-dry condition. The air-dry moisture content obviously varies from one location to another, and also with the season of the year as noted in Section 1.5. In order to compare the strength properties of different woods in the air-dry condition, the measured properties are often "adjusted" to expected values at 12% moisture content, at least in North America. This has been done in the past based on the empirical observations that various mechanical properties increase with decreasing moisture content according to the *exponential equation*

Table 1.7. Mean values for the coefficient B in Eq. (1.28) for several mechanical properties of wood (after USDA Wood Handbook 1955)

Property	B
Static bending	
Fiber stress at proportional limit	0.05
Modulus of rupture	0.04
Modulus of elasticity	0.02
Work to proportional limit	0.08
Work to ultimate load	0.005
Axial compression	
Fiber stress at proportional limit	0.05
Ultimate stress	0.06
Transverse compression	
Stress at proportional limit	0.055
Transverse tension	
Ultimate stress	0.015

$$S_m = S_{12}\{\exp[B(12-M)]\} , \tag{1.28}$$

where S_{12} and S_m are the particular strength properties at moisture content M (%) and at 12%, respectively, and B is a coefficient representing essentially the *fractional increase* in the particular strength property per *percent decrease* in wood moisture content.

For small values of B the *compound interest* formula gives essentially the same results as Eq. (1.28). Thus

$$S_m \simeq S_{12}(1+B)^{(12-M)} . \tag{1.29}$$

Table 1.7 gives approximate mean values of the coefficient B for the mechanical properties of North American woods, based on extensive testing at the USDA Forest Products Laboratory. It should be emphasized that these represent values obtained on clear, straight-grained test samples and not on industrial grades of lumber which may contain various strength-reducing defects. Such defects may be intensified during drying, thus offsetting the increase in strength of the clear wood (Green et al. 1986).

However, for clear straight-grained wood the coefficients listed in Table 1.7 apply, at least in the hygroscopic moisture range above 5% or 6% moisture content. Below this moisture range some mechanical properties may increase less than predicted by Eqs. (1.28) and (1.29), and in some cases may actually decrease. Possibly this is because the wood becomes more rigid and stress concentrations occur which cannot be relieved by transfer of stress to adjacent portions of the wood when it is dry. This effect seems most pronounced for axial tensile strength as is indicated in Fig. 1.34, taken from Gerhards (1982), whereas it appears to be lacking for axial compressive strength. This difference may be related to the fact that stress concentrations may be more readily relieved in compression than in tension modes of failure. It may also be due to the smaller cross-sections used in tensile than in compression test specimens and the greater probability of stress concentrations therefore developing during tension testing. Bodig and Jayne

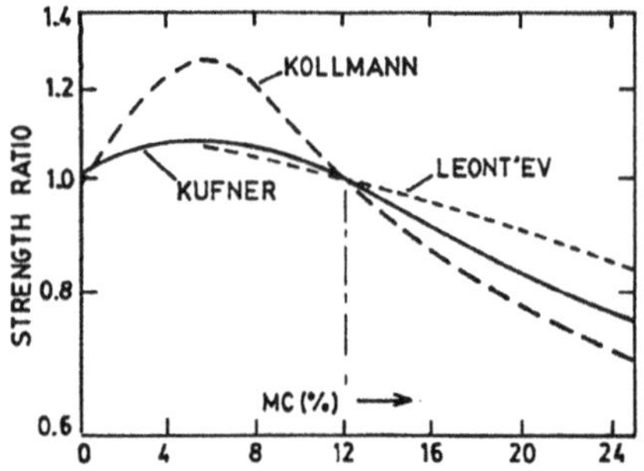

Fig. 1.34. Mean strength ratio (ratio of axial tensile strength at moisture content M to the strength at 12% MC) plotted against M, from three different sources, Kollmann (1940), Kufner (1978), and Leont'ev (1960), all based on ranges given by Gerhards (1982)

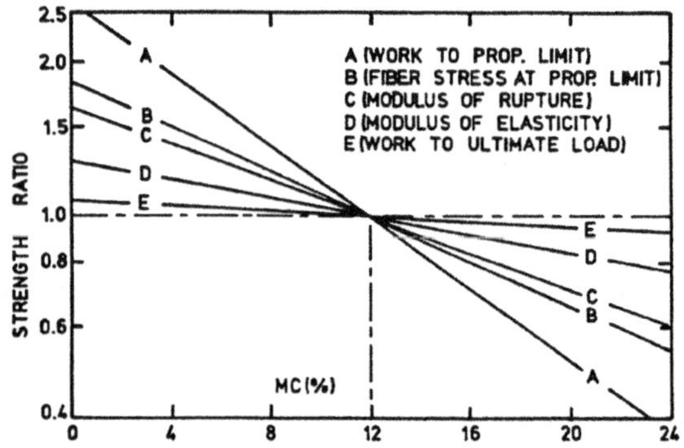

Fig. 1.35. Strength ratio (log scale) curves plotted against wood moisture content, for several flexural properties, relative to strength at 12% moisture content, for defect-free wood

(1982) suggest that cell-wall microfailures due to drying below 5% moisture causes the reduction in strength.

Figure 1.35 shows several curves, similar in format to those given by Gerhards (1982), in which the relative bending strength of wood relative to its value at 12% moisture content, is plotted against moisture content over the moisture range from 0% to 24%, for several flexural properties. The values of B used in each case are those given in Table 1.7 for static bending.

It is of interest to note, both in Table 1.7 and in Fig. 1.35, that the work W_p absorbed by wood in static bending up to the proportional limit is most sensitive to moisture content ($B_{wp} = 0.08$), while that absorbed to ultimate load or failure W_u is least sensitive to moisture ($B_{wu} = 0.005$). It is possible to calculate the coefficient B_{wp} if the coefficients for stress Y_p at proportional limit ($B_{sp} = 0.05$) and for modulus of elasticity E ($B_{ep} = 0.02$) are known. Thus, since $X_p = Y_p/E$, where X_p is the strain at proportional limit, and $W_p = Y_pX_p = Y_p^2/E$, it can be shown that

$$(1+B_{wp}) = (1+B_{sp})^2/(1+B_{ep}) \ . \tag{1.30}$$

According to Table 1.7, $B_{sp} = 0.05$ and $B_{ep} = 0.02$. Substituting these values into Eq. (1.30) yields $B_{wp} = 0.081$, which rounds off to 0.08, in agreement with the value shown in Table 1.7.

The values of B_{wu} cannot be so calculated since there is no proportionability between strain and stress at ultimate or failure load. However, the reason for its very low value is probably due to the greater strain at failure for more moist wood. Thus, although the stress at failure decreases with increasing moisture content, this is essentially compensated for by the increased strain associated with increased pliability of the wood.

Chapter 2

Moisture Sorption Thermodynamics

2.1 Introduction

The wood-water system has been treated by the classical techniques of thermodynamics although the system is not perfectly reversible in the thermodynamic sense. This irreversibility is primarily because of sorption hysteresis, already noted in the discussion of the sorption isotherms for wood. Morton and Hearle (1962, 1975) pointed out that, because hysteresis indicates the existence of states of metastable equilibrium during moisture sorption, application of thermodynamic equations to such systems should be treated cautiously. Nevertheless, it is useful to consider moisture sorption as a reversible process in order to calculate the approximate magnitudes of certain thermodynamic properties.

Before discussing the thermodynamic interactions of water with wood, it is convenient to consider first some of the thermodynamic properties of water itself.

2.2 Thermodynamic Properties of Water

The thermodynamic properties of water which are particularly affected by its interaction with wood are its enthalpy H, entropy S and Gibbs free energy G. The equation relating these parameters at the Kelvin temperature T is

$$H = G + TS \quad . \tag{2.1}$$

All three properties, H, G, and S, vary with temperature.

Equation (2.1) states that the total enthalpy H of a substance is equal to the sum of the Gibbs free energy G and the product of the Kelvin temperature T and entropy S. In the case of water it is possible to calculate the values of H and S at any given temperature relative to their values at zero Kelvin. The relative value of G can be found from Eq. (2.1) written as,

$$G = H - TS \quad . \tag{2.2}$$

The Gibbs free energy per mole of substance is usually expressed as the chemical potential, symbolized by μ. If H and S are also on a molar basis, Eq. (2.2) can be written

$$\mu = H - TS \quad . \tag{2.3}$$

The enthalpy H_i for ice at temperature T may be calculated from

$$H_i = H^0 + \int_0^T (C_p)_i \, dT \quad , \tag{2.4}$$

where H^0 is the molar enthalpy of water (ice) at zero Kelvin, and $(C_p)_i$ is the molar specific heat of ice at temperature T, measured at constant pressure P, generally taken as 1 atm. The molar entropy S_i for ice at temperature T, relative to the entropy S^0, at zero Kelvin, may be calculated from

$$S_i = S^0 + \int_0^T [(C_p)_i/T]\, dT \ . \tag{2.5}$$

The enthalpy H_w for liquid water is equal to that for ice at the melting point T_m, plus the molar heat of melting ΔH_m, plus the additional enthalpy associated with the temperature rise of the water from T_m to a higher temperature T. Thus

$$H_w = (H_i)_m + \Delta H_m + \int_{T_m}^{T} (C_p)_w\, dT \ . \tag{2.6}$$

The molar entropy S_w of liquid water similarly is given by

$$S_w = S_{im} + \Delta H_m/T_m + \int_{T_m}^{T} [(C_p)_w/T]\, dT \ . \tag{2.7}$$

The enthalpy H_v of water vapor above the boiling point T_b can also be calculated. For example, H_v is obtained by adding to $(H_w)_b$ (the value of H_w at T_b) the contributions to H_v due to the heat of vaporization ΔH_0 at T_b and the sensible heat required to raise the vapor temperature above T_b at constant atmospheric pressure. The equation for H_v at temperature T is

$$H_v = (H_w)_b + (\Delta H_0)_b + \int_{T_b}^{T} (C_p)_v\, dT \ , \tag{2.8}$$

where $(C_p)_v$ is the molar heat capacity of the vapor at constant pressure and temperature T. The equivalent equation for S_v is

$$S_v = (S_w)_b + (\Delta H_0)_b/T_b + \int_{T_b}^{T} [(C_p)_v/T]\, dT \ . \tag{2.9}$$

In order to evaluate the enthalpies and entropies in Eqs. (2.4) to (2.9), the molar heat capacity C_p at each temperature, as well as the molar heats of melting ΔH_m and of vaporization ΔH_0, must be known.

The molar heat capacity C_p, measured at constant pressure p, is negligibly different from C_v, the same property measured at constant volume, for ice or liquid water. However, C_p is considerably higher than C_v for water vapor and other gases. For ideal gases, for example, $C_p = C_v + R$, where R is the gas constant.

Figure 2.1 shows how the molar heat capacities C_p of ice and water vary with temperature T(K) at a pressure P of 1 atm. For ice it increases rapidly with temperature, from zero at T = zero (K) to near 9 cal/mol/K° at T_m (273.15 K), the melting point of ice. After melting $(C_p)_w$ approximately doubles to 18 cal/mol/K° for water, remaining nearly constant with temperature to T_b (373.15 K), the boiling point at 1 atm of pressure. At this temperature the value of $(C_p)_v$ for the vapor drops abruptly to 9 cal/mol/K° and decreases slightly with increase in temperature above T_b.

The molar heats of melting, ΔH_m, and of vaporization, ΔH_0, are 1.44 and 9.71 kcal/mol at the respective melting (T_m) and boiling (T_b) points for water.

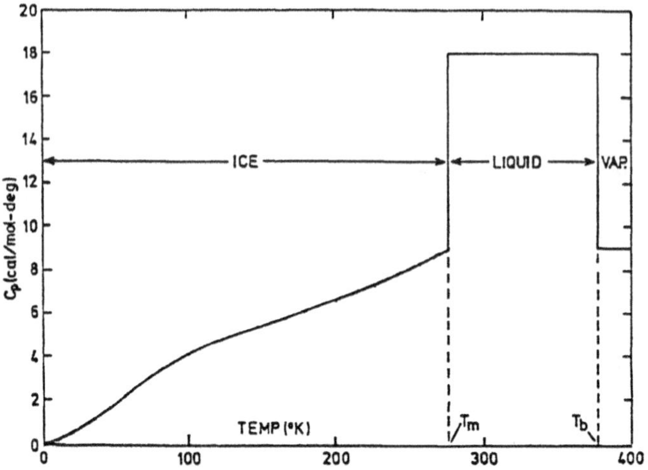

Fig. 2.1. Molar heat capacities (C_p) (cal/mol/K°) of ice, liquid water, and vapor, in relation to Kelvin temperature

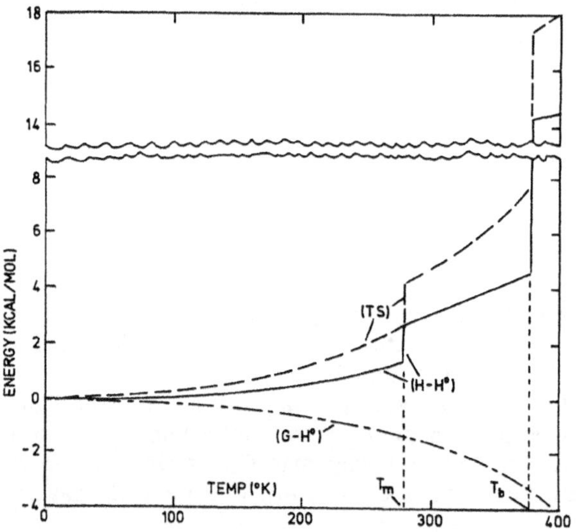

Fig. 2.2. Molar enthalpy (H-H^0), entropy (TS), and free energy (G-H^0) of ice, liquid water, and vapor, in relation to Kelvin temperature. H^0 is enthalpy at zero Kelvin

Figure 2.2 shows how the enthalpy difference, $H - H^0$, increases with temperature, where H^0 is the reference enthalpy, the value at zero Kelvin. The curves are calculated from Eqs. (2.4), (2.6), (2.8).

Figure 2.2 also shows how the entropy for water, expressed as the product TS, increases with temperature. In this case the residual entropy S^0 at zero Kelvin is taken to be zero, in conformity with the Third Law of Thermodynamics. This assumes that the ice is a perfect crystal at °K, although S^0 is actually 0.81 cal/mol/deg, based on statistical mechanical calculations, according to Eisenberg and Kauzmann (1969). The units of TS are the same (kcal/mol) as those of the enthalpy H.

The third curve shown in Fig. 2.2 is that for the molar free energy G, calculated as the difference between H and TS, based on Eq. (2.2). Actually $G-H^0$ is plotted as is the case for the enthalpy H. For example, below the melting point temperature

$$G_i-H^0=H_i-H^0-TS_i\ . \tag{2.10}$$

It is noteworthy that the free energy G decreases with increasing temperature, in contrast to both the enthalpy H and the entropy term TS, because TS increases more rapidly with temperature than does H.

The rate of change of molar free energy with temperature, dG/dT, or dμ/dT, is equal to the negative of the absolute enthalpy S. Thus, differentiating both sides of Eq. (2.2) with respect to temperature T,

$$dG/dT = dH/dT-T(dS/dT)-S$$

$$dG/dT = C_p-T(C_p/T)-S = -S\ , \tag{2.11}$$

where C_p is the molar heat capacity at temperature T.

The temperature region between the melting and normal boiling points of water is of particular interest, since wood is normally exposed to these temperatures during use, and in most cases, during drying. Over this range Eq. (2.6), which gives the enthalpy H_w (cal/mol) of liquid water, may be approximated by

$$H_w \simeq (H_w)_m+(\bar{C}_p)_w(T-T_m)\ , \tag{2.12}$$

where $(H_w)_m$ is sum of the enthalpy of ice $(H_i)_m$ and of the heat of melting (ΔH_m) at the melting point T_m. The values of $(H_i)_m$ and of ΔH_m are 1310 and 1440, and $(H_w)_m$ therefore is 2750 cal/mol. The mean value $(\bar{C}_p)_w$ is close to 18 cal/mol-degree, and Eq. (2.12) may be written

$$H_w \simeq H^0+2750+18(T-273.15)\ \text{(cal/mol)}\ . \tag{2.13}$$

Thus H_w increases linearly from H^0+2750 at 0 °C to H^0+4550 cal/mol at 100 °C, where H^0 is the value of H at zero Kelvin.

The entropy S_w of liquid water can be approximated similarly by use of Eq. (2.6), if $(C_p)_w$ is taken as constant. Thus

$$S_w \simeq (S_w)_m+(\bar{C}_p)_w \ln(T/T_m)\ . \tag{2.14}$$

The entropy $(S_w)_m$ at the melting point T_m is equal to the sum of the entropy of ice $(S_i)_m$ and the entropy change due to melting $(\Delta H_m/T_m)$. These are 9.86 plus 5.28, and therefore $(S_w)_m$ is 15.14 cal/mol/K, and Eq. (2.14) becomes

$$S_w \simeq 15.14+18\ln(T/273.15)\ \text{(cal/mol/K)} \tag{2.15}$$

An equation for the free energy G_w of liquid water is obtained by use of Eqs. (2.2), (2.13), and (2.15). Thus

$$G_w-H^0 \simeq 2750+18(T-273.15)-T[15.14+18\ln(T/273.15)]\ . \tag{2.16}$$

In Fig. 2.3 are plotted curves of H_w-H^0, TS_w and G_w-H^0, for liquid water over the temperature range from 0 °C (273.15 K) to 100 °C (373.15 K), calculated from Eqs. (2.13), (2.15), and (2.16).

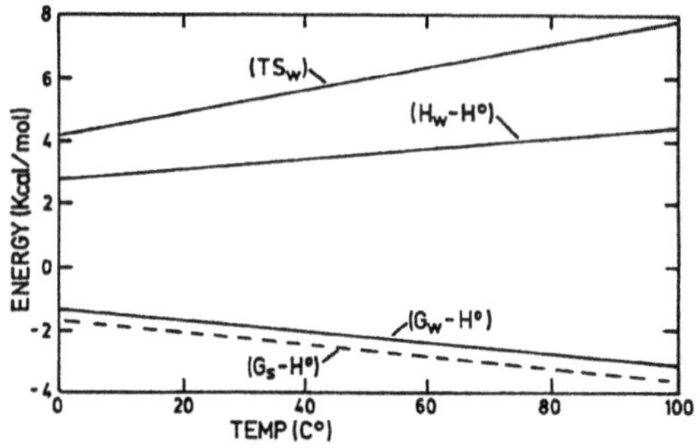

Fig. 2.3. Molar enthalpy (H_w-H^0), entropy (TS_w), and free energy (G_w-H^0) of liquid water, and the free energy of sorbed water (G_s-H^0) at activity of 0.5, in relation to Celsius temperature. H^0 is enthalpy at zero Kelvin

It is well known that sorbed water has a lower vapor pressure p than ordinary liquid water. Its activity h is therefore less than unity. Furthermore, its enthalpy H_s, entropy S_s, and molar free energy G_s or chemical potential μ_s, are all lower or more negative than those of liquid water at the same temperature.

At a given temperature, H_s is lower than H_w by the differential heat of sorption ΔH_s. That is,

$$H_s = H_w - \Delta H_s \ , \tag{2.17}$$

where ΔH_s is the energy (cal/mol water) released when 1 mol of liquid water is sorbed by wood at constant moisture content. The value of ΔH_s is a strong function of wood moisture content (Sect. 2.3.1).

The molar entropy S_s of sorbed water is also lower than S_w. It is given by

$$S_s = S_w - \Delta H_s/T + \Delta G_s/T \ , \tag{2.18}$$

where ΔG_s ($= \mu_w - \mu_s$) is the decrease in molar free energy of liquid water when it is sorbed by wood at a given constant moisture content under isothermal conditions. Since both ΔH_s and ΔG_s vary with wood moisture (Sect. 2.3.1), S_s also varies with this parameter.

Since the molar free energy G, or μ, is a function of both H and S [Eqs. (2.2), (2.3)], G_s or μ_s for sorbed water, obtained from Eqs. (2.17) and (2.18), is

$$G_s = \mu_s = H_s - TS_s = G_w - \Delta G_s \ . \tag{2.19}$$

The decrease in molar free energy ΔG_s associated with the sorption of liquid water by a hygroscopic material such as wood (Sect. 2.4) is given by

$$\Delta G_s = -RT\ln(h) \approx -RT\ln(p/p_0) \ , \tag{2.20}$$

where h is the activity of the sorbed water, essentially equal to the relative vapor pressure (p/p_0) at equilibrium with the sorbed water, and R is the gas constant. Eliminating ΔG_s between Eqs. (2.19) and (2.20) gives

$$G_s = G_w + RT\ln(h) \simeq G_w - RT\ln(p_0/p) \ , \tag{2.21}$$

where the reciprocal p_0/p of equilibrium relative vapor pressure is used. A curve of $G_s - H^0$ is plotted in Fig. 2.3 for wood with an activity of 0.5 for comparison with $G_w - H^0$. It is more negative than $G_w - H^0$ by the term $RT\ln(2)$.

Table 2.1. Values of vapor pressure (p_0) and heats of vaporization per mole (ΔH_0) and per gram (Q_0) at several temperatures, measured calorimetrically and calculated by means of Eqs. (2.22) and 2.25)

Temp. (°C)	p_0 (mmHg)	ΔH_0[a] (kcal mol^{-1})	ΔH_0[b] (kcal mol^{-1})	Q_0[a] (cal g^{-1})	Q_0[b] (cal g^{-1})
0	4.579	10.720	10.756	595.6	597.6
20	17.535	10.528	10.576	584.9	587.6
40	55.324	10.332	10.396	574.0	577.6
60	149.38	10.138	10.216	563.2	567.6
80	355.1	9.931	10.036	551.7	557.6
100	760.0	9.712	9.856	539.6	547.6
Mean	–	10.224	10.306	568.0	572.6

[a] Measured calorimetrically.
[b] Calculated by use of Eqs. (2.22), (2.25).

It has been shown (Fig. 2.2) that the enthalpy of water in each of its three phases, ice, liquid, and vapor, increases steadily with temperature. It also increases abruptly during phase changes, from ice to liquid water ΔH_m, and from liquid water to vapor ΔH_0. A third phase change occurs when ice sublimates directly into water vapor at temperatures below the freezing point. The molar enthalpy change associated with this phase change is the molar heat of sublimation, symbolized here by ΔH_u.

There are two experimental methods for measuring or calculating the enthalpy changes associated with the three phase changes cited above. One is the calorimetric method, in which the phase change energy is measured directly. For example, the molar heat of melting ΔH_m can be obtained by measuring the amount of heat required to melt a mol of ice at the melting point T_m.

A second method is an indirect method based on the *Clausius-Clapeyron* equation. This can be used in the following form to calculate ΔH_0, the molar heat of vaporization of liquid water at Kelvin temperature T

$$\Delta H_0 \approx -R[d\ln(p_0)/d(1/T)] \ , \qquad (2.22)$$

where R is the gas constant and p_0 is the vapor pressure of water at temperature T.

Table 2.1 gives the saturated vapor pressure p_0 of liquid water as a function of temperature from 0° to 100 °C at 20° intervals. An approximate empirical equation which fits these data in Eq. (1.1), repeated here for convenience

$$\ln(p_0) \approx 20.59 - 5205/T \ , \qquad (2.23)$$

where p_0 is given in millimeters of mercury. Based on Eq. (2.23), the slope $d\ln(p_0)/d(1/T)$ is $-5205°$ which, substituted into Eq. (2.22) yields a value for ΔH_0 of $(-1.987)(-5205)$ or 10342 cal/mol-water.

Table 2.1 includes values for ΔH_0 determined calorimetrically. These decrease from 10720 at 0 °C to 9712 cal/mol at 100 °C, with a mean value of 10224, about 1% lower than the mean value of 10342 calculated above.

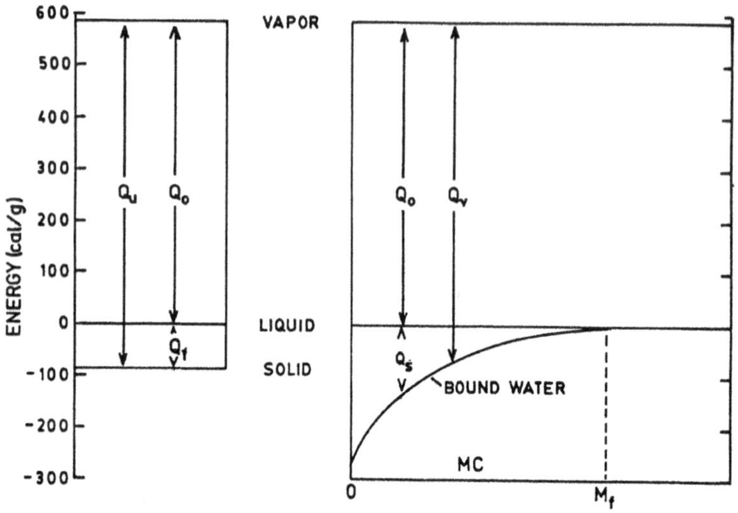

Fig. 2.4. Relative energy levels of water vapor, liquid water, ice and bound water in wood as functions of wood moisture content M, in terms of Q (cal/g)

A better fit to the vapor pressure data listed in Table 2.1 is given by the Kirchoff equation [see Eq. (1.2)]

$$\ln(p_0) \simeq 51.29 - 6651/T - 4.531\ln(T) \ . \tag{2.24}$$

In this case the slope $d\ln(p_0)/d(1/T)$ is given by

$$d\ln(p_0)/d(1/T) \simeq -6651 + 4.531(T) \ . \tag{2.25}$$

The calculated values of ΔH_0 using Eq. (2.25) together with Eq. (2.22) are included in Table 2.1, for comparison with the values obtained calorimetrically. It is apparent that the calculated values of ΔH_0 decrease with increasing temperature to about the same extent as the values measured calorimetrically, but are from 0.3% to 1.5% higher.

Also included in Table 2.1 are values of Q_0, the heat of vaporization per gram of water, defined as $Q_0 = \Delta H_0/18$. It is sometimes more convenient to use a mass rather than a molar basis, as is shown, for example, in Fig. 2.4.

The Clausius-Clapeyron equation can also be used to calculate the molar heat of sublimation ΔH_u if the saturated vapor pressure p_i of ice is known as a function of temperature. Table 2.2 shows the vapor pressure p_i at several subfreezing temperature. Also shown is the vapor pressure p_0 of super-cooled water (water cooled slowly and carefully below T_m to prevent nucleation and resultant ice formation). To calculate ΔH_u, the Clausius-Clapeyron equation can be written

$$\Delta H_u \approx -R[d\ln(p_i)/d(1/T)] \ . \tag{2.26}$$

The slope $d\ln(p_i)/d(1/T)$ for ice, based on the data over the temperature range from 0° to −40°C, is approximately

$$d\ln(p_i)/d(1/T) \approx -6290 + 0.5826(T) \ , \tag{2.27}$$

Table 2.2. Vapor pressures of ice and of supercooled water

Temperature (°C)	p_0 (mmHg)	p_i (mmHg)	p_i/p_0 (ratio)
0	4.579	4.579	1.000
−5	3.163	3.013	0.9526
−10	2.149	1.950	0.9074
−15	1.436	1.241	0.8642
−20	0.9406	0.776	0.8250
−25	0.6053	0.476	0.7864
−30	0.3816	0.2859	0.7492
−35	0.2354	0.1675	0.7116
−40	0.1418	0.0966	0.6812

which, evaluated at $T_m = 273.15$ K, gives a slope of −6131° at the melting point. Substitution of this value into Eq. (2.26) results in $\Delta H_u \approx 12182$ cal/mol of water, and Q_u, defined as $\Delta H_u/18$, is 676.8 cal/g of water at 0 °C. Comparison of these values with ΔH_0 (10756) and Q_0 (597.6), shown in Table 2.1 for liquid water at 0 °C, indicates that the heat of sublimation of ice is substantially greater than the heat of vaporization of liquid water at the melting point, as calculated from the Clausius-Clapeyron equation. The difference (12182 − 10756 = 1426 cal/mol) between ΔH_u and ΔH_0 should be equal to ΔH_m (1385 cal/mol), the heat of melting determined calorimetrically. The difference $\Delta H_u - \Delta H_0$ is approximately 2% higher than ΔH_m in this case.

Equations (2.22) and (2.26) can be combined to give the following equation from which the heat of melting ΔH_m can be calculated directly. Thus

$$\Delta H_m = \Delta H_u - \Delta H_0 = -R\,[d\ln(p_i/p_0)/d(1/T)] \tag{2.28}$$

$$\Delta H_m = +R\,[d\ln(p_0/p_i)/d(1/T)]\ . \tag{2.29}$$

The ratio p_i/p_0 is included in Table 2.2 for each temperature shown. It can be used directly to calculate ΔH_m by use of Eq. (2.28). It is analogous to the relative vapor pressure p/p_0 or activity of water sorbed in the cell wall of wood. The increase of p/p_0 with temperature can also be used in the Clausius-Clapeyron equation to estimate the differential heat of sorption of water by wood.

2.3 Enthalpy Changes During Moisture Sorption

2.3.1 Differential Heat of Sorption

It has been shown that ordinary water exists in three basic forms, solid, liquid and vapor, and that the enthalpy increases during transitions from the solid through the vapor phases. These three phases can coexist in equilibrium only at the triple point (Fig. 1.1).

Moisture in wood can be found in each of these three phases, as well as in a fourth "phase", found only in the cell wall. This is the sorbed water, also referred to as "hygroscopic" or "bound" water. At temperatures above the melting point T_m, only three phases may coexist, sorbed water in the cell walls, capillary water (sometimes referred to as "free" water) in the cell cavities at moisture contents above fiber-saturation, and water vapor in the cell cavities, present at all moisture contents except in fully water-saturated wood. At temperatures slightly below T_m and lower, water is frozen but the other two phases are present.

Sorbed water in the cell wall of wood is analogous to the frozen or solid state of ordinary water in that it has a lower enthalpy than liquid water. However, its enthalpy increases with increasing wood moisture content up to fiber-saturation, above which it is essentially the same as that of liquid water (Fig. 2.4).

The enthalpy of capillary water in the cell cavities is slightly lower than that of ordinary liquid water, for one or both of two reasons. First, the vapor pressure (p_c) of capillary water is lower than (p_0) that of ordinary liquid water because of the *curvature* of the air-water meniscus in a capillary. Second, the presence of *water-soluble extractives* in the capillary water further reduces the vapor pressure.

The lowering of the vapor pressure due to capillary curvature is given by the Kelvin equation which may be written

$$\ln(p_0/p_c) = (2\gamma/r)(\bar{V}/RT) \ , \tag{2.30}$$

where γ (dyne/cm) is the surface tension of the air-water interface, r (cm) is the capillary radius, $\bar{V}$ is the molar volume (18 cc/mol for water), and R is the gas constant in ergs/mol/K. Equation (2.30) is valid for radii as small as approximately 0.01 μm, well below the radius of the smallest cell cavity in wood. This corresponds to a relative vapor pressure p_c/p_0, or activity h, of approximately 0.90 at room temperature, based on Eq. (2.30).

The values of relative vapor pressure p_c/p_0, or activity h, calculated from Eq. (2.30), are approximately 0.9999, 0.999, 0.99, and 0.9 for capillary radii of 10, 1, 0.1, and 0.01 μm, respectively, near room temperature where the surface tension γ is approximately 72 dynes/cm. The cell cavities of wood cells are generally from 1 to perhaps 100 or more μm in radius. The tapering ends may approach 0.1 μm (10^{-5} cm), corresponding to an activity of about 0.99, essentially the same as that for a flat water surface. Therefore, as wood takes up moisture into the cell cavities, one would not anticipate a measurable change in its activity due to filling of the cell cavities, because of the difficulty in measuring such small changes in activity. It is then necessary to use other techniques such as those used to measure the water potential directly by the pressure-plate method as described by Siau (1984) (see also Chap. 1.6).

The enthalpy change ΔH_c associated with capillary sorption can be calculated by use of the Clausius-Clapeyron equation applied to the Kelvin equation. Thus, from Eq. (2.30)

$$\Delta H_c \simeq +R\,[d(\ln(p_0/p_c))/d(1/T)] \tag{2.31}$$

$$\Delta H_c \simeq (2\bar{V}/r)\,[\gamma - T(d\gamma(dT)] \ , \tag{2.32}$$

assuming $\bar{V}$ and r are constant with temperature.

Table 2.3. Calculated values of ΔH_c, ΔG_c, and $T\Delta S_c$, all in calories per mole of water, at several temperatures for a capillary of one micrometer in radius

Temp (°C)	ΔH_c	ΔG_c	$T\Delta S_c$
0	0.98	0.65	0.33
20	1.01	0.63	0.38
40	1.03	0.60	0.43
60	1.06	0.57	0.49
80	1.09	0.54	0.55
100	1.13	0.51	0.62

The values of the surface tension of water are 75.6, 72.75, 69.56, 66.18, 62.6, and 58.9 dynes/cm at temperatures of 0°, 20°, 40°, 60°, 80°, and 100°C, respectively. An empirical equation which fits these data is

$$\gamma \simeq 75.6 - 14.1(\theta/100) - 2.6(\theta/100)^2 \,, \qquad (2.33)$$

where θ is in °C. The rate of change of γ with temperature is given by

$$d\gamma/d\theta = d\gamma/dT \simeq -0.141 - 0.052(\theta/100) \,. \qquad (2.34)$$

Substitution of Eqs. (2.33) and (2.34) into Eq. (2.32) permits evaluation of ΔH_c as a function of temperature T for various values of the capillary radius r. The values of ΔH_c calculated from these equations are in units of ergs/mol. These have been multiplied by 0.2389×10^{-7} cal/erg to obtain the values of ΔH_c (cal/mol) shown in Table 2.3, for r = 1 µm, at each temperature. These are negligibly small compared with the change in enthalpy associated with the sorption of water into the cell wall itself.

Also shown in Table 2.3 are the changes in free energy ΔG_c and the entropy $T\Delta S_c$ in each case. The free energy change was calculated from Eq. (2.20) and the entropy change was taken as the difference between the enthalpy and free energy changes. The corresponding values of ΔH_c, ΔG_c, and $T\Delta S_c$ for a capillary radius of other than 1 µm can be obtained simply by dividing the values shown by that of the radius desired, in µm.

There are two distinct methods of determining the differential heat of sorption. The *isosteric* method, based on the Clausius-Clapeyron equation, is similar to that already described for calculating ΔH_m. It requires sorption isotherms at two or more temperatures. The second is the *calorimetric* method, analogous to that used for measuring the heat of melting directly by calorimetry. The isosteric method will be considered first, since it is an extension of the method just described for calculating ΔH_m.

The vapor pressure p of sorbed water is lower than that of liquid water p_0, since its activity h (= p/p_0) is less than unity. Furthermore it also increases with temperature at constant wood moisture content as Fig. 2.5 shows. Therefore, if the vapor pressure p at a given wood moisture content is known at two or more temperatures, the Clausius-Clapeyron equation can be used to estimate ΔH_v, the heat of vaporization of sorbed water. Thus

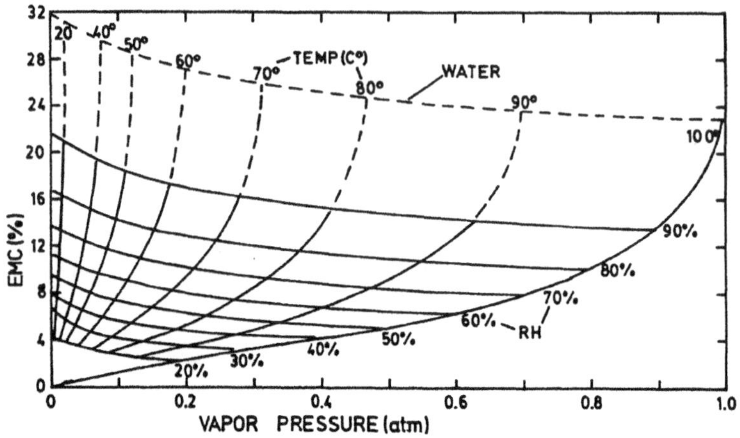

Fig. 2.5. Curves of equilibrium moisture contents as functions of vapor pressure p for wood in relation to temperature and relative humidity. (After Hawley 1931)

$$\Delta H_v \approx -R\,[d\ln(p)/d(1/T)]_m \ , \tag{2.35}$$

provided ln(p) is a linear function of 1/T, as Stamm and Loughborough (1935) first established. They also calculated ΔH_v for wood as a function of wood moisture content, using Eq. (2.35).

The magnitude of ΔH_v is always greater than that of ΔH_0, the heat of vaporization of liquid water, the difference defined as ΔH_s, the differential molar heat of sorption of liquid water by wood. Therefore, ΔH_s is the additional heat energy over and above the molar heat of vaporization ΔH_0 of free water which must be supplied to evaporate 1 mol of sorbed water. The enthalpy H_s of sorbed water is therefore lower than that H_w of free water by ΔH_s, that is

$$\Delta H_s = \Delta H_v - \Delta H_0 = H_v - H_s - (H_v - H_w) \tag{2.36}$$

$$H_s = H_w - \Delta H_s \ , \tag{2.37}$$

Combining Eqs. (2.22) and (2.35) with Eq. (2.36) gives

$$\Delta H_s \simeq +R\,[d\ln(p_0/p)/d(1/T)]_m \ , \tag{2.38}$$

analogous to Eq. (2.29) for calculating the heat of melting ΔH_m of ice.

The Clausius-Clapeyron equation can be written in alternative forms to those used in Eqs. (2.22), (2.26), etc. For example, Eq. (2.38) can be written

$$\Delta H_s \simeq RT^2\,[d\ln(p/p_0)/dT] \ , \tag{2.39}$$

which, if ΔH_s is constant between temperatures T_1 and T_2, can be written

$$\Delta H_s \approx RT_1T_2\,[\ln(h_2/h_1)/(T_2-T_1)]_m \ , \tag{2.40}$$

where h_1 and h_2 are the activities or relative vapor pressures of the wood at Kelvin temperatures T_1 and T_2, respectively, at a constant moisture content.

Equation (2.40) can be used to calculate the value of ΔH_s at a wood moisture content of 10%, based on the two sorption isotherms for wood shown in Fig. 2.6,

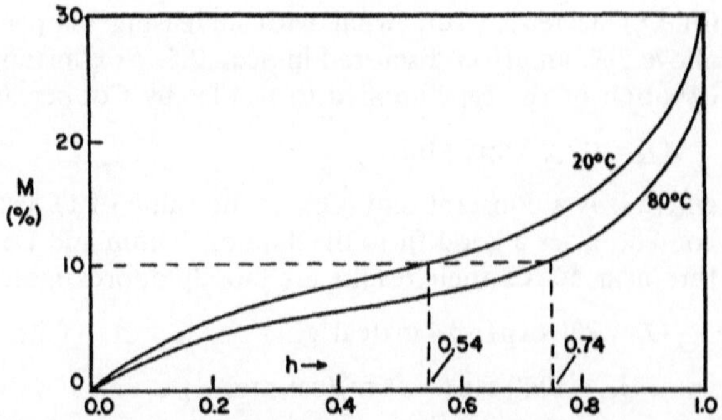

Fig. 2.6. Typical sorption isotherms for wood at 20 °C and 80 °C, showing increase in activity h with increase in temperature for wood at constant moisture content

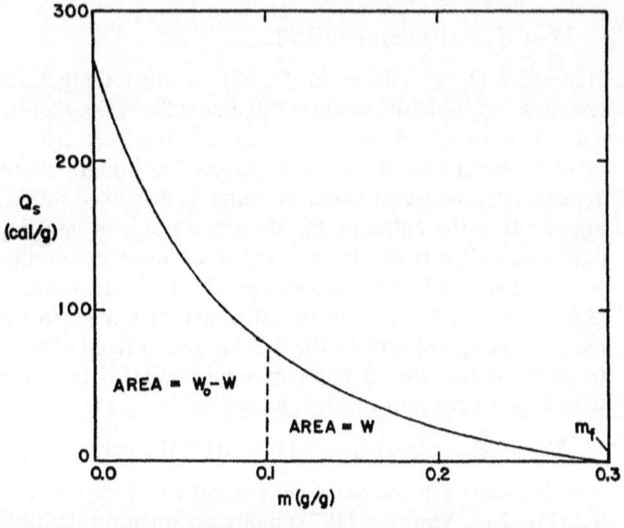

Fig. 2.7. Typical curve of differential heat of sorption Q_s in relation to wood moisture content m. The area under the curve to the left of m = 0.1 is the integral heat of sorption (W_0-W). The area to the right of m = 0.1 is equal to the heat of wetting W, and the total area is equal to the total heat of wetting W_0

at temperatures T_1 and T_2 of 293.15 and 353.15 K, respectively. The corresponding activities h_1 and h_2 are 0.54 and 0.74, which, substituted into Eq. (2.40), with R = 1.987 cal/mol, give $\Delta H_s \simeq 1.987 (293.15) \quad (353.15) [\ln (0.74/0.54)]/60 = 1080$ (cal/mol). Therefore Q_s, defined as $\Delta H_s/18$, is 60 cal/g water at 10% moisture content. Similar calculations made at other moisture contents may be used to obtain a curve of Q_s as a function of wood moisture content, resembling that shown in Fig. 2.7.

Stamm and Loughborough (1935) first applied the Clausius-Clapeyron equation to calculate ΔH_s or Q_s for wood. Using the sorption isotherms for the wood of Sitka spruce over the temperature range from 20° to 100 °C, they obtained curves similar to that shown in Fig. 2.7. They noted that the magnitude of ΔH_s

(or Q_s) decreased somewhat with increasing temperature at moisture contents above 5%, an effect discussed in Sect. 2.5. At constant temperature, an empirical equation of the type applied to textiles by Cooper and Ashpole (1959)

$$Q_s = (Q_s)_0 \exp(-Bm) \ , \tag{2.41}$$

where B is a constant and $(Q_s)_0$ is the value of Q_s extrapolated to zero moisture content, gives a good fit to the data of Stamm and Loughborough. At a temperature near 50 °C, their results are closely approximated by

$$Q_s = 280 \exp(-14\,m)(\text{cal g}^{-1}) \tag{2.42}$$

$$\Delta H_s = 5040 \exp(-14\,m)(\text{cal mol}^{-1}) \ . \tag{2.43}$$

The enthalpy H_s of sorbed water, as defined by Eq. (2.37), can be evaluated by use of Eq. (2.43). Thus

$$H_s = H_w - 5040 \exp(-14\,m) \ . \tag{2.44}$$

Figure 2.8 shows curves of $H_s - H^0$ against temperature for sorbed water at several wood moisture content calculated from Eq. (2.44), using Eq. (2.13) to evaluate the enthalpy H_w of liquid water. It has been assumed for convenience that ΔH_s is independent of temperature, although there is some temperature dependence, the significance of which is discussed in Sect. 2.5. It is evident from Fig. 2.8 that the enthalpy H_s of sorbed water is significantly lower than that of liquid water, particularly at lower wood moisture contents.

The entropy S_s and free energy G_s of sorbed water relative to those (S_w and G_w) of free water can also be calculated at any given wood moisture content by use of Eqs. (2.18) and (2.19) if ΔH_s and ΔG_s are known. For example, the entropy S_s of the sorbed water may be evaluated by use of Eq. (2.18), combined with Eqs. (2.20) and (2.43), giving

$$TS_s = TS_w - 5040 \exp(-14\,m) - RT \ln(h)(\text{cal mol}^{-1}) \ . \tag{2.45}$$

The activity h increases with wood moisture content and temperature above T_m (Fig. 2.6). Simpson (1973) analyzed sorption data of the US Forest Products Laboratory Wood Handbook (1955) over the temperature range from 0° to 100 °C, using several sorption models. For purposes of calculating the activity or relative vapor pressure of wood as a function of its moisture content and temperature, the Bradley model (Chap. 3.2.6) is the most convenient to use here. Based

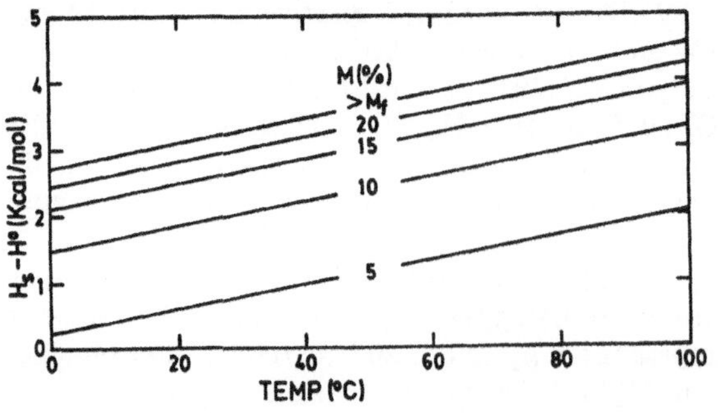

Fig. 2.8. Curves showing enthalpy (H_s-H^0) of sorbed water in relation to temperature for wood at different moisture contents

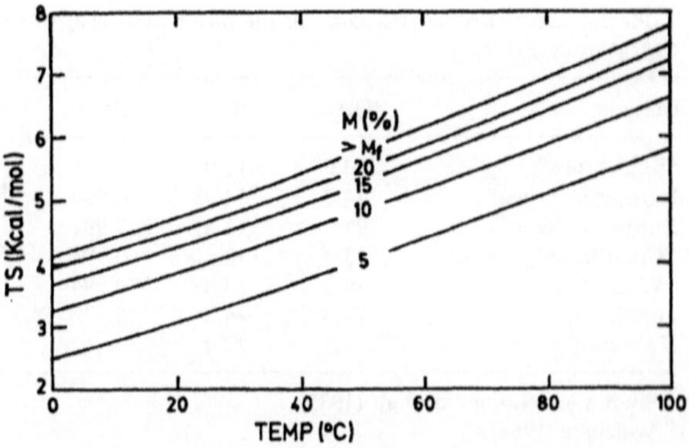

Fig. 2.9. Calculated curves of the entropy term TS_s in relation to temperature for sorbed water at several different wood moisture contents

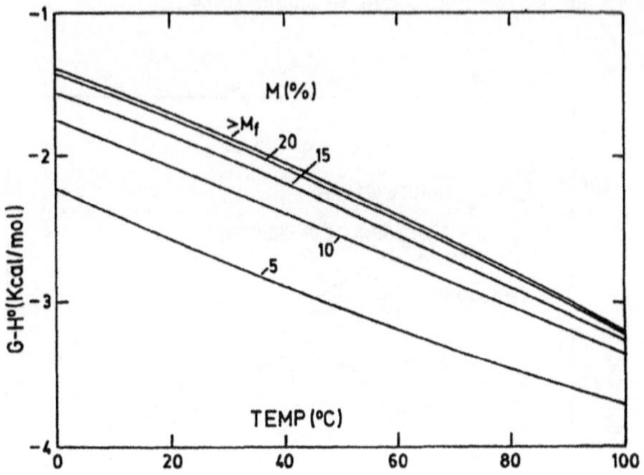

Fig. 2.10. Free energy G_s-H^0 of sorbed water as a function of temperature for several different wood moisture contents

on Simpson's equations giving the three Bradley constants as functions of temperature [Eqs. (3.95)–(3.97)], the activity h of the sorbed water as a function of temperature at 5% moisture content intervals was used to evaluate the term RTln(h) in Eq. (2.45). The entropy S_w of liquid water was evaluated by use of Eq. (2.15). Figure 2.9 shows the calculated curves of TS_s (cal/mol) over the temperature range from 0° to 100 °C for several constant wood moisture contents. It is clear that the entropy S_s of the sorbed water decreases with decrease in wood moisture content at any given temperature.

The free energy G_s of the sorbed water also decreases with decreasing wood moisture content as Eq. (2.21) indicates, because the activity h decreases. Curves of $G_s - H^0$ as functions of temperature are plotted in Fig. 2.10 for the same wood moisture contents shown in Figs. 2.8 and 2.9. Similar curves plotted for wood at one or more constant activities would have somewhat steeper slopes than those at constant wood moisture content because the activity at constant wood moisture content increases with increasing temperature over the temperature range shown.

Weichert (1963) used a vacuum sorption apparatus to measure the sorption isotherms of spruce, beech and "compressed" beech at several temperatures.

Table 2.4. Regression coefficients for the differential heat of sorption Q_s (cal/g) using Eq. (2.41)

Material	$(Q_s)_0$	B	R^2	Reference
Sitka spruce[a]	280.0	14.0	–	a
European spruce[b]	274.0	14.48	0.998	b
European beech[b]	194.2	13.37	0.991	b
"Compressed" beech[b]	283.1	18.86	0.998	b
"Wood fiber"[c]	220.0	13.5	0.994	c
Hinoki cypress[d]	271.4	14.2	0.972	d
Unknown wood[e]	213.7	12.4	–	e

[a] Stamm and Loughborough (1935).
[b] Weichert (1963).
[c] Babbitt (1942), based on data of Katz (1917).
[d] Kajita (1976).
[e] Volbehr (1896), according to Stamm (1964).

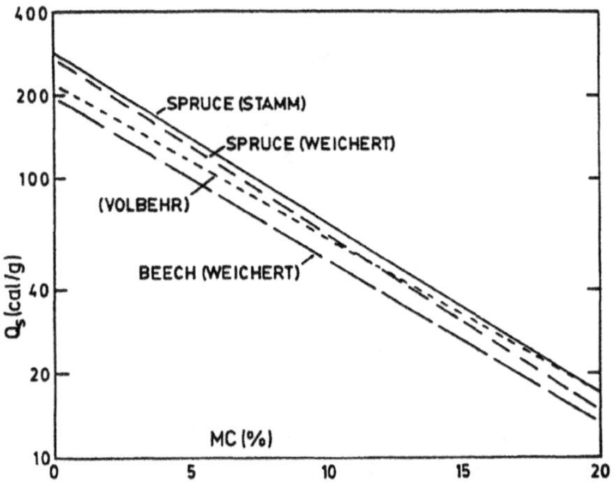

Fig. 2.11. Curves of Q_s (log scale) as functions of wood moisture content obtained from several different sources

Using the Clausius-Clapeyron equation, he calculated Q_s as a function of wood moisture content at 62.5 °C for each wood, with the results shown in Table 2.4. Also shown in the table are the values of $(Q_s)_0$ and B obtained by fitting the data to Eq. (2.41) by a least squares approximation. Figure 2.11 shows plots of Q_s (log scale) against wood moisture content based on Weichert's results, as well as those of Stamm and Loughborough (1935), and of Volbehr (1896).

2.3.2 Heat of Wetting and Integral Heat of Sorption

It is difficult to measure the differential heat of sorption directly by calorimetric means, since this requires the addition of an extremely small amount of liquid water distributed uniformly throughout a finite amount of wood such that its moisture content remains essentially constant. It would be difficult experimentally to add even a small finite increment of moisture to wood, say enough to change its moisture content uniformly by 1%, and measure the heat generated calorimetrically.

Although it is impractical to measure Q_s directly by calorimetry, another related quantity, the heat of wetting, W, can be so measured. It is defined as the heat generated per gram of dry wood when wood at some initial uniform moisture content m is soaked in an excess of liquid water sufficient to wet it well above the fiber-saturation point. The heat of wetting for a given wood sample decreases as its original moisture content increases.

The calorimetric method for measuring the heat of wetting W requires that the wood first be ground into small particles in order to minimize the time required to thoroughly wet all of the wood. Prior to immersing the small wood particles in water, they are equilibrated to some constant moisture content m, and to the initial calorimeter temperature. The wood particles are then immersed in an excess of water, previously introduced into and equilibrated to the calorimeter. The mixture or slurry of wood particles is stirred so as to obtain an uniform moisture and temperature distribution. The heat generated by the reaction is calculated from the temperature rise of the calorimeter and the heat capacities of its various components. The heat generated, divided by the oven-dry weight of the wood, is defined as the heat of wetting W for the wood. It has dimensions of energy per unit *dry mass of the wood*, or calories per gram of oven-dry wood for the units used here.

The heat of wetting for wood which is initially at zero moisture content is the *total* heat of wetting, symbolized here by W_0. The difference between W_0 and W for wood at some higher initial moisture content m, is defined as the *integral heat of sorption*, $W_0 - W$, also having units of calories per gram of dry wood. The terms W and W_0 are related to this differential heat of sorption Q_s as follows, where m_s is the saturation moisture content,

$$W = \int_{m}^{m_s} Q_s \, dm \tag{2.46}$$

$$W_0 = \int_{0}^{m_s} Q_s \, dm \tag{2.47}$$

$$W_0 - W = \int_{0}^{m} Q_s \, dm \; . \tag{2.48}$$

The relationships given by Eqs. (2.46)–(2.48) are shown graphically in Fig. 2.7. For example, the value of W_0, the total heat of wetting, is equal to the total area under the curve. The value of W for an initial moisture content m of 0.1 g/g is equal to the area under the curve to the right of m = 0.1 g/g. Finally, the integral heat of sorption, $W_0 - W$, is the area under the curve between m = 0 and 0.1.

If the heat of wetting is known as a function of moisture content, the differential heat of sorption Q_s can be obtained at any moisture content. Thus Q_s at moisture content m is given by

$$Q_s = d(W_0 - W)/dm = -dW/dm \ , \tag{2.49}$$

since W_0 is a constant.

If Q_s is a simple exponential function of moisture content as given by Eq. (2.41), the relationship between W and Q_s is particularly simple. It can be shown from Eqs. (2.41) and (2.46) that

$$W = W_0 \exp(-Bm) \tag{2.50}$$

$$\ln(W) = \ln(W_0) - Bm = A - Bm \ , \tag{2.51}$$

where $A = \ln(W_0)$, $W_0 = (Q_s)_0/B$, and

$$W = Q_s/B \text{ or } Q_s = BW \ . \tag{2.52}$$

It is also convenient to write Eq. (2.41) in the form

$$\ln(Q_s) = \ln(Q_s)_0 - Bm = C - Bm \ , \tag{2.53}$$

where $C = \ln(Q_s)_0$. Therefore, the empirical constants A, B, and C are related by

$$\ln(B) = C - A \ . \tag{2.54}$$

The earliest heat-of-wetting measurements recorded for wood are those of Volbehr in 1896, according to Stamm (1964). Volbehr's data, fitted to Eq. (2.50) result in

$$W = 17.1 \exp(-12.5\,m) \text{ (cal/g wood)} \ . \tag{2.55}$$

A prediction equation for the differential heat of sorption Q_s based on Eqs. (2.52) and (2.55) therefore is

$$Q_s = 214 \exp(-12.5\,m) \text{ (cal/g water)} \ . \tag{2.56}$$

The curve of Stamm and Loughborough (1935) is approximated by Eq. (2.42) from which the heat of wetting is given by

$$W = 20 \exp(-14\,m) \text{ (cal/g wood)} \ , \tag{2.57}$$

since $W_0 = (Q_s)_0/B = 280/14 = 20$ cal/g wood.

Babbitt (1942) used Eq. (2.49) to calculate Q_s as a function of wood moisture content based on results given by Katz (1917) who measured the heats of wetting on "wood fiber" as a function of initial wood moisture content m. Babbitt (1942) calculated Q_s as a function of m, using the empirical relationship,

$$Q_s = 2.114/(0.0895 + m)^2 \text{ (cal/g)} \ . \tag{2.58}$$

This is approximated by

$$Q_s = 220 \exp(-13.5\,m) \text{ (cal/g)} \ , \tag{2.59}$$

using data calculated from Eq. (2.58) between the limits of m = 0.02 to 0.10 g/g, over which range Eqs. (2.58) and (2.59) give good agreement ($R^2 = 0.994$). The regression coefficients are included in Table 2.4.

2.3.3 Significance of Heats of Wetting and of Sorption

The heat of wetting and differential heat of sorption are interrelated, as has been shown above. However, they have different interpretations in terms of their significance in the wood-water system.

According to Rees, in Hearle and Peters (1960), the differential heat of sorption Q_s is a measure of the excess binding energy of the water molecules to the wood substrate over that between water molecules in the liquid state. The total heat of wetting W_0, on the other hand, is interpreted to be proportional to the total number of sorption sites in the wood available for sorption of water.

2.3.3.1 Heat of Sorption

Stamm (1964) has noted that $(Q_s)_0$, the value of the differential heat of sorption for a number of cellulosic materials including wood, cotton and viscose rayon, is very nearly the same, ranging from 265 to 330 with a mean of 290 (cal/g water) based on data obtained on these materials from ten different sources. He points out that these values (mean for 18 $(Q_s)_0$ of 5.22 kcal/mol of water) are within the range of those of hydrogen bonding. Babbitt (1942) has calculated that the additional bond energy of this magnitude can be accounted for by assuming that two hydrogen bonds connect the oxygen in the water molecule with two oxygen atoms in the hydroxyl groups of the cellulose molecule. Rees (1960) points out that if hydrogen bonding is responsible for the heat of sorption, for every water-water hydrogen bond broken there must be two bonds formed between the water and wood.

Rees (1960) concludes that although the curves of Q_s against moisture content vary considerably with respect to their slopes, the values extrapolated to zero moisture content $[(Q_s)_0]$ are essentially equal, for a wide variety of cellulosic fibers. He cites a mean value for $(Q_s)_0$ of 294 cal/g water and points out that this indicates that the nature of the initial bond between water and dry cellulosic materials is essentially the same for all such materials.

If has been mentioned that curves of Q_s against moisture content are steeper for woods with higher total heats of wetting W_0. This is anticipated since W_0 is equal to the total area under the curve of Q_s against moisture content [Eq. (2.47)]. However, if Q_s is plotted against activity a or relative humidity, the curves for different woods are expected to be much more similar. Babbitt (1942) first demonstrated this effect when he plotted Q_s against relative humidity for several different cellulosic materials, including wood and cotton. Although the curve of Q_s against moisture content for cotton is only half as steep as that for wood, the curve of Q_s against relative humidity is essentially the same for the two materials. This is as expected if the shapes of the sorption isotherms are similar and if the values of $(Q_s)_0$, the differential heat of sorption at zero moisture content, are identical for the two materials. Kelsey and Clarke (1956) confirmed the similarity in the curves relating Q_s and relative humidity for the wood of *Araucaria klinkii* to such curves for other cellulosic materials.

2.3.3.2 Heat of Wetting

There is evidence that the more hygroscopic woods are characterized by a higher value of W_0. This would seem reasonable if woods with larger numbers of sorption sites per unit of dry mass are also more hygroscopic. Such seems to be the case if the hygroscopicity is measured by the magnitude of the fiber-saturation point, M_f, for example. Raczkowski (1963) reported that W_0 was linearly related to the fiber-saturation point such that the ratio W_0/M_f was essentially constant and equal to 0.8. Furthermore, he noted that W_0 for sapwood was up to 30% higher than for heartwood and that alcohol-benzene extraction increased W_0 by as much as 38%. This is probably due to the increase in hygroscopicity resulting from removal of extractives (Higgins 1957) (see also Chap. 4).

Kajita and Mukudai (1971) reported that the total heats of wetting W_0 for *heartwood* ranged from 12.6 to 18.6 cal/g for 16 Japanese woods and 14 woods imported into Japan. In general, those woods with higher extractive content showed lower values of W_0. He noted also that the values for sapwood were generally greater than those for heartwood, probably because of the greater hygroscopicity associated with the lower extractive content of sapwood.

Delgado (1970) found, for 34 different Venezuelan woods segregated into six groups based on their equilibrium moisture contents (EMC) at 67% humidity and room temperature, that W_0 was higher for those groups with higher EMC. Morton and Hearle (1975) show that the integral heat of sorption, W_0-W, for a number of textile materials conditioned to 65% relative humidity, is a linear function of the regain (EMC) at this humidity.

Kajita (1976) also reported a positive relationship between W_0 and EMC calculated at 100% humidity for ten Japanese woods (Fig. 2.12). A linear relationship was observed, described by the regression

$$W_0 = 0.5 + 0.67\, M_f \text{ (cal/g wood)} , \qquad (2.60)$$

with an R^2 of 0.902. The zero intercept in Eq. (2.60) is close to zero, indicating that the ratio W_0/M_f is essentially constant. Kajita (1976) reported that the mean ratio was 0.704 for the ten woods he tested.

Kajita (personal communication) also measured the total heats of wetting W_0 for 22 US hardwoods grown on southern pine sites in the United States. Choong

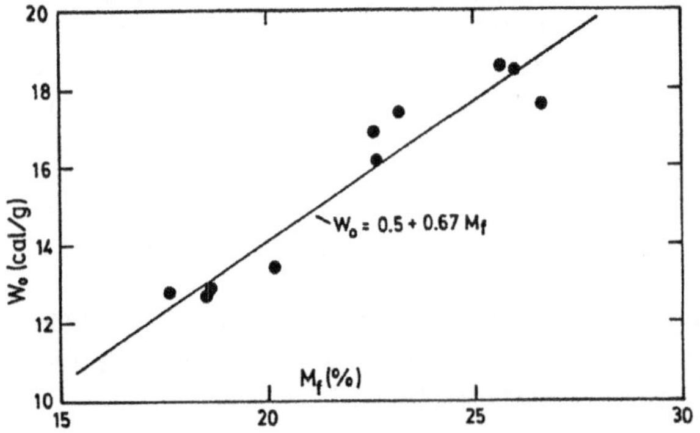

Fig. 2.12. Plotted points and calculated regression curve of total heat of wetting W_0 against estimated fiber saturation moisture content M_f. (Kajita 1976)

and Manwiller (1976) measured the fiber-saturation points by the shrinkage intersection method (Chap. 4) for the same woods. A linear regression of W_0 against M_f resulted in

$$W_0 = 4.64 + 0.480\, M_f \; , \tag{2.61}$$

with an R^2 of 0.55. This was not as good a fit as was the case for the Japanese woods cited above, possibly because the heats of wetting and fiber-saturation point measurements were carried out on different samples.

The various components of wood have been shown (Chap. 1.4.4) to vary with respect to their hygroscopicities, the hemicelluloses being most and lignin the least hygroscopic. Kajita et al. (1977) reported mean values for hemicelluloses, cellulose, and milled wood lignin of *Pinus densiflora* and *Fagus crenata* of about 28, 19 and 14 cal/g dry material, respectively. This is in general agreement with the correlation between the hygroscopicity and total heat of wetting of wood, noted above. Kajita et al. also used the method of mixtures to determine the contribution of each component to the total heat of wetting of the wood.

Kajita (1977) measured the total heats of wetting W_0 for a number of woods, comparing untreated wood with wood treated to remove extractives and to decrease lignin and hemicellulose contents. He reported that W_0 generally increased after extraction with alcohol-benzene, increased with removal of lignin, particularly for short treatment times, and decreased with removal of hemicelluloses from the holocellulose constituent. Raczkowski (1963) also reported a substantial increase in W_0 after alcohol-benzene extraction.

Kajita (1975) found that the total heats of wetting of the inner barks of 22 hardwood species from the United States ranged from 14.16 to 20.45 cal/g, essentially similar to those for wood. However, he noted that values for outer bark were generally considerably lower, indicating the lower hygroscopicity of outer compared with inner bark.

Other factors also influence the total heat of wetting for wood. For example, the measured value of W_0 increases as particle size decreases, according to Kelsey and Clarke (1956). Kajita (1976) reported that W_0 increased from 17.15 to 20.04 cal/g when the fineness of the wood particles was reduced from that passing through 20 to 40 mesh screen to wood flour passing through a 250 mesh screen. It also decreased with mean temperature, an effect discussed in more detail in Section 2.5.

It has been shown that the heat of wetting W for a hygroscopic material decreases with increasing moisture content. However, a hysteresis effect has been observed, in that wood conditioned to a given moisture content by desorption yields a higher value for W than when conditioned to the same moisture content by resorption. Figure 2.13 shows curves obtained by Kajita (1976) with Japanese hinoki (*Chamaecyparis obtusa*) and also with cotton cellulose. The significance of this effect is discussed in Chapter 3.2 under the topic "sorption hysteresis."

It has been shown that the differential heat of sorption Q_s and the heat of wetting W are interrelated. Furthermore, Q_s is obtained by calculation from sorption isotherms using the isosteric method, while W is measured directly by calorimetry. However, Rees (1960) has noted several sources of inaccuracies in calculating Q_s from the sorption isotherms. These are

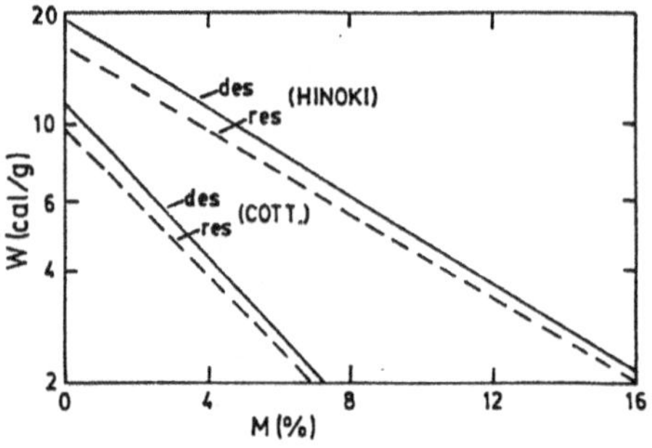

Fig. 2.13. Curves of heat of wetting W (log scale) plotted against initial wood moisture content for Hinoki cypress and cotton, for both resorption and desorption. (Kajita 1976)

1. Plots of sorption isosteres $(\ln(a) \text{ vs. } 1/T)_m$ are not always accurate or linear.
2. The assumption in the Clausius-Clapeyron equation [Eq. (2.40)] that Q_s (or ΔH_s) is constant with temperature is not necessarily correct.
3. The Clausius-Clapeyron equation assumes reversibility in the sorption isotherm, which is not true because of sorption hysteresis.

Based on the inherent inaccuracies in the isosteric method of calculating Q_s (and W), it is concluded that values obtained by classical calorimetric techniques are generally more reliable and reproducible than those based on isosteric calculations from the sorption isotherms.

2.4 Free-Energy and Entropy Changes During Moisture Sorption

It has already been noted that part of the enthalpy decrease when liquid water is taken up by wood is associated with a decrease in the free energy of the water. This free energy decrease ΔG_s of the sorbed water at any given wood moisture content and temperature can be calculated from the sorption isotherm at that temperature. Stamm and Loughborough (1935) first demonstrated this for wood, by use of Eq. (2.20) as follows:

$$\Delta G = (RT/18)\ln(1/h) \ , \qquad (2.62)$$

where ΔG is the free energy *decrease* per gram of liquid water sorbed and 18 (g/mol) is the molecular weight of water.

The other part of the enthalpy decrease is the decrease in entropy $T\Delta S$. It is calculated from $T\Delta S = Q_s - \Delta G$. Thus the differential heat of sorption Q_s consists of two parts, ΔG and $T\Delta S$. The decrease in entropy may be attributed to an increase in order of sorbed water molecules which are attracted to specific sorption sites in the wood, particularly at low wood moisture contents. It is analogous to the decrease in entropy when liquid water freezes into ice, presumably because the ice has a more ordered structure than liquid water. In the latter case, however, all of the heat of fusion is attributable to entropy decrease at T_m, since there is no free energy change, the vapor pressure of ice and liquid water being identical at the freezing point (Fig. 1.1).

Figure 2.14 shows typical curves of Q_s, ΔG and $T\Delta S$ as functions of wood moisture content. Nelson (1983) has shown that $\ln(\Delta G)$ is essentially a linear

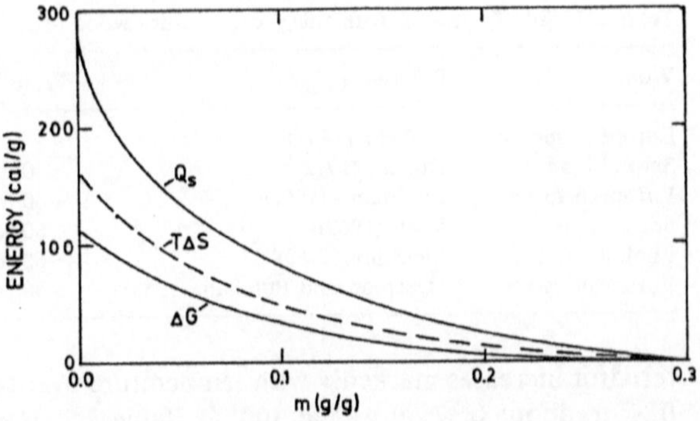

Fig. 2.14. Typical curves of Q_s, ΔG and $T\Delta S$ as related to wood moisture content

function of wood moisture content over most of the relative humidity range (Chap. 3.2.6). Therefore a plot of ΔG against moisture content takes the form

$$\Delta G = (\Delta G)_0 \exp(-Cm) , \tag{2.63}$$

where $(\Delta G)_0$ is the value of ΔG extrapolated to zero moisture content and C is a constant.

A similar equation [Eq. (2.41)] has been applied to data of Q_s against m. Therefore $T\Delta S$ can be evaluated at any temperature and wood moisture content from:

$$T\Delta S = (Q_s)_0 \exp(-Bm) - \Delta G_0 \exp(-Cm) \tag{2.64}$$

if the four constants shown in Eq. (2.64) are known.

Rees, in Hearle and Peters (1960), has pointed out that as the activity approaches zero, 1/h in Eq. (2.62) approaches infinity, and therefore ΔG also approaches infinity. At an activity h near 0.0001, ΔG and Q_s are essentially equal, and below this the absolute value of ΔG is greater than that of Q_s. Therefore the entropy term $T\Delta S$ becomes negative, indicating that the sorbed water molecules are more disordered than those in liquid water, contrary to the case over most of the moisture range. Rees interprets this to mean that water molecules in the wood at very low moisture contents are so far separated from each other that they are less ordered than those in the liquid state. He likens this to the increase in the entropy of a gas during isothermal expansion, resulting from increased spacing of the individual molecules.

2.5 Moisture Sorption and Specific Heat of Wood

The molar heat capacity C of liquid water has been shown (Fig. 2.1) to be essentially constant at 18 cal mol^{-1} over the temperature range from 0° to 100 °C. The specific heat c_w (cal/g) of liquid water is therefore taken as unity over this temperature range.

The specific heat c_0 of dry wood is only about one-third that of liquid water over the same temperature range. Furthermore, in common with most organic ma-

Table 2.5. Specific heat c_0 (cal/g-deg) of oven-dry wood as a function of temperature (°C)

Wood	Reference	Temp. Range °C	Eq. for c_0
European spruce	Volbehr (1896)	–	0.259 + 0.00121 C
Several kinds	Dunlap (1912)	0 to +112	0.266 + 0.00116 C
European spruce	Kühlmann (1962)	−60 to +80	0.267 + 0.00118 C
Spruce pine	Koch (1969)	+60 to +140	0.265 + 0.00104 C
Loblolly pine	McMillin (1969)	+60 to +140	0.271 + 0.00095 C
European beech	Hearmon and Burcham (1955)	+30 to +60	0.28 + 0.001 C

terials it increases markedly with temperature over this range, from about 0.27 at 0 °C to about 0.37 cal g^{-1} at 100 °C. Table 2.5 shows how the specific heats of several kinds of wood measured in the dry condition vary with temperature, based on reported findings from several sources. All of these studies indicate that c_0 increases linearly with temperature over the ranges shown. One might expect a slight nonlinearity, based on the Debye model for the specific heat of materials as a function of temperature, as reported by Ward and Skaar (1963). They also noted a linear increase between −40° and +40 °C in the specific heat of particle board containing about 8% moisture.

Nanassy and Szabo (1978) reported that the specific heat of waferboards increased linearly with temperature over the range from −30° to +20 °C. Their values were somewhat higher than those reported in Table 2.5 for wood, perhaps because of the adhesives used in the waferboards. Nevertheless, the trends were similar to those noted above.

McMillin (1969) found that the specific heats of the principle constituents of dry *Pinus taeda* also increased linearly with temperature between 60° and 140 °C. Lignin yielded somewhat higher values than holocellulose at 40°C. These lignin values increased more slowly with temperature than did the latter and were somewhat lower at higher temperatures than those of holocellulose. The specific heats of extractives were about 35% greater than those of holocellulose and lignin.

The Debye model for explaining the temperature dependence of the specific heat of solids has been applied to wood by Ward and Skaar (1963). For many solids the Debye equation gives an excellent fit to experimental data of atomic heat capacity as a function of temperature.

For example, Fig. 2.15 shows how the constant volume atomic heat capacities C_v of aluminum and carbon (both graphite and diamond) increase with temperature according to the Debye equation. These three experimentally determined curves fit those predicted by the Debye model, as do those for many other solids over the temperature range within which there are no phase transitions.

In the room temperature range (300 K) C_v for aluminum and for many other metals, is nearly constant with temperature, indicating that most of the degrees of freedom associated with interatomic vibrations are effective for absorbing heat. However, for carbon, particularly the strongly crystalline diamond, the strong interatomic bonds are not activated at room temperatures and do not contribute to the heat capacity. Carbon in graphite form with its weaker interatomic bonds has a higher heat capacity than does the strongly bonded diamond.

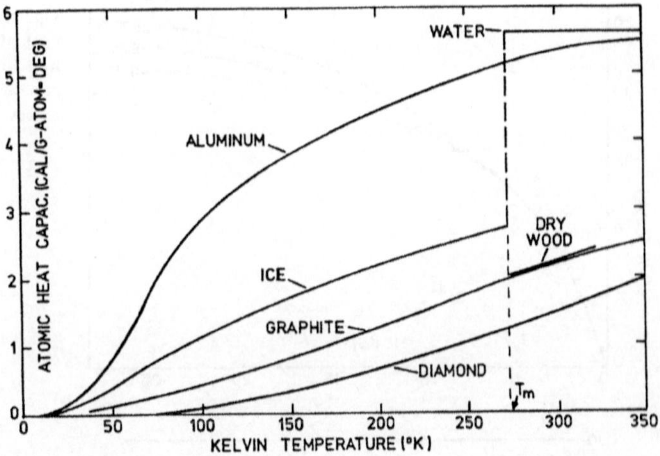

Fig. 2.15. Curves of calculated atomic heat capacities C_v of dry wood, graphite, diamond and of ice and liquid water as functions of temperature

Ward and Skaar (1963) estimated the constant-volume atomic heat capacity for dry wood to be $1.95+0.0085\,\theta$ as a function of Celsius temperature θ, based on Dunlap's (1912) specific heat data. This coincides closely with that of graphite as is shown in Fig. 2.15. Also shown in Fig. 2.15 is the atomic heat capacity of ice and of liquid water as functions of temperature. This was obtained by dividing the molar heat capacities shown in Fig. 2.1 by 3, and multiplying by 0.94, the mean ratio of C_v/C_p (Ward and Skaar 1963).

The specific heat c_m of moist wood is greater than that of dry wood because that of water, c_w, is three or more times greater than that of dry wood. The specific heat c_m has been calculated by the method of mixtures, based on the assumption that the wood and water contribute to c_m in proportion to their weighted isolated heat capacities. Thus:

$$c_m = (c_0 + mc_w)/(1+m)\ , \tag{2.65}$$

where m is the mass of moisture per unit dry mass of wood.

Kelsey and Clarke (1956) have shown that the value of c_m is greater than the value calculated from Eq. (2.65) at any wood moisture content if the *integral heat of wetting*, $W_0 - W$, decreases with increasing temperature. They used the relationship

$$W_0 - W = (q_0 + q_w - q_m)/w_0\ , \tag{2.66}$$

where q_0, q_w and q_m are the heat contents of the dry wood, water, and moist wood, respectively, and w_0 is the dry mass of the wood. The term $(W_0 - W)$ has units of cal/g if heat is in calories and mass in grams.

Differentiating both sides of Eq. (2.66) with respect to temperature T gives:

$$d(W_0 - W)/dT = c_0 + mc_w - (1+m)c_m \tag{2.67}$$

when the following identities are used,

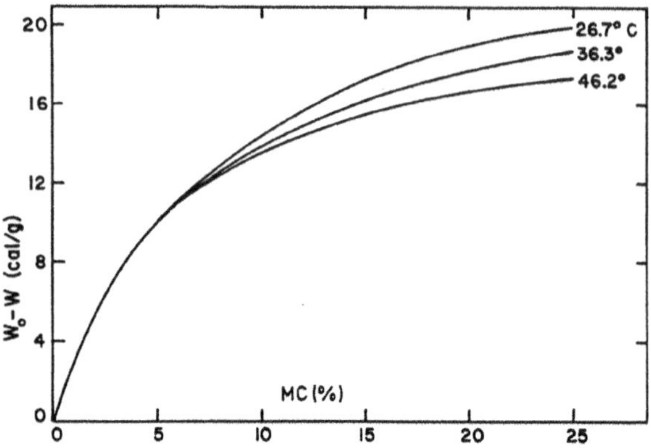

Fig. 2.16. Curves of integral heat of sorption W_0-W in relation to wood moisture content at several temperatures. (After Kelsey and Clarke 1956)

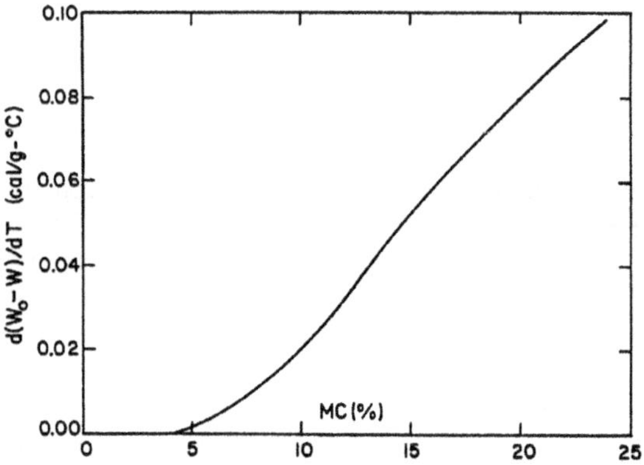

Fig. 2.17. Calculated curve of $d(W_0\text{-}W)/dT$ as a function of wood moisture content. (After Kelsey and Clarke 1956)

$$c_0 = (1/w_0))(dq_0/dT) \tag{2.68}$$

$$c_w = (1/w_w)(dq_w/dT) = (1/mw_0(dq_w/dT) \tag{2.69}$$

$$c_m = (1/w_m)(dq_m/dT) = (1/[w_0(1+m)(dq_m/dT)] \ . \tag{2.70}$$

Equation (2.67) can be rearranged to give c_m. Thus

$$c_m = [c_0 + mc_w - d(W_0 - W)/dT]/(1+m) \ , \tag{2.71}$$

which differs from Eq. (2.65) in that it contains an extra term in the numerator to account for the variation in the integral heat of sorption with temperature.

Kelsey and Clarke (1956) measured the heat of wetting W of klinki pine (*Araucaria klinkii*) for different initial moisture contents from 0% to 24%, at three different temperatures. Their results (Fig. 2.16) showed that $(W_0 - W)$ decreased with increasing wood temperature at moisture contents above $m = 0.05$ g/g. They calculated $d(W_0 - W)/dT$ as a function of wood moisture content, with the results shown in Fig. 2.17.

The increase Δc_m in the specific heat of moist wood [Eq. (2.71)] over that obtained by Eq. (2.65), the simple method of mixtures equation, is given by

$$\Delta c_m = -(1/m)d(W_0 - W)/dT \ . \tag{2.72}$$

Kelsey and Clarke (1956) attributed the increase in c_m to an increase in the specific heat of the sorbed water over that of liquid water c_w. An alternative interpretation of their findings is that the increase in the specific heat of moist wood is due to an increase in the specific heat of the dry wood associated with the opening up to the cell-wall structure by the sorbed water. This permits more freedom for the wood components to absorb additional heat energy as temperature increases. As Morton and Hearle (1975) point out, the specific heat of sorbed water itself should be lower than that of liquid water. This is because it behaves more like ice, having more restricted vibrational modes than liquid water and therefore a lower heat capacity, just as ice does compared with liquid water (Fig. 2.1).

This interpretation of the results of Kelsey and Clarke (1956) leads to the relation that

$$\Delta c_0 = -d(W_0 - W)/dT \,, \tag{2.73}$$

assuming there is no change in the heat capacity of the sorbed water. If the sorbed water has a lower heat capacity than liquid water, c_w is lower than unity, and Δc_0 may be even greater than Eq. (2.73) indicates.

The values of Δc_m obtained by Kelsey and Clarke (1956) on klinki pine using Eq. (2.72) are shown in Table 2.6. Hearmon and Burcham (1955) measured both the heat of wetting and specific heat of beech (*Fagus* sp.) wood over the moisture and temperature range shown in Table 2.6. Their study generally confirmed the experimental results of Kelsey and Clarke.

The curves of Stamm and Loughborough (1935) of Q_s against moisture content, obtained by the isosteric or thermodynamic method, also show a variation of Q_s with temperature above 5%. Analysis of these curves, which can be converted into terms of $W_0 - W$ by use of Eq. (2.48), gives qualitative agreement with the findings of Kelsey and Clarke (1956). However, the excess specific heat Δc_m is about half the values found by Kelsey and Clarke. This may be because of the difficulty in obtaining precise results using the thermodynamic or isosteric method, as Kelsey and Clarke have pointed out.

Table 2.6. Apparent increase Δc_m in specific heat of moist wood

Source	Wood condition		Δc_m (cal g^{-1}/°C)	
	m (g/g)	Temp (°C)	Meas.	Eq. (2.72)
Beech	0.107	30	0.02	0.02
(Hearmon and Burcham 1955)		60	0.04	0.05
	0.140	30	0.02	0.02
		60	0.06	0.05
	0.216	30	0.03	0.03
		60	0.06	0.07
	0.310	30	0.04	0.05
		60	0.09	0.09
Klinki pine	0.05	27–46	–	0.00
(Kelsey and Clarke 1956)	0.10	27–46	–	0.02
	0.16	27–46	–	0.05
	0.24	27–46	–	0.08

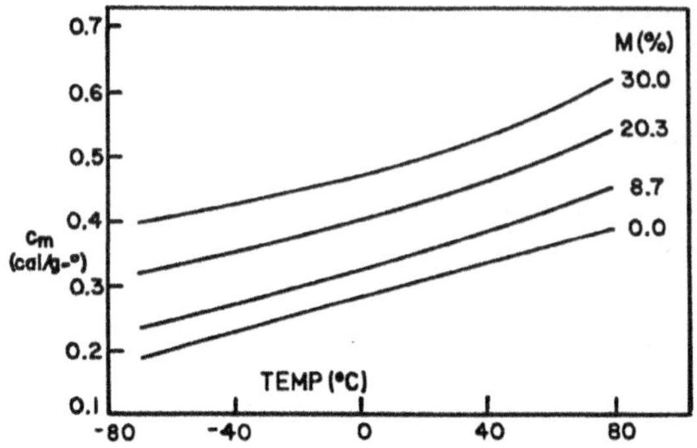

Fig. 2.18. Curves of specific heat of moist wood in relation to temperature. (After Kühlmann 1962)

Kühlmann (1962) measured the specific heats of wood and of particle board using a quasi-steady-state thermal conductivity apparatus over the moisture range from 0% to 30% at temperatures between $-60°$ and $+80\,°C$. His results for moist wood are shown in Fig. 2.18 and those for dry wood in Table 2.5.

2.6 Swelling Pressure of Wood

When wood takes up water, it swells to an extent which depends on the conditions under which the water is absorbed. One should expect, therefore, that wood restrained from swelling should exert a stress in the direction in which it is restrained. This is analogous to the swelling pressure exerted by an osmotic solution against a restraining piston (Fig. 2.19).

When an aqueous solution such as sugar in water is separated from pure water by a semipermeable membrane, a driving force tends to cause water to move into the solution from the pure water. This driving force is known as the osmotic pressure and can be calculated from

$$\pi = (RT/\bar{V})\ln(1/h) = (RT/18)\ln(1/h) \ , \qquad (2.74)$$

where the osmotic pressure π is in dynes/cm^2 if R is in cgs units and 18 cc/mol is the partial molar volume $\bar{V}$ of the water.

Water can be prevented from migrating through the membrane into the solution by applying a force F on the piston (Fig. 2.19) such that F/A, where A is the cross-sectional area of the piston, is equal to π. This provides one method for

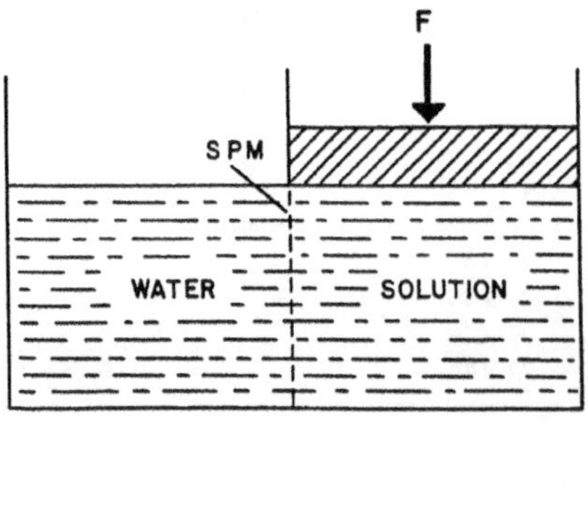

Fig. 2.19. Osmotic pressure diagram showing force F required to prevent water from moving into an aqueous solution through a semipermeable membrane (SPM)

measuring the osmotic pressure of a solution. If a pressure greater than π is applied water may be forced to migrate from the solution into the pure water reservoir. This principle is used in the "reverse osmosis" technique for purifying water (Merten 1966).

In the case of wood, an analogy exists to an aqueous solution. The wood can be thought of as being partially dissolved in the sorbed water. As with a true solution this water is held at a vapor pressure p, lower than the saturated vapor pressure p_0 of pure water. In fact, wood is a material of the general class called gels which are similar to solutions in some respects but different in others, as noted below.

Barkas (1949) lists six characteristics of gels which are important for understanding the swelling pressure of wood as well as the relationship between the stress applied to wood and its equilibrium moisture content. These are:

1. *Gels are hygroscopic.* They hold moisture at a vapor pressure p at equilibrium with the surrounding atmosphere.
2. *Gels swell* when they take up water, to a greater or lesser extent than the volume of water taken up. Swelling may be anisotropic.
3. *Gels exert forces if restrained from swelling* when exposed to vapor pressures higher than their equilibrium vapor pressures. These forces may be different in different directions. Gels also sorb less moisture at a given vapor pressure when so restrained.
4. *Gels possess some rigidity* and can therefore withstand static shear stresses, unlike solutions which cannot.
5. *Gels show limited sorption and swelling* in a saturated atmosphere. Thus they differ again from solutions, which swell to infinite dilution in a water-vapor-saturated atmosphere.
6. *Gels show sorption hysteresis* in contrast to solutions, which do not.

Thus gels are similar to solutions in the first three characteristics but differ in the last three, since they exhibit some rigidity, limited sorption and swelling, and sorption hysteresis.

A hypothetical sequence of development of the cell wall of wood, shown schematically in Fig. 2.20, illustrates the difference between solutions and the gel which makes up the cell wall of wood. Sugar molecules shown in a are in solution

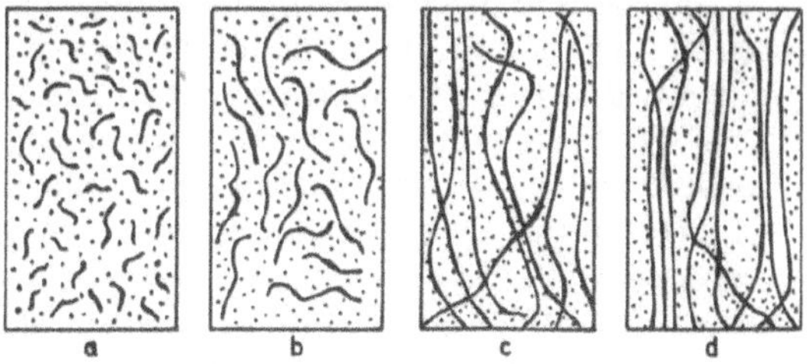

Fig. 2.20. Illustration of a hypothetical sequence in the development of the cell wall showing: **a** Sugar molecules in solution. **b** Sugar molecules partially polymerized into randomly oriented cellulose molecules. **c** Cellulose molecules completely polymerized and partially oriented. **d** Cellulose molecules in final oriented configuration

in the cell lumen. In Fig. 2.20 b they have begun to polymerize into cellulose molecules. As the molecules continue to polymerize they may attain the configuration shown in Fig. 2.20 c, in which some of the cellulose molecules become aligned. Finally they become further polymerized and aligned into the gel system of cellulose and water which constitutes the cell wall of wood, shown in Fig. 2.20 d. Lignin and other cell-wall components are omitted in this simple schematic diagram for reasons of clarity.

The three essential differences between solutions and gels are apparent in contrasting the solution of sugar molecules in water shown in Fig. 2.20 a with those of the cell wall shown in d. The rigidity of the gel is caused by the cellulose network which resists distortion. The limited sorption and swelling is a result of the same network. Finally, sorption hysteresis in the cell wall is also caused by inelastic deformation in the gel network (Sect. 2.8), according to Barkas.

In order to measure the swelling pressure of a gel such as wood it is not necessary to provide a semipermeable membrane between the water and the gel. This is because the solute molecules, cell-wall molecules in this case, are essentially immobile and cannot migrate into the pure water. However, water molecules can migrate through the wood or gel structure and thereby pass into the gel. The swelling pressure of the gel can be measured by exposing the gel to liquid water while it is restrained from swelling.

The swelling pressure of the cell wall or gel material must be distinguished from that of the gross wood, including the cell cavities or air spaces. The swelling pressure of the gross wood is limited by the wood's crushing strength, which is a function of its density, moisture content, and temperature. Most experimental measurements of the swelling pressure of wood have been carried out on the gross wood, usually in one or more of the three principal structural directions. Perkitny (1960) has reviewed most of the important studies of the swelling pressure of gross wood and has proposed standardized procedures for making and reporting such measurements. The swelling pressure of the gross wood will be considered first, followed by that of the cell wall.

The swelling pressure of wood is usually measured by first drying the wood to a low moisture content followed by exposure to a humid atmosphere or to liquid water under conditions such that the wood is mechanically restrained from swelling in one or more directions. Usually the wood is restrained from swelling in one dimension only (Fig. 2.21).

Simpson and Skaar (1968), for example, reported curves of swelling pressure against time for red oak, of the form shown in Fig. 2.22. These samples were initially dried to near 1% moisture content and then exposed to an atmosphere of 50% or 80% humidity at either of two temperatures, 80° or 130°F, under conditions where they were restrained from swelling either tangentially or radially.

It is clear that the swelling pressure in each case reached a peak value and then decreased with time. This may be caused by the simultaneous occurrence of swelling and relaxation of force with time. As the wood swells initially, it exerts an increasing pressure against the restraining clamps. However, there is also a strong relaxation of stress because of the large rheological flow component in the wood. The initial rapid swelling causes the strong initial increase in swelling pressure, but the cumulative effects of stress relaxation later dominate the process and the

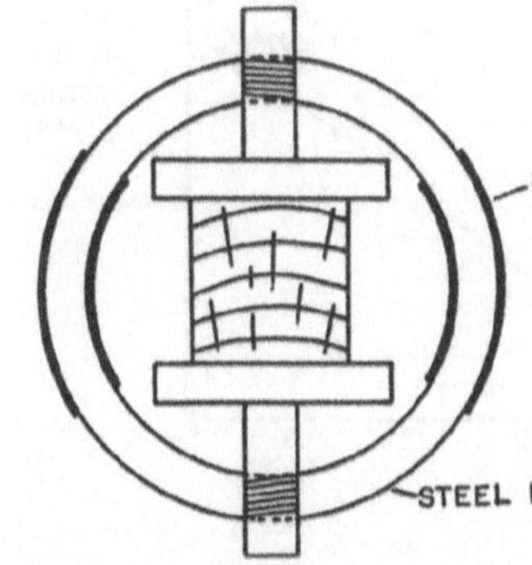

Fig. 2.21. Apparatus for measuring unidirectional swelling stress in wood, showing strain gages attached to steel ring. These measure the force exerted by the wood as it is restrained from swelling by the two plates. (Simpson 1966)

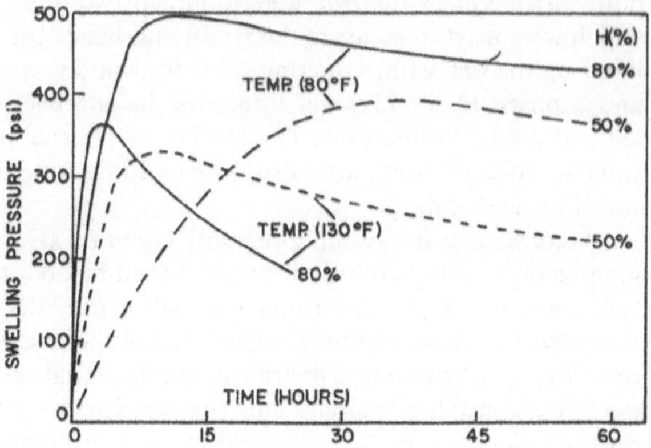

Fig. 2.22. Tangential swelling pressure versus time for dry red oak samples exposed to humidities of 50% and 80% at temperatures of 80 °F and 130 °F. (Simpson and Skaar 1968)

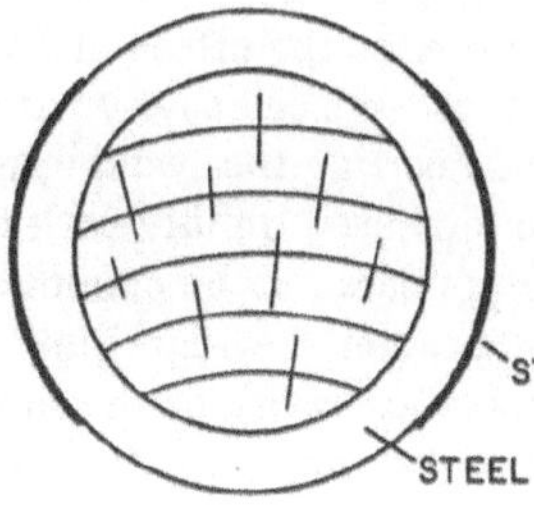

Fig. 2.23. Steel restraining ring with attached strain gages of the type used to measure transverse swelling pressure in wood. (Tarkow and Turner 1958)

overall swelling pressure decreases with time. Similar effects have been observed by Keylwerth (1962, 1964) for wood restrained in any one of the three principal structural directions.

In the experiments described above, the swelling pressure was measured in a single direction in any given sample. In this case the maximum pressure which can be developed is limited by the compressive strength of the wood in the direction of restraint.

Tarkow and Turner (1958) were able to measure the mean transverse swelling in two dimensions, using the arrangement shown in Fig. 2.23. The restraining

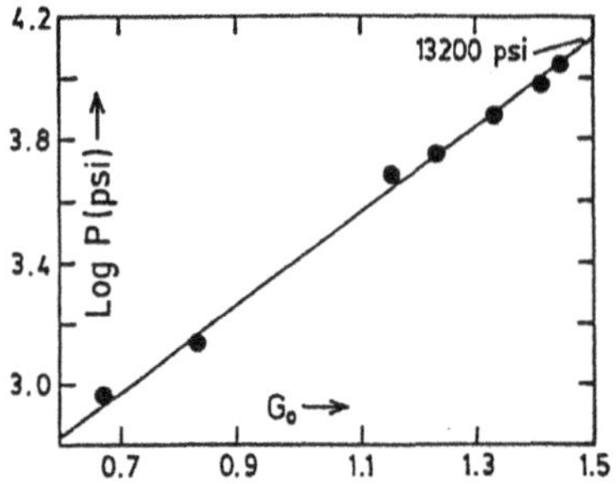

Fig. 2.24. Logarithm of swelling pressure in relation to dry wood specific gravity, together with experimental points obtained by Tarkow and Turner (1958)

force was exerted by a steel ring into which the dry sample was inserted. On the outer perimeter of the ring were mounted two electrical resistance strain gauges which were used to measure the strain and hence the stress that the wood exerted in trying to swell within the ring. The dry wood was force-fitted into the steel ring and exposed to liquid water after first having been conditioned to equilibrium with 30% relative humidity. The swelling or internal pressure on the ring was measured at different times after exposure to liquid water in order to trace the pressure variation with time.

According to the swelling-pressure equation [Eq. (2.74)] the maximum swelling pressure in the cell wall of wood should be about 1560 atm at room temperature when wood at equilibrium with 30% humidity is exposed to saturated air. However, the values obtained experimentally were always considerably lower than this. This is attributed primarily to the fact that when wood is restrained from swelling externally, it swells into the cell cavities since these provide space for swelling. Tarkow and Turner could measure no appreciable swelling in the grain direction of the samples.

In order to minimize the effect of swelling into the cell cavities, Tarkow and Turner (1958) reduced the cell cavity volume by densifying the wood to different degrees prior to measuring the swelling pressures. They found that the measured swelling pressure increased rapidly with the extent of densification of the wood. The relationship appeared to be exponential as shown in Fig. 2.24, which gives the relationship between the logarithm of the maximum swelling pressure observed and the specific gravity of the samples, based on dry weight and volume. It appears that the maximum apparent swelling pressure, as measured, increased exponentially with increasing specific gravity. The maximum value measured was 11000 psi, about 750 atm, obtained for a sample of 1.44 specific gravity. It therefore contained some void space into which swelling could take place. Extrapolation of the curve to specific gravity of 1.5, that of the dry cell wall, gives a swelling pressure of approximately 900 atm (13200 psi).

The discrepancy between theoretical (1560 atm) and measured (900 atm) swelling pressures of the cell wall may be partially accounted for by the fact that the cell wall itself is somewhat compressible and some of the stress was reduced in compressing the wood. Barkas (1949), in his discussion of stresses associated with moisture sorption, uses the compressibility of the gel to modify Eq. (2.74), which does not account for the properties of the gel itself, but only those of the water.

The approach of Barkas to swelling pressure and associated phenomena in wood is treated next.

2.7 Mechanical Stress and Sorption (Hygroelastic Effect)

When wood is restrained from swelling and exerts a swelling pressure against the restraining structure, it also comes to a lower EMC than wood which is unrestrained from swelling. This effect also operates in the reverse sense; that is, if a mechanical tension stress is applied to wood, its EMC is increased compared with that of an unstressed sample.

Barkas (1949) had treated this effect quantitatively, considering wood to be a hygroscopic gel and using the generalized Porter equation in the form

$$(\partial V/\partial m)_p dP_m = v\,dp \ , \tag{2.75}$$

where V is the specific volume (cc/g) of the gel in terms of the swollen volume of the gel per unit dry mass (g), m (g/g) is the mass of water per unit mass of dry gel, dP_m is the increase in hydrostatic pressure required to raise the vapor pressure of the gel by the increment dp, and v (cc/g) is the specific volume of the water vapor. The term dV/dm actually represents the apparent specific volume (cc/g) of the sorbed water.

If the Porter equation is integrated, it reduces to the osmotic-pressure equation, provided certain assumptions are made. Thus, letting ϱ represent $\partial m/\partial V$, the apparent density (reciprocal of specific volume) of the sorbed water, and integrating between the limits p_1 to p_2, the Porter equation becomes

$$P_m = \int_{p_1}^{p_2} \varrho v\,dp \ . \tag{2.76}$$

If the ideal gas law is now assumed to hold, and if the apparent density ϱ of the sorbed water is considered to be constant, Eq. (2.76) becomes the familiar osmotic-pressure equation

$$P_m = (RT/18)\ln(p_2/p_1) \ . \tag{2.77}$$

Barkas cites four objections to integrating the Porter equation and applying the resulting simplified osmotic pressure equation to gels. First, the integrated equation neglects the rigidity of gels. Second, it contains no provision for data concerning the gel itself, but only of the water in the gel. Third, it does not distinguish between the swelling pressure at constant moisture content from that at constant volume. And fourth, the vapor is presumed to obey the ideal gas law.

Because of these factors Barkas proposed a more generalized osmotic theory based on the Porter equation [Eq. (2.75)], one that interrelated hydrostatic pressure P, specific volume V, moisture content m, and vapor pressure p. Figure 2.25 shows such a diagram calculated by Barkas from his generalized equation for the wood substance of Sitka spruce, using data of Stamm and Seborg (1935) on the specific volumes of this material over the full range of its sorption isotherm. This diagram shows that the moisture content m and the specific volume V of the

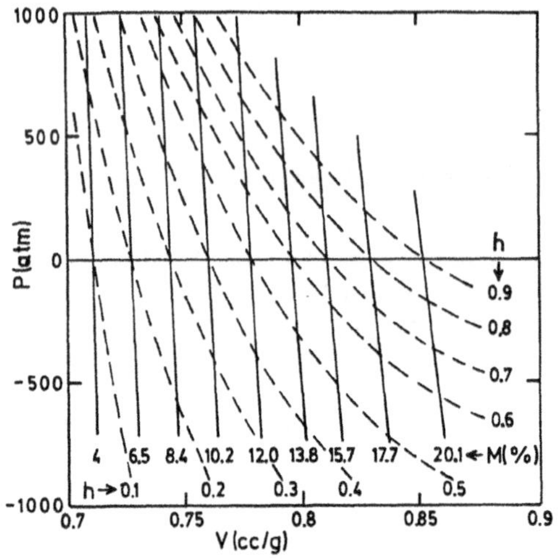

Fig. 2.25. Hydrostatic pressure P as a function of specific volume V for the cell wall of Sitka spruce at several constant moisture contents and relative vapor pressures h. (After Barkas 1949)

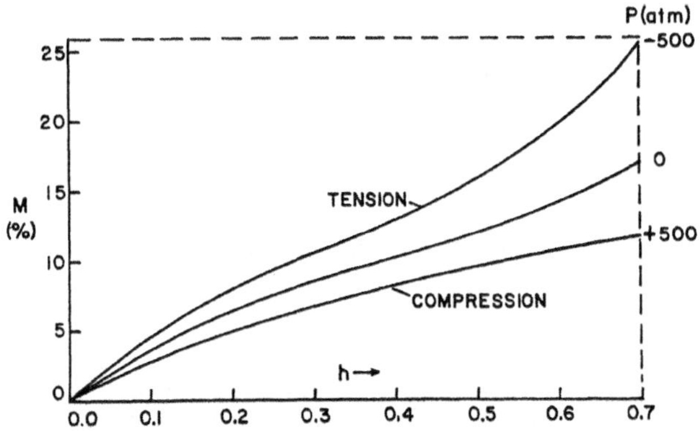

Fig. 2.26. Sorption isotherms for the cell wall of Sitka spruce in relation to hydrostatic stress P as calculated from curves of Fig. 2.25

cell wall are dependent not only on the vapor pressure p, or on h, but also on the external hydrostatic pressure P.

The curves shown in Fig. 2.25 can be replotted in the typical sorption isotherm form to show the effect of hydrostatic pressure or stress P on the sorption isotherms. For example, Fig. 2.26 shows such curves – the one for the stress-free condition, one for a positive hydrostatic pressure of 5×10^8 dynes/cm^2 (approximately 500 atm), and one for a negative hydrostatic pressure (tension) of 5×10^8 dynes/cm^2. It is clear from these three curves that a hydrostatic pressure decreases the equilibrium moisture content at a given vapor pressure and that a hydrostatic tension increases the equilibrium moisture content.

Barkas has pointed out that even in the absence of external restraint there is an internal restraint to swelling because of the nature of the cell-wall structure. This internal restraint, caused by at least two factors, effectively reduces the hygroscopicity of the wood in the same way as does an external pressure or stress. One of these internal restraining factors is attributed to the differences in fibril

Fig. 2.27. Small portion of the pressure-volume diagram of Fig. 2.25 greatly enlarged. (Barkas 1949)

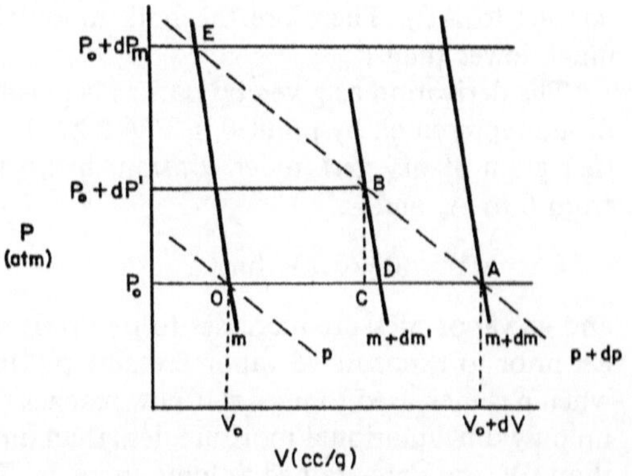

orientation in the S1 and S3 layers of the cell wall compared with the S2 layer (Fig. 4.6). These fibrils have their long axes nearly perpendicular to those of the bulky S2 layer and therefore reduce the external swelling and sorption. A second restraining factor is the presence of interfibrillar bonds which limit the swelling between fibrils. According to the fringe-micellar theory, for example, these bonds would consist of the cellulose chain molecules which interweave among the crystalline regions of the cell wall, thus restricting the degree of swelling in the intercrystalline or amorphous regions.

Figure 2.27, taken from Barkas (1949), shows a small portion of the P, V, m, p diagram, such as shown in Fig. 2.25, but for a gel which is assumed for convenience to be homogeneous, isotropic, and perfectly elastic. Before formally deriving the equations used by Barkas it is instructive first to consider the principles involved as illustrated in Fig. 2.27. The gel is at a fractional moisture content m (g/g), at equilibrium with vapor pressure p, and having a specific volume V_0 (cc/g) based on swollen gel volume and dry gel mass, at the reference hydrostatic pressure P_0, which may be atmospheric pressure, for example. This condition is defined by point 0 in the diagram.

If the gel is now exposed to a higher vapor pressure, p+dp, and allowed to swell freely at constant hydrostatic pressure P_0, it swells to a new specific volume V_0+dV as it adsorbs an additional moisture increment dm. The new equilibrium condition corresponds to point A on the diagram.

If an increment of hydrostatic pressure or stress dP′ is now applied, maintaining a constant vapor pressure p+dp, the moisture content reduces to m+dm′, lower than m+dm, and the specific volume V decreases. This new stress-induced equilibrium corresponds to point B.

In order to reduce the moisture content to m at the vapor pressure p+dp a total stress increment dP_m must be applied. At this equilibrium condition (point E) the specific volume is lower than at point 0, even though m is the same. The difference is attributed to the compressibility, or to its reciprocal, the bulk modulus k_m of the gel at constant moisture content, which is directly proportional to the slope $(\partial P/\partial V)_m$ of the constant m line 0E. It should be noted that this is much steeper than the slope $(\partial P/\partial V)_p$ of the line of constant p (essentially parallel to AE). Therefore the bulk modulus k_p at constant vapor pressure is much lower than k_m.

The derivation as given by Barkas begins by considering the equilibrium condition represented by point 0 in Fig. 2.27. If the gel absorbs dm grams of water (per gram of dry gel) under constant hydrostatic stress P_0, its volume increases from 0 to A, and

$$0A = dV = (\partial V/\partial m)_P dm \tag{2.78}$$

and its vapor pressure increases to p+dp. If a partial restraint is imposed on the gel prior to exposure to vapor pressure p+dp, the stress P_0 increases to P_0+dP when p is increased to p+dp. It now reaches equilibrium at point B but has taken up only dm′ additional moisture, less than dm. Its volume increase dV′ is smaller than dV, the unrestrained volume increase. Thus

$$OC = dV' = (\partial V/\partial m)'dm' \ . \tag{2.79}$$

The volumetric compression due to the stress dP′ is 0D−0C or CD, where 0D is the volume increase the gel would have exhibited if it had taken up dm′ grams of moisture at constant stress P_0.

The volumetric strain associated with the increment of moisture dm′ under this stress increment dP′ is CD/V, or

$$CD/V = [(\partial V/\partial m)'dm' - (\partial V/\partial m)dm']/V \ . \tag{2.80}$$

The volumetric or hydraulic stress dP′, associated with restrained sorption, can also be calculated from the differences in the moisture sorption increments dm and dm′ if it is assumed that the sorption increment difference (dm−dm′) is directly proportional to the volume difference (dV−dV′). Thus, in Fig. 2.27 this means that the distances 0D and 0A are proportional to dm′ and dm, or that

$$(0A-0D)/0A = DA/0A = (dm-dm')/dm \ . \tag{2.81}$$

Furthermore, it is assumed that the triangles 0EA and DBA are similar, which means that the volume strain for a given vapor-pressure increment dp is directly proportional to the hydraulic stress applied in restraining the swelling associated with the increment of vapor pressure. Thus, by the law of similar triangles

$$DA/0A = dP'/dP_m \ , \tag{2.82}$$

and from the Eq. (2.81)

$$dP'/dP_m = (dm-dm')/dm \tag{2.83}$$

or

$$dP' = [1-(dm'/dm)]dP_m \ . \tag{2.84}$$

But, from the Porter equation, $dP_m = (v\,dp)/(\partial V/\partial m)_p$, and $dP' = v\,dp\,[1-(dm'/dm)]/(\partial V/\partial m)_p$ or

$$dP' = \frac{v\,[(\partial p/\partial m)' - (\partial p/\partial m)]dm'}{(\partial V/\partial m)_p} \ , \tag{2.85}$$

which is the stress associated with restrained sorption. The bulk modulus k_m is equal to the hydraulic stress in Eq. (2.85) divided by the volume strain in Eq. (2.80). Thus, making this substitution and rearranging, k_m becomes

$$k_m = \frac{-Vv\,[(\partial p/\partial m)' - (\partial p/\partial m)]}{(\partial V/\partial m)_p\,[(\partial V/\partial m)' - (\partial V/\partial m)]} \,. \tag{2.86}$$

Equation (2.86) can be applied to isotropic gels or to solutions. It relates the bulk modulus k_m at constant moisture content to the difference in the slopes of the sorption isotherm for a gel in the restrained $(\partial p/\partial m)'$ and unrestrained $(\partial p/\partial m)$ condition and to the corresponding differences in the differential volume change for the restrained $(\partial V/\partial m)'$ and unrestrained $(\partial V/\partial m)$ conditions.

For an anisotropic gel such as wood there are shear strains involved even when it is subjected to hydrostatic pressures which are uniform on all surfaces. This is both true at the macroscopic or gross level, such as in a block of wood, and also at the microscopic or cell-wall level, as Barkas points out. Therefore, Eq. (2.86) cannot be applied rigorously. However, it is possible to apply it in a modified form for mechanical stresses resulting from unidirectional restraints applied normal to the principal orthotropic axes of wood such as in the radial or tangential directions during swelling. In this case, if Poisson effects and shear stresses are neglected, and if $(\partial V/\partial m)$ and $(\partial V/\partial m)'$ are taken to be equivalent to $(V/x)(\partial x/\partial m)$, and $(V/x)(\partial x/\partial m)'$, respectively, Eq. (2.86) can be modified to

$$E_m = \frac{-v\,[(\partial p/\partial m)' - (\partial p/\partial m)]}{(V/x)(\partial x/\partial m)_p\,[(1/x)(\partial x/\partial m)' - (1/x)(\partial x/\partial m)]} \tag{2.87}$$

where E_m is Young's modulus in the direction x of swelling, $(1/x)(\partial x/\partial m)$ is the unrestrained rate of swelling in the x direction and $(1/x)(\partial x/\partial m)'$ is the restrained swelling in the x direction, the other terms having the same meaning as in Eq. (2.86).

In order to apply Eqs. (2.86) or (2.87) to calculate the bulk modulus k_m or Young's modulus E_m, it is necessary to have the slope of the sorption isotherm is the unrestrained $(\partial p/\partial m)$ and in the restrained condition $(\partial p/\partial m)'$, and also the volumetric or linear swelling in the unrestrained $(\partial V/\partial m)$ or $(1/x)(\partial x/\partial m)$ and restrained $(\partial V/\partial m)'$ or $(1/x)(\partial x/\partial m)'$ conditions. Barkas (1945) gives data obtained from measurements on spruce cubes which were restrained from swelling in the tangential directions during adsorption from equilibrium with h at 0.258 to equilibrium with h at 0.538. These data, together with those obtained on unrestrained samples exposed to the same conditions, are given in Table 2.7. It should be noted that the terms $(\partial p/\partial m)$ and $(\partial p/\partial m)'$ in Eqs. (2.86) or (2.87) can be written in terms of relative vapor pressure h, in place of p, in which case the term $p_0 v$ is substituted for v.

Substitution of the data into Eq. (2.87) gives the following

$$E_m = \frac{-(3.41\times 10^9)\,[(0.279/0.0382) - (0.279/0.0396)]}{(2.581)(0.386)(0.120 - 0.386)}$$

$$= 3.32\times 10^9 \text{ dynes/cm}^2, \text{ or } 48\,000 \text{ psi} \,.$$

Table 2.7. Data obtained by Barkas (1945) on Sitka spruce for calculating Young's modulus in tangential direction

Mean value of h	0.398
Δh	0.279
p_0v (dyne-cm/g)	3.41×10^9
mean value of V (cc/g dry wood)	2.581
Δm unrestrained (g/g)	0.0396
Δm' restrained (g/g)	0.0382
(1/x) (dx/dm) unrestrained $(g/g)^{-1}$	0.386
(1/x) (dx/dm)' restrained $(g/g)^{-1}$	0.120
Young's modulus E_m (dynes/cm²)	3.32×10^9

This value agreed reasonably well with measurements made on a mechanical testing machine using matched material conditioned to equilibrium with h = 0.40.

Bello (1968) had calculated Young's moduli in the radial and tangential directions from measurements on several hardwood species which were first dried and then restrained from transverse swelling by use of steel rings into which samples were inserted (Fig. 2.23) prior to exposure to a higher humidity. In this case the values of E obtained were in the right order or magnitude. He also found that the EMC was 1.44% lower on the average for restrained than for unrestrained samples when dry samples were exposed to 87% humidity, and 0.51% lower when exposed to 58% humidity, both at 77 °F.

Equations (2.86) and (2.87) relate the elastic moduli of gels to the differences in swelling and water sorption between samples which are partially restrained from swelling compared with those which are unrestrained. It has been indicated during the development of these equations that the application of a compressive mechanical stress decreases sorption while a tension stress increases sorption. This is clear from Figs. 2.25, 2.26, and 2.27, and also from the Porter equation itself [Eq. (2.75)].

It is more convenient experimentally to measure the effect of a constant mechanical stress, applied continually in one direction only, on the equilibrium moisture sorption at constant vapor pressure. For this case the Porter equation can be written in the form

$$(\partial V/\partial m)_p = v(\partial p/\partial P)_m \,, \tag{2.88}$$

or, more conveniently for an uniaxial stress σ applied in the x direction only, as

$$(V/x)(\partial x/\partial m)_\sigma = v(\partial p/\partial \sigma)_m \,. \tag{2.89}$$

From the rules of partial differentiation it can be shown that

$$(\partial p/\partial \sigma)_m = -(\partial m/\partial \sigma)_p(\partial p/\partial m)_\sigma \,. \tag{2.90}$$

Substituting this into the preceding equation and rearranging gives

$$(\partial m/\partial \sigma)_p = -(V/v)(1/x)(\partial x/\partial p)_\sigma \,. \tag{2.91}$$

Equation (2.91) can also be written in the form

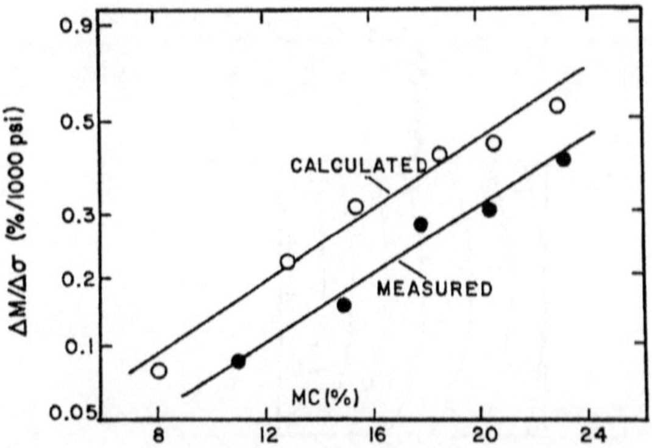

Fig. 2.28. Curves of $\Delta M/\Delta\sigma$ (log scale) against wood moisture content M calculated from Eq. (2.29) and also measured directly. (Simpson 1971 b)

$$\left(\frac{\partial m}{\partial \sigma}\right)_p = -\left[\frac{(1+m)}{\varrho v p_0}\right]\left[\left(\frac{1}{x}\right)\left(\frac{\partial x}{\partial h}\right)\right]_\sigma \tag{2.92}$$

by recalling that $V = (1+m)/\varrho$, where ϱ is the density at moisture content m.

Equation (2.92) is in a convenient form for comparing experimental results with those of theory. It states that the change in moisture content m per unit of applied stress is directly proportional to the moisture content m and to the humidity expansion coefficient $[(1/x)(dx/dh)]$ [Eqs. (4.41), (4.42)], and inversely proportional to the wood density.

Simpson (1971 b) had carried out experiments in which he measured the effect $(\partial m/\partial \sigma)_p$ for compression and tension stresses on red oak in both the tangential and radial directions over a range of wood moisture contents. He also calculated the anticipated value from Eq. (2.92). A typical pair of curves showing both experimental and calculated values are shown in Fig. 2.28. The experimental values were always smaller than the calculated values, but the variation with wood moisture content was similar.

2.8 Mechanical Stress and Sorption Hysteresis

Barkas (1949) extended his theory as given above to explain sorption hysteresis (Chap. 1.4.2). To illustrate the principles involved, Barkas used the example of a wood specimen partially restrained from swelling in one dimension by a linear elastic spring as shown in Fig. 2.29. In this case, the pressure is the unidirectional stress P_x in the x direction, and the change in volume is represented primarily by the change in the x-dimension. For simplicity of concept, the stresses and dimensional changes in other directions are not considered.

Following Barkas (1949), we will consider two cases. In the first case the wood is assumed to be perfectly elastic with no plastic flow when a stress is applied, illustrated by Fig. 2.30 a. In the second case the wood will exhibit rheological properties, as it in fact does, resulting in some irreversible plastic flow (Fig. 2.30 b).

In the first, reversible case, shown in Fig. 2.30 a, the wood is perfectly elastic. It is initially at a moisture content M_0, at equilibrium with a relative humidity

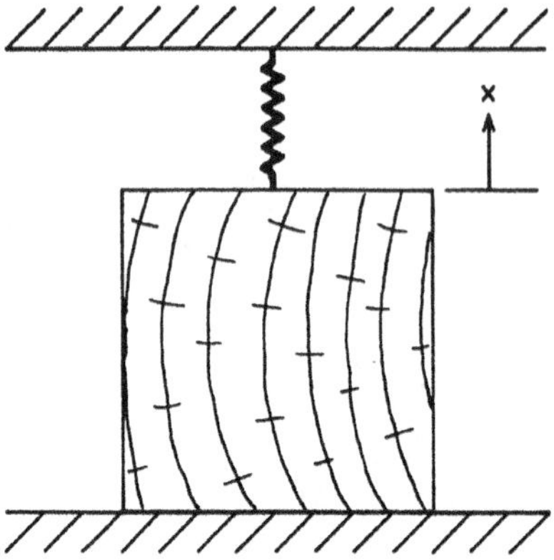

Fig. 2.29. Schematic diagram showing a wood sample swelling and shrinking against a linear spring

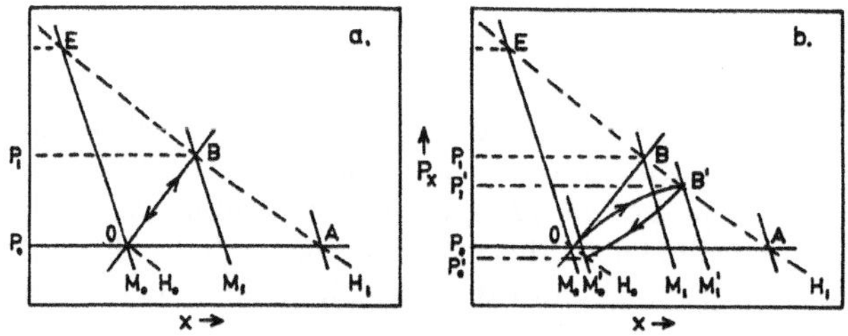

Fig. 2.30. Diagrams showing unidirectional stress P_x against displacement X for wood during moisture sorption: **a** For the elastic reversible case. **b** For the inelastic irreversible case. (After Barkas 1949)

H_0 under the initial stress P_0 exerted by the elastic spring. If the humidity increases from H_0 to H_1, the wood will adsorb moisture and swell against the spring until some new equilibrium point is reached at stress P_1 and moisture content M_1. Thus, it will move from point 0 to point B on the diagram. The actual magnitude of M_1 is inversely related to the increase in stress $(P_1 - P_0)$. If the humidity H is now decreased back to the initial value H_0, M_1 will decrease to M_0 and P_1 to P_0, thus returning exactly to the starting point 0.

It should be noted that the actual location of point B, and therefore the values of P_1 and M_1, depend on the stiffness of the restraining spring shown in Fig. 2.29. If this spring becomes more stiff, the equilibrium point B moves toward the point E with a greater compressive stress P_1 and reduced moisture content M_1. The converse is true if the spring is weaker or less stiff. However, for the reversible, perfectly elastic case, the equilibrium point returns to 0 when H is reduced to H_0, regardless of the location of B along the line H_1.

In the reversible perfectly elastic case cited above sorption hysteresis would not occur. According to Barkas (1949), this in fact is the situation one would expect to find in aqueous solutions where shear stresses do not occur at equilibrium.

In the second case such as for the wood, where irreversible plastic flow occurs, the location of point B after increasing the humidity from H_0 to H_1, is closer to

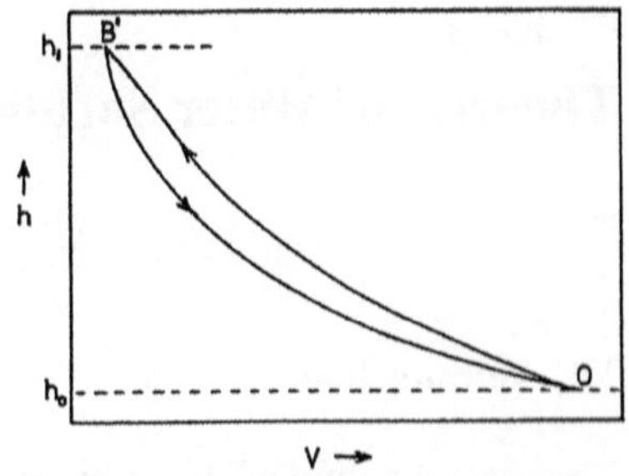

Fig. 2.31. Diagram showing hysteresis in the activity h versus vapor volume phase V. (After Barkas 1949)

the point A than for the perfectly elastic case. This is because the wood relaxes as it approaches the stress P_1 and never quite reaches that level. Rather it moves toward the point B′ as shown in Fig. 2.30b. When equilibrium is finally attained at point B′ the stress P_1' is lower than the elastic value P_1, and the equilibrium moisture content M_1' is higher than M_1.

Lowering the humidity to the original value of H_0 results in a reduction in moisture content and stress. The final equilibrium value P_0' will be lower than P_0 and that of M_0' will be higher than the original value M_0. The area within the loop represents the difference in the work done by the wood on the spring during resorption and the work recovered by the wood during desorption. There is therefore a net energy loss in the cycle. As Barkas points out, the humidity must be lowered below H_0 for the wood to attain its original moisture content M_0, since it is now in tension.

Barkas (1949) has used the example of the wood subjected to swelling against an external spring (Fig. 2.29) to illustrate a principle. Actually there are restraining forces within the cell wall itself which act similarly to the external spring in resisting swelling. It is difficult or impossible to reproduce the pressure-volume diagram of Fig. 2.30b for the cell wall. However, Barkas suggests using the corresponding pressure-volume diagram for the water vapor itself to calculate the loss in energy or work during a sorption cycle. This is possible in an isothermal system if it is assumed that the same amount of energy is required to evaporate water at a given moisture content during desorption as is recovered during condensation at the same moisture content.

The vapor pressure-volume diagram corresponding to the wood pressure-volume diagram of Fig. 2.30b is shown in Fig. 2.31. It can be calculated from the resorption and desorption isotherms at any given temperature. The area between the resorption and desorption curves in Fig. 2.31 gives the work lost during sorption hysteresis. It can be calculated easily and should correspond to the area between the curves in Fig. 2.30b. Barkas (1949) has calculated that this area represents approximately 17% of the total work of desorption for spruce wood. If this area is proportional to the mean difference between the resorption and desorption isotherms, the R/D ratio can be estimated as $1-0.17=0.83$, close to the R/D ratios given in Chapter 1 (Table 1.3).

Chapter 3

Theories of Water Sorption by Wood

3.1 Introduction

The general nature of the sorption isotherm and the thermodynamics of moisture sorption by wood have been discussed in Chapters 1 and 2, respectively. In the present chapter, we will consider some of the equations, both empirical and theoretical, which have been used to quantify the sorption isotherms for wood.

As noted elsewhere, wood is a natural material, produced by biological processes in the living tree. For this reason; it has moisture sorption properties similar to those of most biologically produced materials. Those produced by animals include wool, silk, hair, keratin, etc. These are protein materials but have sorption isotherms similar to plant-produced materials, including carbohydrate fibers such as cotton, jute, bamboo, rice straw, etc., as well as food materials such as wheat, corn, rice, nuts, etc. The latter may include both carbohydrates and proteins.

Many equations have been proposed and tested for describing the moisture sorption isotherms of biological materials. Some of these are purely empirical, while others are based on theoretical considerations, and some are semi-empirical.

Van den Berg and Bruin (1981) have reviewed the literature and compiled a list of 77 isotherm equations which have been applied to biological materials, including wood and other fibrous materials. They grouped the equations into four general categories, based on the physical models assumed in their derivations. These are: localized monolayer sorption models, multilayer sorption models, sorption models used in polymer science, and empirical models. A further breakdown of these four categories is shown in Table 3.1.

A few of the 77 isotherm equations mentioned above have been applied to the sorption isotherms for wood, primarily by Simpson (1973), who used sorption data from the USDA Forest Products Laboratory Wood Handbook (1955). Several of these equations gave quite satisfactory fits to the experimental data, but the thermodynamic constants, particularly those related to the heat of sorption, often were not of the expected magnitude, according to a later paper by Simpson (1979).

Venkateswaran (1970) lists 18 sorption isotherm equations which have been proposed to explain the mechanism of sorption on cellulosic materials. He points out that in the derivation of most of these equations, one or more common assumptions are made. In terms of water sorption by wood these assumptions can be stated as follows: the energy of interaction with the dry wood of the first or primary water molecules taken up is generally higher than the interaction energy of secondary water molecules taken up by the more moist material; the interaction energy of the secondary water molecules with the moist wood is essentially equal to the heat of condensation of liquid water and is constant; there is no energy of

Table 3.1. Overview of moisture sorption models

A.	Localized monolayer sorption models
B.	Multilayer sorption models
	1. Homogeneous sorption
	2. Polarized sorption layers
	3. Liquid film
	4. Condensation (capillary)
	5. Other multilayer models
C.	Polymer science sorption models
	1. Solution models
	2. Localized sorption and solution
	3. Miscellaneous
D.	Empirical sorption models
	1. Partially theoretical
	2. Fully empirical

interaction between adjacent sorbed water molecules. Some of the sorption equations are derived from classical thermodynamic considerations, others from statistical considerations, and still others from combinations of these or other treatments. As noted above, some are empirical, in part or in entirety.

Many of the 77 equations listed by Van den Berg and Bruin (1981) are mathematically equivalent, according to Boquet et al. (1980). In other words, isotherm equations derived by different workers using different physical and/or mathematical models can be rearranged and shown to have the same mathematical formulations in many cases. For example, several theories produce an empirical isotherm of the form first derived by Hailwood and Horrobin (1946)

$$h/m = A + Bh - Ch^2 \,, \tag{3.1}$$

where A, B, and C are empirical constants, related to various fundamental constants, the nature of which depends on the physical model assumed for sorption. Other forms of empirical equations, similar to one another, are produced by several different sorption theories.

Only a few representative isotherm equations will be derived here, including the Dent (1977) variation of the BET equation, the Hailwood-Horrobin (1946) equation, and the Peirce (1929) equation. Several other equations will be listed or partially derived. Interpretation of the significance of some of the constants and their interrelationships are given in several cases.

Before proceeding with the formal derivations, it is helpful to consider some of the common factors involved in most sorption theories. It has already been noted that bound water in wood is generally considered to consist of two components, one strongly and the other weakly bound. Two general approaches have been taken in developing most theoretical sorption isotherms. In one approach, sorption is considered to be a surface phenomenon, and in the other a solution phenomenon (Table 3.1). White and Eyring (1947) have pointed out that in both cases the existence of strong sorption sites is assumed. These sites may represent either a primary surface layer (surface theories), or sites distributed throughout the volume of the sorbate (solution theories) but equally accessible to water.

Thus, according to White and Eyring, "the equations for multilayer surface sorption and for extremely imperfect solutions in which mixing terms are negligible have the same form."

Another important factor, both in fitting sorption isotherm models and in interpreting thermodynamic data, is the accuracy and reproducibility of *sorption equilibrium*. This varies considerably among different workers. In some cases, an arbitrary time after exposure to new conditions is allowed, at which time equilibrium is assumed. In other cases, the precision of measurement determines the attainment of equilibrium, that is, when the measuring apparatus is not sufficiently sensitive to detect small changes in weight. In some cases, equilibrium is assumed to have been attained if weight changes between successive weighing periods are less than some arbitrary value.

Sometimes more serious factors affect the reliability of sorption data, such as the precise measurement of equilibrium relative humidity and temperature. The difficulties involved in measuring and controlling relative humidity have been discussed in Chapter 1 under *Hygrometry* (Chap. 1.2.2). These need not be repeated here except to point out that it is particularly difficult to control and measure relative humidity levels above 90%.

Finally, it should be noted that isotherms measured at high temperatures may be affected by irreversible changes in the wood brought about by thermal degradation. These are not ordinary hysteresis effects but those caused by permanent changes in the basic cell-wall structure. These generally tend to reduce the hygroscopicity of wood and similar materials.

In summary, it may be stated that the sorption isotherm data may be at fault as much as theories used to explain them, particularly if isotherms are measured over a wide temperature range. This may be so for the sorption data accumulated over many years and published in the USDA Wood Handbook (1955). These data were obtained by a number of techniques and may not be the best data bank to use for testing the various sorption theories given here. Nevertheless, it has been a source of "standard" isotherms and has been used by Simpson (1973, 1979) to test a number of sorption theories. It provides the widest range of sorption data currently available for wood.

3.2 Moisture Sorption Theories and Equations

Several moisture sorption theories have been proposed which postulate that moisture is taken up by wood on internal sorption sites attractive to water. The best known of these theories is that of Brunauer et al. (1938), the well-known Brunauer-Emmett-Teller or BET theory, which assumes that from one to many layers of water molecules may be sorbed on these sorption sites. The BET theory is an extension of an earlier theory of Langmuir (1918) which assumes that only a single layer of vapor molecules could be condensed on these internal surfaces.

The BET sorption theory has been successfully applied to wood and other hygroscopic solids at water vapor activities below approximately 0.4. However, it does not describe the sorption isotherm satisfactorily at higher activities. A modification of the BET theory proposed by Dent (1977) gives a more satisfactory description of the moisture sorption isotherm for wood. It will be discussed first, followed by consideration of the better-known BET theory.

3.2.1 Dent Sorption Theory

In both the Dent and BET surface sorption theories, water is presumed to be sorbed in two forms. These are "primary" molecules sorbed directly on primary sorption sites within the cell wall, and "secondary" molecules sorbed on secondary sites. The primary sites are those of high binding energy, such as accessible hydroxyl groups. The secondary sites exhibit lower binding energies and may be considered to be superposed on primary sites or other secondary sites (Fig. 3.1).

The model described above is identical for the Dent and BET theories. The two theories differ in that the BET model assumes that the thermodynamic properties of the secondary water are identical with those of ordinary liquid water, while the Dent model assumes that they are different. Both models assume that the properties of water in the various secondary layers, that is, the second, third, fourth, etc. layers, are identical.

The internal wood surface is presumed to have a total area A available for sorption of water molecules. At a given moisture content a portion of this total area, A_0, is free of water. The area A_1 is covered with a monolayer, A_2 with two layers, A_3 with three layers, A_i with i layers, etc. (Fig. 3.2).

When the internal wood surface is at equilibrium with water vapor in the surrounding atmosphere, there exists an equilibrium between the rates of condensation and evaporation of water at any layer. The rate of condensation on A_0 is presumed to be equal to the rate of evaporation from A_1. Also, the condensation rate on A_0 is taken to be proportional to the vapor activity h, at constant temperature. The rate of evaporation from area A_1 is proportional to A_1. Thus, the equilibrium between rate of condensation on A_0 and of evaporation on A_1 can be written

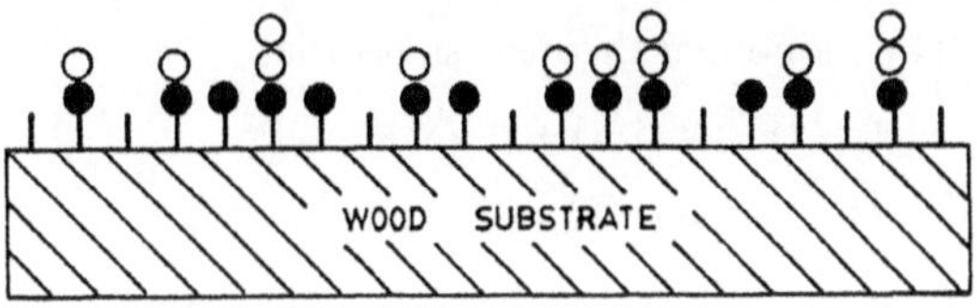

Fig. 3.1. Schematic diagram showing primary sorption sites (*vertical lines*) in the cell wall, some of which are unoccupied and others occupied by primary water molecules (*dark circles*) and sometimes by secondary sorbed molecules (*open circles*)

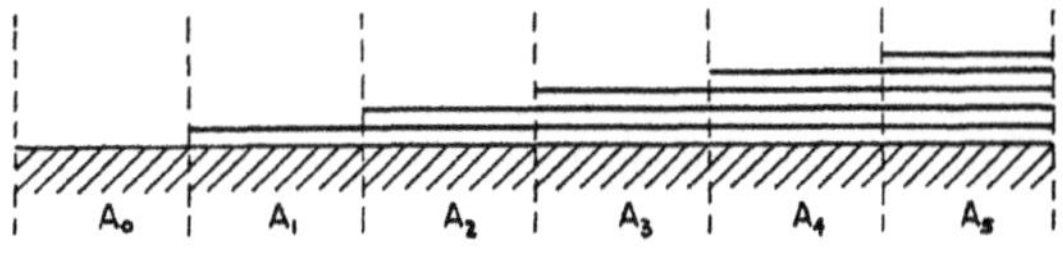

Fig. 3.2. Surface sorption model, showing the surfaces A_0, A_1, A_2, etc. covered by 0, 1, 2, etc. layers of water molecules

$$A_1 = b_1 h A_0 \ , \tag{3.2}$$

where b_1 is the constant of proportionality.

Similar equilibrium relationships are presumed to exist between other pairs of adjacent areas, as follows:

$$A_2 = b_2 h A_1 : A_3 = b_3 h A_2 : A_4 = b_4 h A_3 \tag{3.3}$$

$$A_i = b_i h A_{i-i} \ . \tag{3.4}$$

The total internal surface area A is equal to

$$A = A_0 + A_1 + A_2 + A_3 + \ldots A_i \ . \tag{3.5}$$

The total volume V_0 possible in the monolayer is given by

$$V_0 = d_0 A = d_0 (A_0 + A_1 + A_2 + \ldots A_i) \ , \tag{3.6}$$

where d_0 is the thickness of a single layer of water. The total volume V in all layers is

$$V = d_0 (A_1 + 2A_2 + 3A_2 + \ldots iA_i) \ . \tag{3.7}$$

An important assumption is now made to the effect that the proportionality coefficients b_i are all equal to each other for all cases except for $i = 1$. In other words,

$$b_2 = b_3 = b_4 = \ldots b_i \quad (i \neq 1) \ . \tag{3.8}$$

With this assumption, the ratio

$$A_i = b_2 h A_{i-1} \quad (i \neq 1) \ , \tag{3.9}$$

and Eqs. (3.6) and (3.7) can be written as

$$V_0 = d_0 [A_0 + A_1 (1 + x + x^2 + x^3 + \ldots)] \tag{3.10}$$

$$V = d_0 A_1 (1 + 2x + 3x^2 + 4x^3 + \ldots) \ , \tag{3.11}$$

where $x = b_2 h$.

The bionomial theorem states that

$$(1-x)^{-1} = (1 + x + x^2 + x^3 + \ldots) \tag{3.12}$$

$$(1-x)^{-2} = (1 + 2x + 3x^2 + 4x^3 + \ldots) \ . \tag{3.13}$$

Combining Eqs. (3.10)–(3.13), the ratio V/V_0 can be written

$$V/V_0 = [A_1/(1-x)^2]/[A_0 + A_1/(1-x)] \ . \tag{3.14}$$

Eliminating A_1 by use of Eq. (3.2) and substituting $b_2 h$ for x, and rearranging, Eq. (3.14) becomes

$$V/V_0 = b_1 h/[(1 - b_2 h)(1 - b_2 h + b_1 h)] \ . \tag{3.15}$$

Equation (3.15) can be written in terms of wood moisture content M or m if it is assumed that the ratios M/M_0 or $m/m_0 = V/V_0$, where M_0 and m_0 are the moisture contents corresponding to complete monolayer coverage of all available sorption sites. Thus,

$$m/m_0 = M/M_0 = b_1h/[(1-b_2h)(1-b_2h+b_1h)] \quad . \tag{3.16}$$

Equations (3.15) and (3.16) give the ratio of the total moisture content of the wood to that of the completely filled monolayer. This total moisture consists of two components, primary and secondary water. An equation for the primary water can be derived using the following relationships for the volume V_1 of primary water

$$V_1 = d_0(A_1+A_2+A_3\ldots) \; , \tag{3.17}$$

which, divided by V_0, given by Eq. (3.6), can be shown to yield

$$V_1/V_0 = M_1/M_0 = m_1/m_0 = b_1h/(1-b_2h+b_1h) \quad . \tag{3.18}$$

The secondary water M_2 of m_2 can be obtained by subtracting Eq. (3.18) from Eq. (3.16). The resulting equation reduces to

$$M_2/M_0 = m_2/m_0 = b_1b_2h^2/[(1-b_2h)(1-b_2h+b_1h)] \quad . \tag{3.19}$$

Equations (3.18) and (3.19) give the primary and secondary moisture contents m_1 and m_2 as functions of the three constants b_1, b_2, and m_0 and of the activity h. The sum of m_1 and m_2 is equal to the total moisture content m.

If the constants b_1, b_2, and m_0 are known, the sorption isotherm can be predicted. However, the sorption isotherm is usually measured experimentally, from which the constants b_1, b_2, and m_0 can be evaluated. The most convenient relationship for doing so is obtained by rearranging Eq. (3.16) into the following form

$$h/m = (1-b_2h+b_1h)(1-b_2h)/m_0b_1 \quad . \tag{3.20}$$

This can be rearranged to

$$h/m = (1/m_0b_1)+[(b_1-b_2)/m_0b_1]h-[(b_1b_2-b_2^2)/m_0b_1]h^2 \; , \tag{3.21}$$

a parabola of the form similar to Eq. (3.1)

$$h/m = A+Bh-Ch^2 \; , \tag{3.22}$$

where

$$A = 1/m_0b_1 : B = (b_1-b_2)/m_0b_1 : C = (b_1b_2-b_2^2)/m_0b_1 \quad . \tag{3.23}$$

In practice, the measured sorption isotherm is plotted as h/m or H/M against h, where H = 100 h. A parabola of the form shown in Eq. (3.22) is then fitted to the plotted points and the values of the empirical constants A, B, and C are calculated using linear regression techniques. Figure 3.3 shows plotted points and fitted curves of H/M against h at 25 °C for wood and bark samples, taken from Okoh and Skaar (1980). Table 3.2 gives the sorption data and the calculated values of A, B, and C. It also lists values for the constants b_1, b_2, and M_0 calculated from the empirical constants A, B, and C, using the following relationships, derivable from Eq. (3.23),

$$b_2 = [-B+(B^2+4AC)^{0.5}]/(2A) \tag{3.24}$$

$$b_1 = 2b_2+B/A \tag{3.25}$$

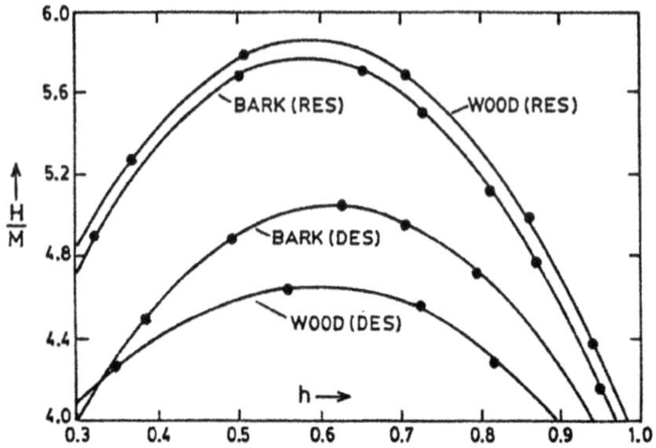

Fig. 3.3. Plotted points and fitted curves of the ratio H/M against relative vapor pressure for resorption and desorption based on the mean values for the woods and barks of ten species (Okoh and Skaar 1980)

Table 3.2. Resorption and desorption humidities and mean equilibrium moisture contents for woods and barks for ten species at +25 °C. Also shown are calculated values of A, B, C, b_1, b_2, and M_0 (Okoh and Skaar 1980)

	Wood-res.		Bark-res.		Wood-des.		Bark-des.	
	H(%)	M(%)	H(%)	M(%)	H(%)	M(%)	H(%)	M(%)
	37.0	7.01	32.3	6.59	35.0	8.20	38.5	8.47
	51.4	8.95	50.6	8.91	56.4	12.15	49.8	10.17
	71.2	12.55	65.9	11.61	72.8	15.95	63.3	12.46
	86.2	17.13	73.3	13.31	81.8	19.11	71.1	14.13
	94.1	21.74	81.8	15.90			80.0	16.97
			87.3	18.53				
			95.0	23.02				
A	1.64		1.48		2.23		1.18	
B	14.3		14.6		8.3		12.5	
C	12.1		12.4		7.0		10.1	
b_1	10.26		11.48		5.15		11.5	
b_2	0.777		0.789		0.712		0.7	
M_0	5.94		5.91		8.70		6.94	

$$M_0 = 100\,m_0 = 100/(A b_1) \ . \tag{3.26}$$

Figure 3.4 shows the calculated sorption isotherms and the experimental points for the same samples for which the curves shown in Fig. 3.3 were calculated. They were calculated from Eq. (3.22) written in the form

$$M = H/(A + Bh - Ch^2) \ , \tag{3.27}$$

where $H = 100h$. It is of interest to note that the mean resorption isotherm for the barks of the ten hardwoods are nearly identical with those for the woods. Furthermore, the values of M_0 corresponding to complete monolayer coverage are also similar for resorption. Hysteresis, however, appears to be much less pronounced for the barks than for the woods, the desorption isotherms being appreciably higher for the woods. This difference is also reflected in M_0, which is

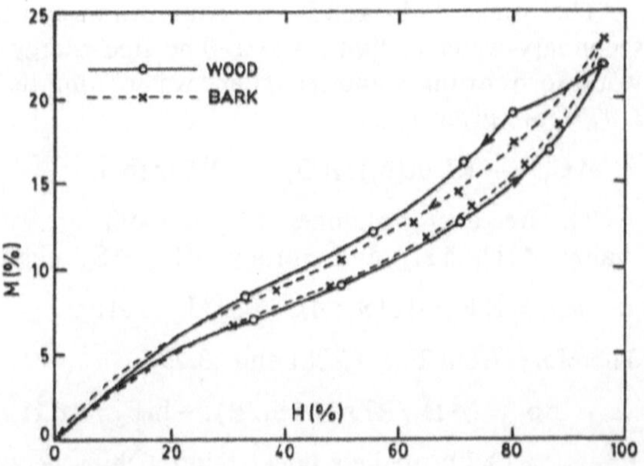

Fig. 3.4. Plotted points and fitted resorption and desorption isotherm curves based on the mean values for the woods and barks of ten species (Okoh and Skaar 1980)

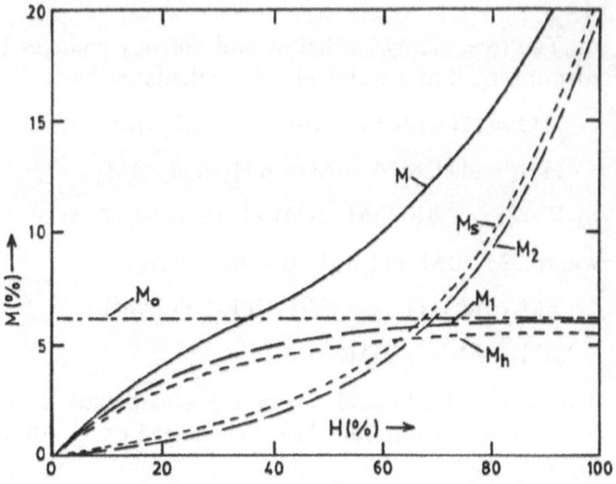

Fig. 3.5. Mean resorption isotherms for the woods of ten species and of the curves of M_h, M_s, M_1, and M_2, as functions of H. Also shown is the constant M_0

considerably greater for desorption for the woods (8.70%) than for the barks (6.94%), according to Table 3.2

The two components of the sorption isotherm, primary water M_1 and secondary water M_2, calculated by use of Eqs. (3.18) and (3.19), respectively, are plotted against relative humidity H in Fig. 3.5, for the mean adsorption isotherms for the ten hardwoods. Also shown are curves labeled M_h and M_s which relate to the Hailwood-Horrobin model discussed in Section 3.2.3. It is apparent that the curve for primary water is typical of the type 1 isotherm discussed in Chapter 1, characteristic of sorption restricted to a monolayer. The secondary isotherm conforms to the type 3 classification associated with multilayered sorption. As noted in Chapter 1, the total isotherm is of the type 2 classification, typical of hygroscopic materials such as wood.

The horizontal line in Figure 3.5 labeled M_0 is the moisture content of the fully saturated monolayer. It is the limiting moisture content for M_1, which does not attain this value even at saturation humidity. This is presumably because some sorption sites are vacant even under these conditions, as Eq. (3.18) indicates.

The constants b_1 and b_2 are equilibrium constants relating the primary and secondary water to liquid water. The free energy changes ΔG_1 and ΔG_2 when wood sorbs primary and secondary water from the liquid state, according to Dent (1977), are given by

$$\Delta G_1 = -RT\ln(b_1) : \Delta G_2 = -RT\ln(b_2) \ . \tag{3.28}$$

The free energy changes ΔG_1 and ΔG_2 are related to the corresponding enthalpy (ΔH_1, ΔH_2) and entropy (ΔS_1, ΔS_2) changes as follows

$$\Delta G_1 = \Delta H_1 - T\Delta S_1 : \Delta G_2 = \Delta H_2 - T\Delta S_2 \ . \tag{3.29}$$

Therefore, from Eqs. (3.28) and (3.29)

$$-\ln b_1 = (\Delta H_1/RT) - (\Delta S_1/R) : -\ln b_2 = (\Delta H_2/RT) - (\Delta S_2/R) \ . \tag{3.30}$$

Equation (3.30) predicts linear relationships between plots of $\ln(b_1)$ and $\ln(b_2)$ against $1/T$, from which ΔH_1 and ΔH_2 can be calculated, as well as ΔS_1 and ΔS_2.

The free energy, enthalpy, and entropy changes (ΔG_s, ΔH_2, and $T\Delta S_2$) for sorption of liquid water may be calculated by

$$\Delta G_s = \Delta G_1(\partial M_1/\partial M) + \Delta G_2(\partial M_2/\partial M) \tag{3.31}$$

$$\Delta H_s = \Delta H_1(\partial M_1/\partial M) + \Delta H_2(\partial M_2/\partial M) \tag{3.32}$$

$$T\Delta S_s = T\Delta S_1(\partial M_1/\partial M) + T\Delta S_2(\partial M_2/\partial M) \ , \tag{3.33}$$

where $\partial M_1/\partial M$ and $\partial M_2/\partial M$ are given by

$$\partial M_1/\partial M = (1 - b_2 h)^2/(1 + b_1 b_2 h^2 - b_2^2 h^2) \tag{3.34}$$

$$\partial M_2/\partial M = 1 - \partial M_1/\partial M \ . \tag{3.35}$$

The values of ΔG_s, ΔH_s, and $T\Delta S_s$ calculated from these equations should be identical with those calculated by use of Eqs. (2.20), (2.38), and (2.45), respectively, in Chapter 2.

In order to test the ability of the Dent model to predict the values of ΔG_s, ΔH_s, and $T\Delta S_s$ using Eqs. (3.31), (3.32), and (3.33), the same sorption data must be used to calculate the values directly by means of Eqs. (2.20) and (2.39) or (2.40). The reason for this is that the values of these parameters calculated from different sets of sorption data may differ considerably. Furthermore, data at two or more temperatures are required to evaluate ΔH_s and $T\Delta S_s$.

The procedure described above will be applied here to sorption data for wood tabulated from several sources by Wengert and Mitchell (1979). These isotherm data are given for only three relative humidities, 30%, 50%, and 80%, as shown in Table 3.3. This makes it convenient to fit the Dent model, since a minimum of three points are required. Furthermore, the values are in the center of the humidity range where the model is most applicable. Data at two temperatures, 21 °C and 49 °C, are shown in Table 3.3. Those at the third temperature of 82 °C are not included here, since only two are necessary and data at the highest temperature are more likely to be in error.

Each set of sorption data was fitted to Eq. (3.27) to calculate the empirical coefficients A, B, and C. From these the Dent coefficients M_0, b_1, and b_2 were

Table 3.3. Summary of equilibrium moisture content data on wood (Wengert and Mitchell 1979)

Temp.	21 °C			49 °C		
Rel. hum.	30%	50%	80%	30%	50%	80%
Data 1[a]	6.1	9.2	16.0	5.2	8.0	14.1
Data 2[a]	6.2	9.6	17.3	4.7	7.9	14.9
Data 3[a]	5.9	8.9	16.4	5.2	8.1	14.3
Data 4[a]	7.4	10.8	17.7	5.8	8.8	15.1
Mean	6.4	9.6	16.8	5.2	8.2	14.6

[a] 1. USDA Wood Handbook (1974). 2. Cech and Paff (1977). 3. Hildebrand (1970). 4. Pratt (1974).

Table 3.4. Summary of Dent model coefficients based on data shown in Table 3.3, and data for sheep wool given by Dent (1977)

Source[a]	Temp (°C)	M_0 (%)	b_1	b_2	ΔG_1 (cal/g)[b]	ΔG_2 (cal/g)
1	21	7.5	4.53	0.723	−49	+10
1	49	6.8	3.93	0.719	−49	+12
2	21	8.0	4.00	0.741	−45	+10
2	49	8.2	2.20	0.703	−28	+12
3	21	6.6	5.54	0.787	−56	+8
3	49	7.2	3.51	0.708	−45	+12
4	21	9.0	4.99	0.678	−52	+13
4	49	7.6	4.06	0.699	−50	+13
5	25	8.1	9.2	0.75	−72	+9
5	40	8.3	6.3	0.71	−64	+12

[a] 1. USDA Wood Handbook (1974). 2. Cech and Pfaff (1977). 3. Hildebrand (1970). 4. Pratt (1974). 5. Dent (1977).
[b] Note that ΔG_1 and ΔG_2 are in cal g^{-1} rather than cal mol^{-1} as in Eqs. (3.28) and (3.29).

evaluated by use of Eqs. (3.24 to 3.26). They are tabulated in Table 3.4, together with the values of ΔG_1 and ΔG_2, calculated by means of Eq. (3.28). Also shown are the values of these coefficients as given by Dent (1977) based on sorption data for sheep wool, at 20° and 40 °C. Wool isotherms are comparable in shapes and magnitudes to those for wood.

Table 3.5 lists the mean values of ΔG_1 and ΔG_2 for each set of isotherm data shown in Table 3.4, as well as the mean values of Q_1, Q_2, $T\Delta S_1$, and $T\Delta S_2$ calculated by use of Eqs. (3.29) and (3.30) using data for b_1 and b_2 at the two temperatures in each case. For example, using the sorption isotherm values of b_1 and ΔG_1 shown in Table 3.3 for sorption data of Cech and Pfaff (1977)

$$Q_1 = \{RT_1T_2/[18(T_2-T_1)]\}\ln(2.20/4.00) = -224 \text{ cal/g}$$

$$TS_1 = Q_1 - \Delta G_1 = -224-(-45-28)/2 = -188 \text{ cal/g} ,$$

where T_1 and T_2 are 294 and 322 K, respectively.

It is evident from Tables 3.4 and 3.5 that there are considerable discrepancies among the values of the parameters shown for the different sets of sorption data. It is anticipated for example that ΔG_1 should approximate $(\Delta G)_0$, the value of

Table 3.5. Mean values of the thermodynamic coefficients (cal/g) at the mean temperature indicated (Sources as in Table 3.4)[a]

Source	Temp (°C)	ΔG_1	ΔG_2	Q_1	Q_2	$T\Delta S_1$	$T\Delta S_2$
1	35	−49	+11	−53	−2	−4	−13
2	35	−36	+11	−224	−20	−188	−31
3	35	−50	+10	−170	−40	−120	−50
4	35	−51	+13	−78	−66	−27	−79
5	32	−68	+10	−260	−38	−192	−49

[a] Note that ΔG and $T\Delta S$ are in cal/g rather than in cal/mol as in Eqs. (3.28−3.33).

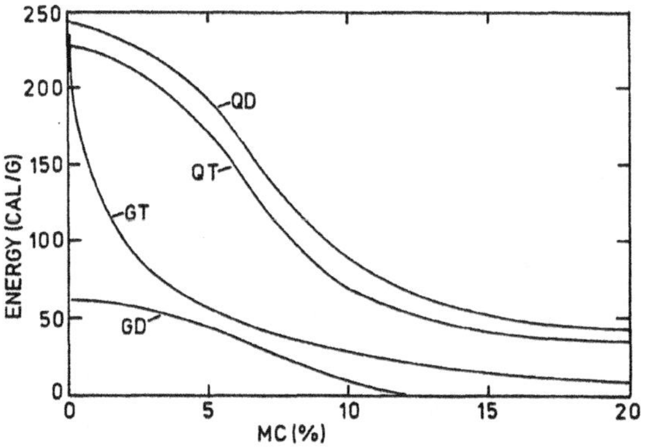

Fig. 3.6. Curves of ΔG_s predicted (GD) by the Dent model using Eq. (3.31) and calculated thermodynamically (GT) by Eq. (2.20), and of Q_s predicted (QD) by the Dent model using Eq. (3.32) and calculated thermodynamically (QT) by Eq. (2.40), all plotted against moisture content, for sheep wool

the free energy decrease extrapolated to zero moisture content by use of Eq. (2.63), approximately 100 cal/g. The values shown in Table 3.5 are approximately half this value. Likewise the values of Q_1 should be close to $(Q_s)_0$, the value of the differential heat of sorption extrapolated to zero moisture content (Table 2.4), in the order of 250 cal/g. The values shown for wood range in magnitude from 53 to 224 with a mean of 131, about half the magnitude anticipated. The value of Q_1 for sheep wool shown in Table 3.5 however is 260 cal/g, of the magnitude expected.

It is instructive also to compare curves of the free energy ΔG_s and enthalpy changes Q_s predicted by the Dent model with those calculated from the same data using the standard thermodynamic equations. Figure 3.6 shows four curves, calculated from the constants for sheep wool isotherms given by Dent (1977). Two of these curves are plots of the free energy change ΔG_s, and two of the enthalpy changes Q_s, all as functions of moisture content. It is apparent that the magnitudes of the predicted free energy decrease ΔG_s are much lower than those calculated from Eq. (2.20) applied to the same data. In fact the predicted values become negative (the free energy increases during sorption) above about 12% moisture content. This is due to the positive value of ΔG_2 and the strong rate of increase dM_2/dM of secondary water at the higher moisture contents.

The experimental (QT) and predicted (QD) curves of the decrease in enthalpy or differential heat of sorption Q_s, on the other hand, are in reasonable agree-

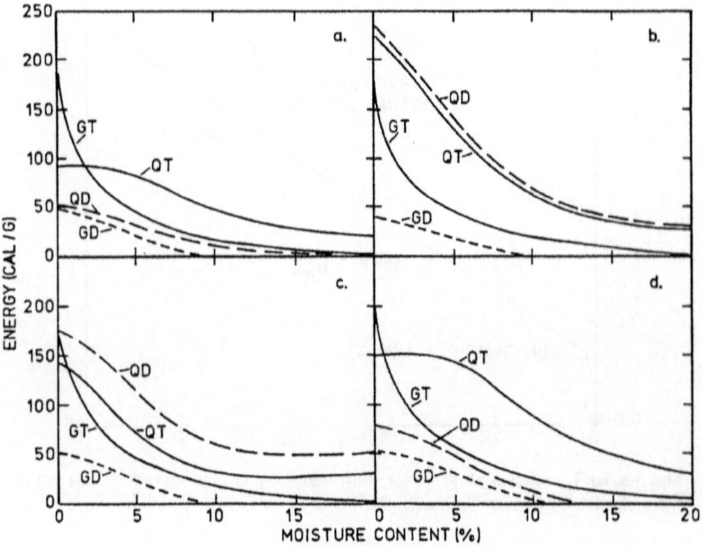

Fig. 3.7. Curves similar to those of Fig. 3.6, but calculated from sorption isotherms for wood, using data of: **a** = USDA Wood Handbook (1974). **b** = Cech and Pfaff (1977). **c** = Hildebrand (1970). **d** = Pratt (1974)

ment based on these data. Furthermore, they decrease rapidly with increasing moisture content, as described by Eq. (2.53).

Four sets of curves, each including the same four parameters shown in Fig. 3.6 are plotted in Fig. 3.7. Each set of curves refers to one of the four sets of wood isotherm coefficients listed in Tables 3.3 and 3.4. The curves calculated from the data of Hildebrand (1970) and of Cech and Pfaff (1977) give better agreement between experimental (QT) and theoretical (QD) values of Q_s than do those of the other two sets of data. The reverse is true for the curves of ΔG_s although, as noted above, all of the values of ΔG_s determined from the model (GD) are lower than those obtained (GT) by application of Eq. (2.20) to the sorption data directly.

The entropy change term $T\Delta S_s$ is negative in all cases since both $T\Delta S_1$ and $T\Delta S_2$ are negative (Table 3.4). The values of $T\Delta S_s$ calculated from Eq. (3.33) are not plotted here as a function of moisture content. They will always be negative, as anticipated, but their magnitudes do not always decrease with increasing moisture content as one would expect. Again it is not known if this is a limitation in the sorption data or in the sorption theory.

3.2.2 BET Sorption Theory

The Brunauer-Emmett-Teller (BET) theory, first proposed by Brunauer et al. (1938) is the most widely used model for describing the sorption of gases on surfaces. It has been particularly useful in calculating surface areas and the energies

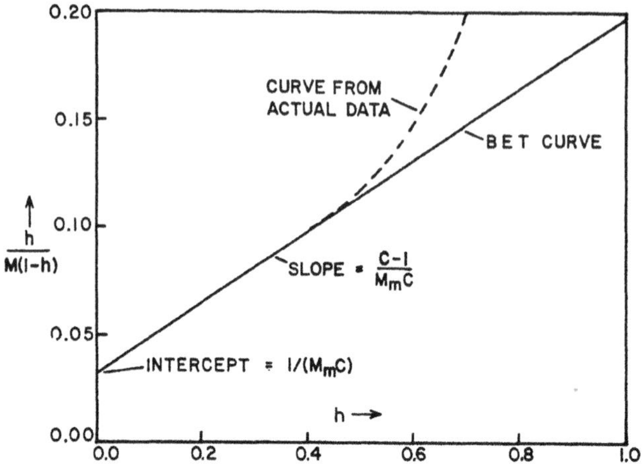

Fig. 3.8. BET isotherm plot of data from USDA Wood Handbook (1955) at 110 °F, showing deviation from the BET curve at high activity h

of sorption. As noted earlier, it fits the moisture sorption isotherm for hygroscopic materials such as wood at relative vapor pressures below 0.4 quite well. However, at higher humidities, the fit is not satisfactory, although a modification limiting the number of allowable layers improves the fit considerably, as discussed below.

The BET sorption model differs from the Dent model in that the former assumes that the thermodynamic properties of the secondary layers of water are essentially the same as those of liquid water. This means that the term b_2 in the Dent model is taken as unity. With this assumption Eq. (3.16) can be written

$$M/M_m = Ch/[(1-h)(1-h+Ch)] \ , \tag{3.36}$$

where the constant C replaces b_1 in the Dent model, and M_m replaces M_0. Equation (3.36) can be rearranged into the commonly used form of the BET equation given here as

$$h/[M(1-h)] = (1/M_mC)+(C-1)h/M_mC \ . \tag{3.37}$$

Equation (3.37) predicts that a plot of the sorption data in the form $h/[M(1-h)]$ against h is a straight line with a slope of $(C-1)/M_mC$ and an intercept $(1/M_mC)$ as shown in Fig. 3.8. The two BET constants M_m and C can be calculated if the theoretical relationship predicted by Eq. (3.37) is obeyed. The BET constants may be slightly different from the corresponding constants M_0 and b_1 of the Dent theory because of the influence of the constant b_2 in the latter. They have the same interpretation, however.

The BET equation was modified by Brunauer et al. (1983) to improve the fit to sorption at relative vapor pressures above 0.4. This was done by restricting the maximum number of layers permissible to n layers, thus modifying Eq. (3.37) to

$$\frac{h}{M(1-h)} = \frac{1-h+Ch(1-h^n)}{M_mC[1-(n+1-nh)h^n]} \ . \tag{3.38}$$

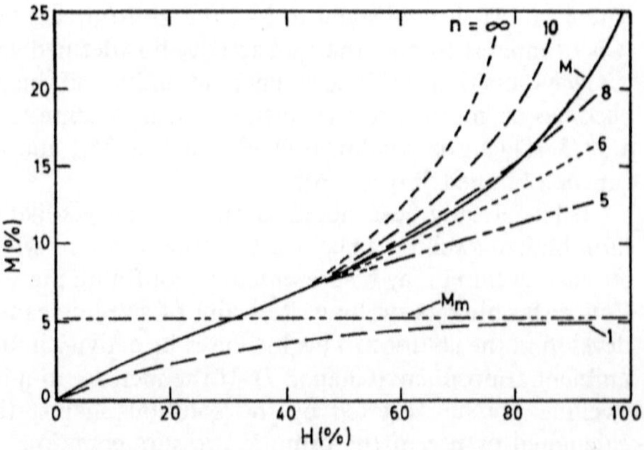

Fig. 3.9. Comparison of predicted BET sorption isotherms (*broken lines*), for several values of n, with the actual isotherm, using same data as in Fig. 3.8. The *horizontal line* represents M_m, the fully complete monolayer

It reduces to Eq. (3.37) when n = infinity, that is when the number of layers is unrestricted. When n = 1, Eq. (3.38) reduces to the simple Langmuir (1918) equation in the form

$$M_1/M_m = h/[(1/C)+h] \ , \tag{3.39}$$

which predicts a simple type 1 isotherm, characteristic of sorption in a monolayer.

Equation (3.38) can be used to calculate the sorption isotherm if C and M_m are evaluated by use of the linear portion of Eq. (3.37), the unmodified BET equation which applies only below an h near 0.4. At least two data points below this activity are required to define the straight line. These values of C and M_m, together with assumed values of n, are then substituted into Eq. (3.38) rearranged into a form suitable for calculating moisture content M as a function of h. Figure 3.9 shows several such curves for different values of n, together with the actual sorption isotherm used to evaluate M_m and C from the curves shown in Fig. 3.8.

It is apparent from Fig. 3.9 that the sorption curve is fitted best for values of n of 6, 7, or 8. The value of n used in each case is the *maximum* number of layers permitted on any sorption site. The *average* number of layers at any relative vapor pressure may be estimated at any value of h as the ratio M/M_1, where M is the total moisture content and M_1 that of the monolayer, obtained from Eq. (3.39), shown as the lowest curve in Fig. 3.9. These ratios are approximately 1.5, 1.8, 2.5, and 3.5 for h = 0.3, 0.5, 0.7, and 0.9, respectively.

The unmodified BET equation [Eq. (3.28)], has also been derived by the methods of statistical thermodynamics (White and Eyring 1947, G. King 1960). Two equilibrium equations are obtained by these methods. The first, derived from the condition that ΔG_s as a function of the amount of primary water M_1 is a minimum, i.e., $\partial(\Delta G_s)/\partial M_1$ is zero, may be written

$$C(M-M_1)(M_m-M_1) = (M_1)^2 \ . \tag{3.40}$$

The second, based on the assumption that the chemical potentials of the sorbed moisture μ_s and of its vapor, μ_v, are identical at equilibrium, may be written

$$h = (M-M_1)/M = M_2/M \ , \tag{3.41}$$

where $M - M_1$ is equivalent to M_2, the moisture content of the secondary layers. It is of interest to note that the activity h is defined as the ratio M_2/M. Since M_2 is always less than M, h never becomes unity and the predicted sorption isotherm becomes asymptotic to the vertical axis as h approaches unity. Equations (3.40) and (3.41) can be combined by eliminating M_1, and the standard BET equation can be obtained [Eq. (3.36)].

It has already been noted (Sect. 2.7) that gel-like materials such as wood exhibit limited swelling, in contrast to true solutions which have no limit. In the latter case swelling may be prevented by confining the solution to a limited volume, thus restraining its swelling. Restraint of swelling causes a hydrostatic pressure to develop in the solution. This increases its activity h until it is equal to that of the ambient atmosphere (Chap. 2.7). If the increase in h is known, the hydrostatic or swelling pressure exerted by the solution against the confining walls can be calculated by use of the osmotic pressure equation.

Within the cell wall of wood the restraint to swelling is inherent in the intertwining and cross-linking of the long cellulosic molecules, as well as in the layered structure of the microfibrils. Work must be done by the sorbed water to overcome the resistance of this network to swelling. The magnitude of this work or free energy loss due to swelling can be calculated if the change in activity h of the sorbed water due to the swelling stress is known.

It will be assumed here that the unrestrained sorption is predicted by the BET isotherm, using the constants M_m and C calculated from the sorption isotherm below $h = 0.4$. This isotherm will then be compared with the isotherm calculated by fitting the Dent isotherm model of the experimental sorption data, thus giving a convenient representation of the experimental isotherm. The highest curve in Fig. 3.9, that for n = infinity, is of the form predicted by the simple BET isotherm [Eq. (3.35)]. Figure 3.10 shows a replot of this curve, as well as the Dent isotherm, assumed here to be equivalent to the actual isotherm. The two horizontal lines indicate the magnitudes of the BET and Dent constants M_m and M_0, respectively. Curves of M_1 and M_2, calculated by Eqs. (3.40) and (3.41), are not shown but are similar to those in Fig. 3.5.

At a given moisture content the activity h of the experimental isotherm is always greater than h′, that predicted by the BET model. The difference can be

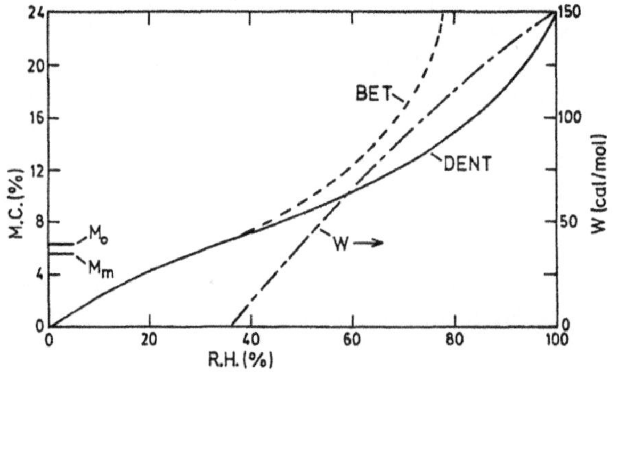

Fig. 3.10. Comparison of the Dent and BET (assuming n = infinity) isotherms, including values of M_0 and M_m. Also shown is a curve giving the calculated value, based on Eq. (3.44), of work W required to swell the wood, as a function of h

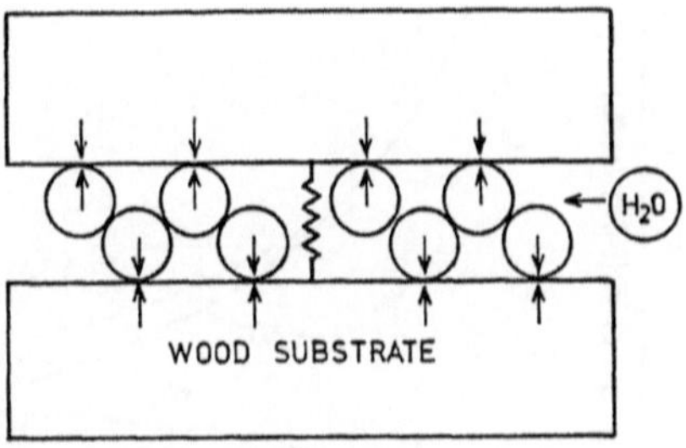

Fig. 3.11. Schematic diagram showing the work done by water molecules (*circles*) against the resistant spring (representing the cohesive elastic strength of the internal wood bonds) during water sorption and swelling

accounted for by assuming that the wood resists swelling and exerts a hydrostatic pressure on the sorbed water, thereby increasing its activity from h′ to h. The pressure presumably arises from the resistance to deformation of bonds in the cell wall as the water molecules force themselves in, as Fig. 3.11 illustrates.

Equation (3.41) gives h′ based on the BET model. Taking the logarithm of both sides and multiplying by RT, gives

$$RT\ln(M_2/M) = RT\ln(h') \ , \tag{3.42}$$

equivalent to the free energy change $\Delta G'_s$ in the sorbed water under stress-free conditions. If the wood resists swelling, work must be done by the water to overcome this resistance. Equation (3.42) is then modified to account for this additional work. Thus

$$RT\ln(M_2/M) = -W + RT\ln(h) \ , \tag{3.43}$$

where W is the work done by the water in swelling the wood. The magnitude of h is therefore greater than h′, the value predicted by the BET equation.

Equations (3.42) and (3.43) can be combined and rearranged to solve for W. Thus

$$W = RT\ln(h/h') \ . \tag{3.44}$$

The work W required to swell the wood is plotted as a function of relative humidity H in Fig. 3.10. It increases with increasing relative humidity H or activity h as anticipated, since more energy is expended in swelling the wood at higher moisture contents.

The molar free energy change ΔG_s of sorption is given by Eq. (2.20), repeated here as

$$\Delta G_s = RT\ln(h) \ . \tag{3.45}$$

That predicted from the BET equation is

$$\Delta G'_s = RT\ln(h') \ . \tag{3.46}$$

Combining Eqs. (3.42)–(3.46) gives

$$\Delta G_s = W + \Delta G'_s \ . \tag{3.47}$$

It should be noted that the free energy changes ΔG_s and $\Delta G'_s$ are negative while the work W is positive. Therefore, the absolute magnitude of ΔG_s is smaller than that of $\Delta G'_s$, that is, it is less negative than $\Delta G'_s$.

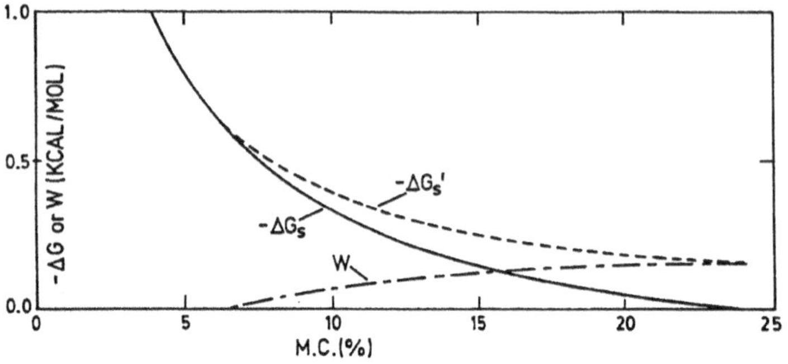

Fig. 3.12. Curves of the calculated free energy changes ($-\Delta G_s$) and of the work of swelling (W), plotted against moisture content

Figure 3.12 shows plots of $-\Delta G_s$, W and $-\Delta G_s'$ against wood moisture content. It is evident that the work of swelling W becomes more important at higher moisture contents. In other words, based on this model, the work done by the wood on the sorbed water increases its free energy content to a greater extent at high than at low wood moisture contents. Presumably this is because the swelling of the wood increases with increasing moisture content.

According to this model, the swelling at low moisture contents does not involve significant swelling stress. This may be interpreted to indicate that the take-up of primary water is not accompanied by sufficient bond deformation in the wood structure to resist the swelling. Although there is some wood swelling at low moisture contents, the wall structure may be sufficiently porous and flexible to permit reduced and/or stress-free swelling. This would partially explain the reduced hygroexpansion coefficients of wood at low moisture contents (Fig. 4.10).

Incorporating the work of swelling into the BET model as discussed above, Eqs. (3.40) and (3.41) may be modified to

$$C(M-M_1)(M_m-M_1) = (M_1)^2 \tag{3.48}$$

$$h = X(M-M_1)/M \ , \tag{3.49}$$

where $X = \exp(W/RT)$. These may be combined to obtain a modified BET equation similar to Eq. (3.36) but including the swelling stress term X. Thus

$$M/M_m = Ch/\{(X-h)[1+(Ch-h)/X]\} \ . \tag{3.50}$$

When the work of swelling W is zero, X = unity and Eq. (3.50) reduces to Eq. (3.36). Since W is positive if swelling restraint is present, the value of X is greater than unity.

G. King (1960) derived an equation similar to Eq. (3.50) for constrained sorption in textiles. He also notes that restricting the number of layers in the BET model [Eq. (3.38)] has an effect on the sorption isotherm similar to that of constrained sorption, that is, sorption restricted to a finite number of layers.

White and Eyring (1947), using statistical thermodynamics, derived sorption equations similar to Eqs. (3.48) and (3.49). They calculated the magnitude of W based on a model for the bond deformation within the swelling fiber matrix for textile fibres.

3.2.3 Hailwood-Horrobin Sorption Theory

The Hailwood and Horrobin (1946) moisture sorption theory predicts the sigmoid type 2 isotherm normally found for wood, as given by Eq. (3.1). It has been applied to explain the sorption isotherms for textile materials as well as for wood. Although it has been criticized because of some of the assumptions used in its derivation, it provides an estimate of certain fundamental sorption parameters similar to those obtained from the Dent model (Sect. 3.2.1).

The Hailwood-Horrobin (HH) model considers that part of the sorbed water forms a hydrate with the wood and the balance forms a solid solution in the cell wall. In the model derived here only a single hydrate is considered, although the authors also discuss a model with two distinct kinds of hydrate. The cell wall is presumed to consist of three chemical "species"; dry wood, hydrated wood and dissolved water, treated as an ideal solution.

If X_0, X_h, and X_d are the numbers of moles of dry wood, hydrated wood (also water of hydration) and dissolved water, respectively, the total number of moles of all three species is $X_0+X_h+X_d$. Assuming ideal solution behavior, the activities A_0, A_h, and A_d are equal to their mole fractions in the solution, or

$$A_0 = X_0/(X_0+X_h+X_d) \tag{3.51}$$

$$A_h = X_h/(X_0+X_h+X_d) \tag{3.52}$$

$$A_d = X_d/(X_0+X_h+X_d) \ . \tag{3.53}$$

At equilibrium, the equilibrium constant K_h, defined as the ratio of the activity A_h of the hydrate to the product A_0A_d of the activities of the two reactants, dissolved water and unhydrated wood, is, using Eqs. (3.51) and (3.52)

$$K_h = A_h/A_0A_d = X_h/X_0A_d$$

or

$$X_h = K_hX_0A_d \ . \tag{3.54}$$

The constant K_d, defined as the ratio A_d/h is the equilibrium constant between the dissolved water and the vapor in the atmosphere. Thus,

$$A_d = K_dh \ . \tag{3.55}$$

The ratio $X_h/(X_h+X_0)$ is equal to the moles X_h of hydrated wood, and of water of hydration, per mole of dry wood, since this is given by X_h+X_0. Combining this with Eqs. (3.54) and (3.55) gives

$$X_h/(X_h+X_0) = K_hK_dh/(K_hK_dh+1) \ . \tag{3.56}$$

The ratio $X_d/(X_h+X_0)$ is equal to the moles of dissolved water X_0 per mole of dry wood, obtained by inverting Eq. (3.53) and rearranging to

$$(X_h+X_0)/X_d = (1/A_d)-1 \ , \tag{3.57}$$

which, inverted and combined with Eq. (3.55), yields

$$X_d(X_h+X_0) = K_d h/(1-K_d h) \; . \tag{3.58}$$

The terms X_h and X_d are equivalent to

$$X_h = m_h w_0/18 : X_d = m_d w_0/18 \; , \tag{3.59}$$

where m_h and m_d are the moisture contents corresponding to the hydrated and dissolved water, respectively, w_0 is the dry weight of the wood and 18 is the molecular weight of the water. The sum X_h+X_0 is the number of moles of dry wood, which may be expressed as

$$X_h+X_0 = w_0/W \; , \tag{3.60}$$

where W is the apparent molecular weight of the wood per mole of sorption or hydration sites. The sum of Eqs. (3.56) and (3.58) gives the total moles of water per mole of dry wood. Combining this with Eqs. (3.59) and (3.60) gives

$$(X_h+X_d)/(X_h+X_0) = (W/18)(m_h+m_d) = (W/18)m \; , \tag{3.61}$$

since the total moisture content $m = m_h+m_d$.

The term 18/W is equivalent to m_0, the moisture content corresponding to complete hydration of all of the available sorption sites in the wood. It is analogous to, and will be shown to be equal to, the moisture content m_0 corresponding to complete monolayer coverage in the Dent model (Sect. 3.2.1).

Equation (3.61) can be rearranged, substituting m_0 for 18/W, and combined with Eqs. (3.56) and (3.58) to give

$$m = m_h+m_d = m_0 K_h K_d h/(1+K_h K_d h)+m_0 K_d h/(1-K_d h) \; . \tag{3.62}$$

The first term corresponds to m_h, the moisture content of the hydrated water and the second to m_d, that of the dissolved water according to this model.

According to Eq. (3.62), the two equilibrium constants K_h and K_d, together with m_0, determine the sorption isotherm. They also determine the magnitude of m_h and m_d at any vapor activity h.

Equation (3.62) can be rearranged into a form identical with Eq. (3.22), obtained from the Dent model. Thus

$$h/m = A+Bh-Ch^2 \; , \tag{3.63}$$

where

$$A = 1/[m_0 K_d(K_h+1)] : B = (K_h-1)/[m_0(K_h+1)] : C = K_h K_d/[m_0(K_h+1)] \; . \tag{3.64}$$

Thus both the Hailwood-Horrobin (1946) and Dent (1977) sorption theories predict the same empirical parameters A, B, and C and also three fundamental constants; K_h, K_d, and m_0, or b_1, b_2, and m_0 which can be calculated from A, B, and C. It can be shown from Eqs. (3.22) and (3.64) that m_0 is identical for the two models, as are b_2 and K_d. However, b_1 and K_h are slightly different. For example, the values of K_h included in Table 3.6 are somewhat larger than the corresponding values of b_1.

Although the water of hydration m_h of the Hailwood-Horrobin model and the primary water m_1 of the Dent model are analogous, the latter is slightly greater at any total moisture content (Fig. 3.5). Therefore, m_2 is slightly lower than m_s since $m_1+m_2=m=m_h+m_d$, as Fig. 3.5 shows.

Hailwood and Horrobin (1946) proposed a more general solution model than the single hydrate model already described. In the more general model, hydrates of several kinds are proposed, each with a different equilibrium constant K_1 relating it and the dissolved water. For two hydrates, m_h in Eq. (3.62) is replaced by

$$m_h=\frac{m_0(K_hK_dh+2K_hK_h'K_d^2h^2)}{(1+K_hK_dh+K_hK_h'K_d^2h^2)}, \tag{3.65}$$

where K_h and K_h' are the respective equilibrium constants between the first and second hydrates and the dissolved water, and K_d is that between the dissolved water and its vapor, as in Eq. (3.62). The equation for m_d remains the same as in Eq. (3.62). Equation (3.65) reduces to Eq. (3.62) if $K_h'=0$, that is, if the second hydrate does not exist.

It should be noted that the values of the constants K_h, K_d, and m_0 will be different in general for the two-hydrate than for the single-hydrate model. The calculated values of m, m_h, and m_d will also be different, in general. Figure 3.13, adapted from Simpson (1979), shows the total sorption isotherm at 40 °C, based on USDA Wood Handbook (1955) data, and also curves of M_h and M_d for both the single and two-hydrate models. The ratio of M_h/M_d is somewhat greater for the two-hydrate model. The total isotherm is also slightly different for the two models although this is not apparent from the figure.

Spalt (1958) was the first to analyze moisture sorption isotherms for wood by means of the Hailwood-Horrobin single-hydrate model, which he applied to his own resorption and desorption data on eight hardwoods and eight softwoods at 90 °F. His calculated values of A, B, C from Eq. (3.63), as well as the constants M_0, K_h, and K_d, are listed in Table 3.6, for both resorption and desorption isotherms for three of the woods. Also included in the table are calculated values

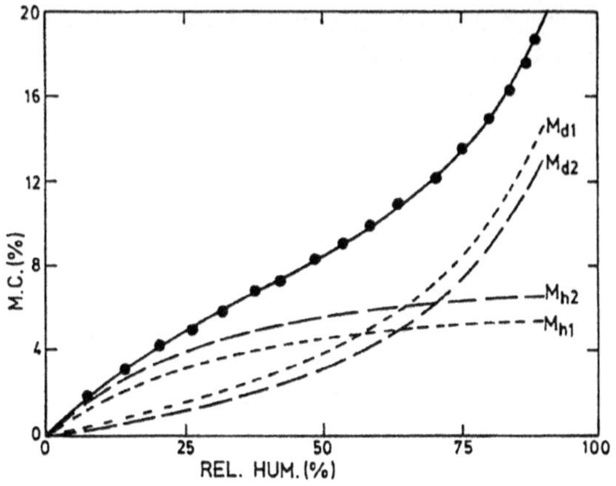

Fig. 3.13. Total sorption isotherm at 40 °C, together with curves of hydrated M_h and dissolved M_d water for both the single-hydrate and two-hydrate Hailwood-Horrobin sorption models. (After Simpson 1979)

Table 3.6. Hailwood-Horrobin and Dent coefficients for various woods at 90 °F (Spalt 1958)

Kind of wood	A	B	C	M_0	K_d (b_2)	K_h	b_1
Redwood (D)	1.774	9.85	7.25	8.22	0.659	9.43	6.87
Redwood (R)	2.267	13.84	11.72	5.79	0.754	9.09	7.61
Basswood (D)	2.800	8.84	8.30	7.63	0.758	5.17	4.68
Basswood (R)	2.321	16.37	15.20	7.49	0.830	9.49	8.71
W. spruce (D)	1.388	10.30	8.39	8.11	0.741	11.02	8.91
W. spruce (R)	2.209	12.13	10.33	6.47	0.750	8.32	6.99
W. spruce (D)[a]	0.377	13.20	14.66	3.57	0.712	2.79	2.70
W. spruce (R)[a]	11.744	19.69	20.68	2.71	0.731	3.29	3.14

[a] Acetylated W. spruce.

Table 3.7. List of equivalent parameters for sorption models giving a parabolic relationship between $h\,m^{-1}$ and h (after Boquet et al. 1980)

Model[a]	A	B	C	b_1[b]	M_0
1	$1/[(b_1+1)b_2m_0]$	$Ab_2(b_1-1)$	$Ab_1b_2^2$	$(B/Ab_2)+1$	$1/[Ab_2(b_1+1)]$
2	$1/(b_1b_2m_0)$	$Ab_2(b_1-2)$	$Ab_2^2(b_1-1)$	$(B/Ab_2)+2$	$1/(Ab_1b_2)$
3	$1/(b_1m_0)$	$A(b_1-b_2-1)$	$Ab_2(b_1-1)$	$(B/A)+b_2+1$	$1/(Ab_1)$
4	$1/(b_1m_0)$	$A(b_1-2b_2)$	$Ab_2(b_1-b_2)$	$(B/A)+2b_2$	$1/(Ab_1)$
5	$1/(b_1b_2m_0)$	$A(b_1-2b_2)$	$Ab_2(b_1-b_2)$	$(B/A)+2b_2$	$1/(Ab_1b_2)$
6	$1/(b_1m_0)$	$A(b_1-2b_2)$	$Ab_2(b_1-b_2)$	$(B/A)+2b_2$	$1/(Ab_1)$

[a] 1. Hailwood and Horrobin 1946. 2., 3. Anderson 1946. 4. R.D. Anderson and Hall 1948. 5. Gascoyne and Pethig 1977. 6. Dent 1977.
[b] The constant b_2, the same for all models, is given in Eq. (3.24) $b_2=[-B+(B^2+4AC)^{0.5}]/(2A)$.

for the Dent coefficient b_1. The other Dent coefficients b_2 and M_0 are identical with K_d and M_0, respectively.

Two important observations may be made from the data in Table 3.6. First, the value for M_0 is always lower for resorption from the dry condition compared with desorption. After rewetting, these sites again become available possibly because of the swelling forces of the water. The great reduction in M_0 for acetylated spruce may also be anticipated since acetylation replaces some of the hydroxyl groups which are strong sorption sites. The second observation is that different woods vary considerably with respect to their hygroscopicities, particularly in the values of M_0. Some of these differences are probably due to the presence of extractives. For example, Wangaard and Granados (1967) noted an increase in M_0 in many cases after woods with high extractive contents were extracted with water, presumably because removal of the extractives freed additional sorption sites.

Several other isotherm equations can be rearranged into the form of Eqs. (3.22) and (3.63), according to Boquet et al. (1980). The details of the theories on which these equations are based are not given here, but the constants b_1, b_2, and m_0 are essentially similar to those of the Dent model as given in Eq. (3.23). The relationships of the empirical constants A, B, and C of Eqs. (3.22) and (3.63) to b_1, b_2, and M_0 are shown in Table 3.7.

It should be noted that, in general, the constants b_1 and M_0 have different values for the different models, for a given set of empirical constants A, B, and C. If these are determined from empirical sorption data the constants b_1, b_2, and M_0 can be derived from each of the models in the manner shown by Eqs. (3.25) to (3.27). The constant b_2 is identical for all six models shown. It is obtained by use of Eq. (3.24). It is clear from Table 3.7 that the constants b_1, b_2, and M_0 of the Anderson-Hall model are identical with those of the Dent model.

The Hailwood-Horrobin constants K_h and K_d may be used to calculate the free energy changes ΔG_h and ΔG_d, for the hydrated and dissolved water, respectively, compared with liquid water. Thus

$$\Delta G_h = -RT\ln(K_h) : \Delta G_d = -RT\ln(K_d) \ . \tag{3.66}$$

Likewise, by analogy with Eqs. (3.29) and (3.30)

$$\Delta H_h = d[\ln(K_h)]/d(1/T) : \Delta H_d = d[\ln(K_d)]/d(1/T) \tag{3.67}$$

$$T\Delta S_h = \Delta H_h - \Delta G_h : T\Delta S_d = \Delta H_d - \Delta G_d \ . \tag{3.68}$$

Equations analogous to Eqs. (3.31) to (3.33) may be used to estimate ΔG_s, ΔH_s, and $T\Delta S_s$. These give results essentially similar to those given by the Dent model shown in Figs. 3.6 and 3.7.

Simpson (1979) has evaluated ΔH_h and ΔH_d, based on the temperature dependence of K_h and K_d of the Hailwood-Horrobin model, using sorption data of the USDA Wood Handbook (1955) over the temperature range from zero to 100 °C. He also evaluated similar constants based on other sorption theories which he tested. He concluded that the Hailwood-Horrobin model did not predict satisfactory values of ΔH_h and ΔH_d based on the sorption data he used. Some of the other models such as those of Peirce (1929) and of G. King (1960) gave more satisfactory values. Simpson noted that the simplifying assumptions on which some of the sorption theories are based may preclude their predicting the heats of sorption accurately. He also pointed out that the experimental data might be at fault, an important factor in evaluating sorption theories.

3.2.4 Peirce Sorption Theory

One of the earliest moisture sorption theories applied to textiles is that of Peirce (1929). He separated sorbed water into two components, designated here as alpha water m_a and beta water m_b, a precedent followed in many subsequent sorption theories including those already discussed. The essential elements of his theory are given here.

Peirce postulated the existence of a finite number of sorption sites in the textile fiber to which water molecules could be attached, from one to several molecules per site. We will designate by N, N_a, and N_b, the total number, the number of alpha molecules, and the number of beta molecules, respectively, per sorption site. Thus,

$$N = N_a + N_b \ . \tag{3.69}$$

The maximum possible value of N_a is unity, since no more than one molecule of alpha or directly bound water is allowed on any site. The term N_b, however, may be greater than unity since there may be more than one beta or indirectly bound molecule on a given site. Beta molecules can only be sorbed on sites already occupied by alpha molecules or other beta molecules. It also follows that N may be greater than unity.

The model assumes that the rate of increase of N_a with N is equivalent to the fraction $(1-N_a)$ of sites which are unoccupied, that is

$$dN_a/dN = (1-N_a) \ . \tag{3.70}$$

This can be integrated and rearranged to give

$$N_a = 1-\exp(-N) \ , \tag{3.71}$$

and N_b, using Eq. (3.69) is

$$N_b = N-1+\exp(-N) \ . \tag{3.72}$$

In order to relate vapor activity h to N it is assumed that the amount of beta water is most important in determining h. The relationship used is

$$dh/dN_b = B(1-h) \ , \tag{3.73}$$

which states that the rate of change of h with respect to N_b is proportional to $(1-h)$, that is the increase in vapor pressure per unit increase in beta water decreases linearly with increasing h. The proportionality constant B is included because only a fraction $1/B$ of the sites is presumed to be available for indirect sorption. The remaining sites, $(1-(1/B))$, can only sorb alpha water, because of steric hindrance factors. Equation (3.73) can be integrated to give

$$h = 1-\exp(-BN_b) \ . \tag{3.74}$$

An additional term is added to account for evaporation of alpha or directly attached water molecules not covered by beta molecules. This is presumed to be proportional to $N_a(1-h)$, or, from Eq. (3.74), $N_a\exp(-BN_b)$. Thus, Eq. (3.74) is modified to

$$h = 1-\exp(-BN_b)+KN_a\exp(-BN_b) \ , \tag{3.75}$$

which can be rearranged to

$$1-h = (1-KN_a)\exp(-BN_b) \ , \tag{3.76}$$

where K is a constant, determined by the relative effect on h of the alpha and beta water molecules.

The total isotherm in terms of N can be obtained by use of Eqs. (3.71) and (3.72) and Eq. (3.76) becomes

$$1-h = \{1-K[1-\exp(-N)]\}\exp\{-B[N-1+\exp(-N)]\} \ . \tag{3.77}$$

The term N is equivalent to the ratio M/M_0 or m/m_0, in which M_0 or m_0 correspond to the moisture content of a completely full monolayer in the BET or Dent models (Sects. 3.2.2, 3.2.3). The terms N_a and N_b are equivalent to the ratios M_a/M_0 or m_a/m_0, and M_b/M_0 or m_b/m_0, respectively. Thus

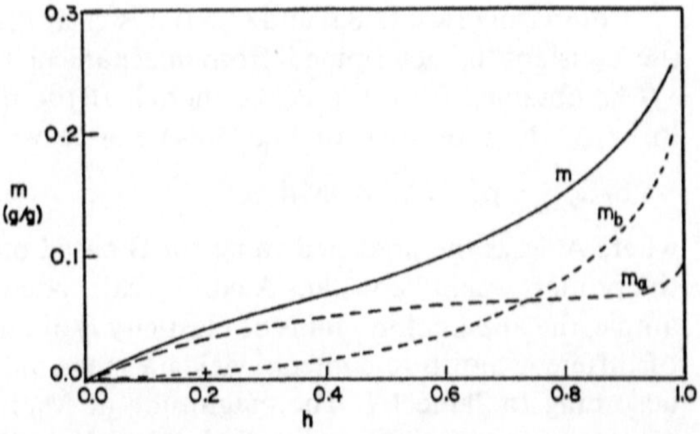

Fig. 3.14. Total sorption isotherm at 40 °C, together with curves of alpha m_a and beta m_b water calculated from the Peirce (1929) model

$$N_a = M_a/M_0 : N_b = M_b/M_0 : N = M/M_0 \; . \tag{3.78}$$

The complete sorption isotherm relating activity h and moisture content m, from Eqs. (3.77) and (3.78), is

$$1-h = \{1-K[1-\exp(-M/M_0)]\}\exp\{-B[(M/M_0)-1+\exp(-M/M_0)]\} \tag{3.79}$$

In order to evaluate the sorption isotherm, three constants must be known; B, K, and M_0. In general these must be evaluated from equilibrium sorption data, using nonlinear regression analysis. Simpson (1973) has done so for the USDA Wood Handbook (1955) sorption data at temperatures ranging from 0 °C to 100 °C. Fitting regression equations of B, K, and M_0 as linear functions of temperature in degrees C, using data from 20° to 70 °C, gave the following results

$$K = 0.246 - 0.000851\,(C) \tag{3.80}$$

$$B = 3.227 - 0.0318 \;\, (C) \tag{3.81}$$

$$M_0 = 100\, m_0 = 15.5 - 0.144\,(C) \tag{3.82}$$

Figure 3.14 shows curves of m, m_a, and m_b as functions of h calculated from the values of K, B, and m_0 determined at 40 °C. Note that the alpha and beta moisture curves are similar to those predicted by the Dent and Hailwood-Horrobin sorption isotherm models (Fig. 3.5), although the value of m_0 is greater.

Peirce (1929) proposed that the modulus of rigidity of textile fibers should be inversely related to the proportion of alpha water in cotton fibers. This has been corroborated for a number of natural fibers according to Meredith (1957). Equation (3.71), written in the form

$$(M_0 - M_a)/M_0 = \exp(-M/M_0) \tag{3.83}$$

is similar to the exponential relationship often found to relate moisture content and several of the mechanical properties of wood. This is Eq. (1.28), written in the form

$$S/S_0 = \exp(-BM) \; , \tag{3.84}$$

where S and S_0 are the strengths at moisture content M and zero moisture content, respectively, and B is a constant of proportionality. Equation (3.84) applies only in the hygroscopic moisture range.

Comparing Eqs. (3.83) and (3.84) it is clear that if Peirce's assumption is valid, the constant B, determined from mechanical tests should be proportional to $1/M_0$, obtained from the Peirce model. If the proportionality constant is taken to be A, the ratio S/S_0 of Eq. (3.84) can be written

$$S/S_0 = \exp[-(A/M_0)M)] , \tag{3.85}$$

where $A/M_0$is the predicted value for B based on the Peirce model. The value of the proportionality constant A can be calculated if B and M_0 are known. For example, the apparent modulus of elasticity E obtained from bending tests on wood of different moisture contents yields a mean value for B of approximately 0.02, according to Table 1.7. The magnitude of M_0 is in the order of 8% to 10% at room temperature. Since $A = BM_0$ from Eqs. (3.84) and (3.85), the value of A ranges from 0.16 to 0.2 based on the values of B and M_0 given above. This is interpreted to mean that increasing the content of alpha water in the wood is approximately 16% to 20% effective in reducing the modulus of elasticity in bending.

Other mechanical properties are more sensitive to moisture change. For example (Table 1.7), the values of B for the modulus of rupture, fiber stress at proportional limit and work to proportional limit are 0.04, 0.05 and 0.08, respectively. The corresponding values for A are therefore 0.32 to 0.4, 0.4 to 0.5, and 0.64 to 0.8. The relationship of mechanical strength to alpha moisture according to the Peirce model is based on the assumption that these strength properties are determined by the number of hydrogen bonds crosslinking the cellulose and other constituents in the noncrystalline regions of the cell wall. Sorption of alpha water presumably involves rupturing these hydrogen bonds in proportion to the amount of alpha water.

Increasing temperature reduces the mechanical properties of wood. It also reduces the value of M_0, thus having an effect similar to that of increasing moisture content M. Thus the Peirce model also predicts at least some of the reduction of mechanical strength with increase of temperature.

It has already been noted that the water of hydration M_h of the Hailwood-Horrobin model and the primary water M_1 of the Dent model are analogous to alpha water. Furthermore, they increase with total moisture content in much the same manner as does alpha water. The discussion of the role of alpha water in reducing wood strength can also be applied to these and similar sorption models.

3.2.5 Enderby-King Sorption Equations

G. King (1960) derived an equation, using statistical thermodynamics, of the following form

$$m = m_0\{[ak_1p/(1+k_1p)]+[bk_2p/(1-k_2p)]\} , \tag{3.86}$$

where m_0 is the moisture content of the saturated monolayer. The terms a and b are constants, proportional to the numbers of two kinds of sorption sites. Those associated with a allow only one water molecule to be sorbed (Fig. 3.15a), the site then being saturated. The b sites allow for additional water molecules to be sorb-

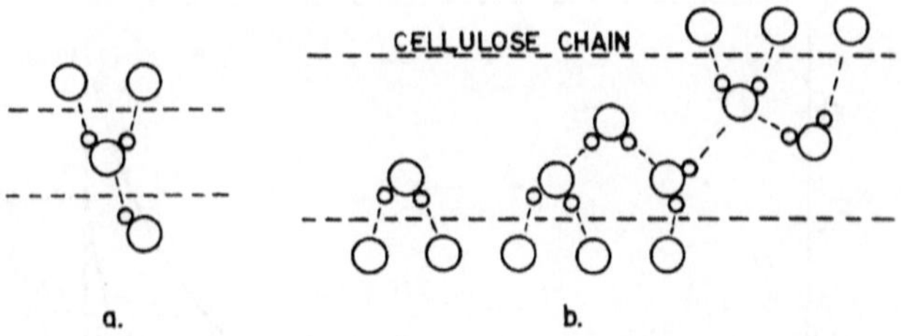

Fig. 3.15. Schematic diagram illustrating how water molecules may be sorbed between cellulosic chains in the case of: **a** = Primary sites. **b** = Secondary sites. (After Enderby 1955)

ed (Fig. 3.15b). The constants k_1 and k_2 are related to the sorption energies for the two kinds of sites. Equation (3.86) is similar to one proposed by Enderby (1955), based on the model shown in Fig. 3.15.

The Enderby-King equation can be written in terms of activity h and in a form similar to the Hailwood-Horrobin by making the following substitutions

$$k_1 = K_1 K_2/p_0 = K_1 k_2 : k_2 = K_2/p_0 \tag{3.87}$$

Thus, letting $p = hp_0$, Eq. (3.86) becomes

$$m = m_0\{[aK_1 K_2 h/(1+K_1 K_2 h)] + [bK_2 h/(1-K_2 h)]\} \tag{3.88}$$

This is identical with the Hailwood-Horrobin equation, Eq. (3.62), if a and b are unity and K_1 and K_2 are equivalent to K_h and K_d, respectively. Simpson (1973) has tabulated values of a, b, k_1, k_2, and M_p ($= 18/m_0$) from USDA Wood Handbook (1955) sorption data over the temperature range from 0° to 100 °C. The values of a and b are in the order of 2 and 0.5, respectively, with some temperature variation [Eqs. (3.89), (3.90)]. This indicates that there are approximately four times as many a as b types of sorption sites. The terms k_1 and k_2 decrease with temperature, approximately to the same extent as p_0 increases, such that the products $k_1 p_0$ ($= K_1$) and $k_2 p_0$ ($= K_2$) vary to only a small extent, much as K_h and K_d of the Hailwood-Horrobin model. The term M_0 decreases in the same manner with increase in temperature as in the Hailwood-Horrobin model.

The following regression equations were obtained for the terms a, b, K_1, K_2, and M_0 based on data reported by Simpson (1973), but limited to the temperature range from 20 °C to 70 °C.

$$a = 1.039 + 0.0212(C) \tag{3.89}$$

$$b = 0.587 - 0.000609(C) \tag{3.90}$$

$$K_1 = 2.71 + 0.00915(C) - 0.000423(C)^2 \tag{3.91}$$

$$K_2 = 0.827 + 0.000607(C) \tag{3.92}$$

$$M_0 = 7.75 - 0.0234(C) - 0.000234(C)^2 \tag{3.93}$$

It is possible to calculate ΔG_1, and ΔG_2 at any temperature from the values of K_1 and K_2 using equations similar to Eq. (3.28). Also ΔH_1 and ΔH_2 may be calculated from the variation of K_1 and K_2 with temperature from equations analogous to Eqs. (3.30) and (3.67).

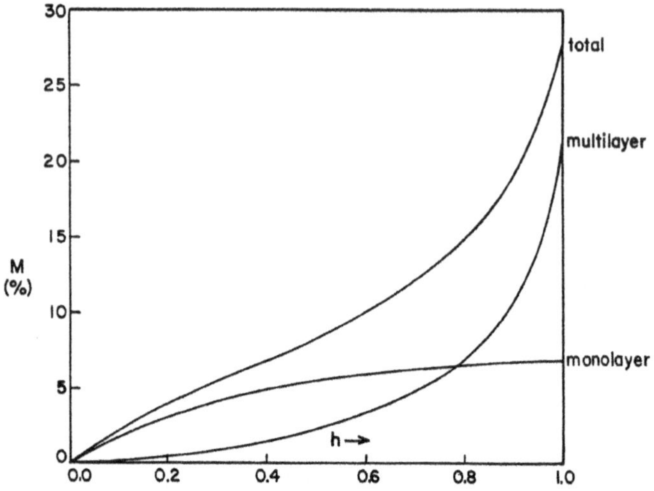

Fig. 3.16. Total sorption isotherm at 110 °F (43.3 °C) taken from Simpson (1973), together with the calculated curves for the moisture contents of the a sites M_a and of the b sites M_b of the Enderby-King model (G. King 1960)

Figure 3.16 shows the sorption isotherm for wood from Simpson's (1973) values at 43.3 °C, together with the calculated values for the two kinds of water, the a and b components of Eq. (3.86). These are similar to those for M_h and M_d of the Hailwood-Horrobin model.

3.2.6 Bradley and Related Sorption Equations

Bradley (1936) proposed a sorption model which has been applied to the sorption isotherms of textiles (Morton and Hearle 1962). It is based on the reduction in dipole attraction between successive layers of sorbed water molecules. The final equation is

$$\ln(1/h) = K_2 K_1^M + K_3 \ , \qquad (3.94)$$

where K_1, K_2, and K_3 are constants whose values depend on the electric field strength of the sorption surface, the dipole characteristics of the sorbed water, and the temperature.

Simpson (1973) has evaluated the constants K_1, K_2, and K_3 at various temperatures from the USDA Wood Handbook (1955) sorption data. Calculation of these constants as functions of temperature between 20° and 70 °C give

$$K_1 = 0.838 - 0.000182(C) - 0.0466(C)^2 \qquad (3.95)$$

$$K_2 = 3.65 + 0.00442(C) - 0.000191(C)^2 \qquad (3.96)$$

$$K_3 = 0.278 - 0.00065(C) \ . \qquad (3.97)$$

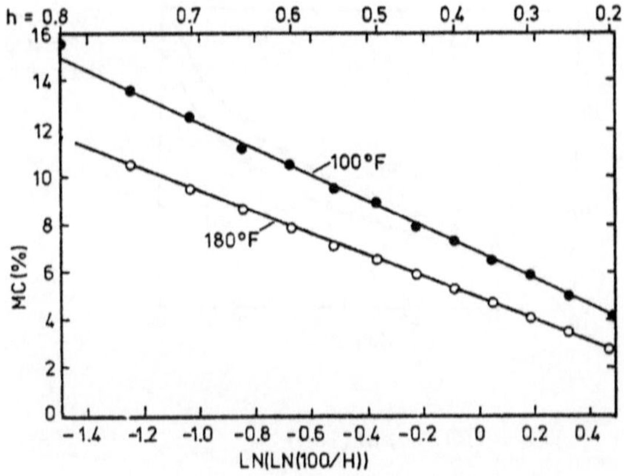

Fig. 3.17. Curve showing the linear relationship between moisture content and ln [ln (1/h)] as predicted by Eq. (3.99)

It is clear that, except for large values of M, K_3 is small compared with the product $K_2K_1^M$ in Eq. (3.94), which may therefore be approximated by

$$\ln(1/h) \simeq K_2K_1^M \tag{3.98}$$

This may be written in two alternate linear forms,

$$M \simeq \{\ln(K_2) - \ln[\ln(1/h)]\}/[-\ln(K_1)] \tag{3.99}$$

$$h = \exp\{-K_2 \exp[M \ln(K_1)]\} \ . \tag{3.100}$$

Figure 3.17 shows a plot of M against ln [ln (1/h)] at 100 °F, based on Eq. (3.99) and using data calculated from Eqs. (3.95) and (3.96). It is evident that the equation fits the data well over the range from h = 0.2 to 0.8.

N. T. Anderson and McCarthy (1963) used classical thermodynamics to obtain the following equation

$$h = \exp[(-18AB/RT)\exp(AM)] \ , \tag{3.101}$$

where A and B are constants. They used the empirical observation that the logarithm of Q_s, the differential heat of sorption, decreased linearly with increasing M. However, they assumed that the entropy change $T\Delta S_s$ was zero, so that Q_s was taken to be equal to the free energy change ΔG_s. Equation (3.101) reduces to Eq. (3.100) if $A = \ln(K_1)$ and $18AB/RT = K_2$.

Nelson (1983) recognized that the free energy change (ΔG_s) should be used rather than the heat of sorption in deriving Eq. (3.101). He therefore derived an equation similar to that of N. T. Anderson and McCarthy (1963), but based on the observed empirical linear relationship between the logarithm of the decrease of free energy associated with sorption, and wood moisture content, given by Eq. (2.63). This relationship may be written

$$\Delta G_s = D \exp(-CM) \ , \tag{3.102}$$

where C and D are empirical constants, the latter equivalent to ΔG_0, the value of ΔG_s extrapolated to zero moisture content. From Eq. (2.62), ΔG_s is dependent on the activity h as follows,

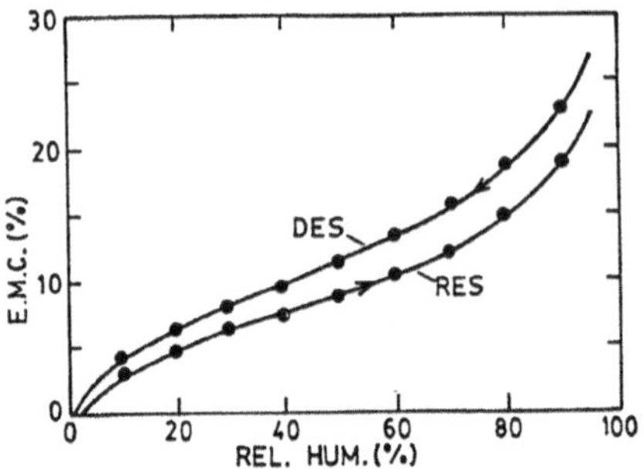

Fig. 3.18. Desorption and resorption isotherms calculated by Nelson (1983) using sorption data of Kelsey (1957) at 25 °C

$$\Delta G_s = (RT/18) \ln (1/h) \ . \tag{3.103}$$

Combining Eqs. (3.102) and (3.103) to eliminate ΔG_s, and rearranging, yields either of the following

$$M = \{\ln(18D/RT) - \ln[\ln(1/h)]\}/C \tag{3.104}$$

$$h = \exp[(-18D/RT)\exp(-CM)] \ . \tag{3.105}$$

These are identical with Eqs. (3.99) and (3.100), respectively, if $C = -\ln(K_1)$ and $D = K_2RT/18$. Also $C = -A$, and D is equal to the product AB of Eq. (3.101).

Nelson (1983) applied his model to sorption data from several materials, including wood. He concluded that the sorption isotherm for wood is adequately described by his model over the activity range from 0.05 to 0.90. Figure 3.18 shows desorption and resorption isotherms calculated by Nelson (1983) and also the experimental data points reported by Kelsey (1957). It should be noted that the calculated isotherms become negative as the activity h approaches zero. This is because the term 1/h in Eq. (3.104) approaches infinity as h approaches zero.

3.2.7 Capillary Condensation and Sorption

Most of the sorption models discussed tend to underestimate the actual sorption values obtained at high humidities. This may be due to capillary condensation at these high humidities, since the relative vapor pressure h over a concave air-water meniscus, say of radius r, is less than unity. Therefore, condensation in such a capillary may take place even if the ambient atmosphere is not fully saturated.

The Kelvin equation gives the relationship between the radius r and the equilibrium relative vapor pressure h. This may be written for water as

$$r = 2V_wS/[RT\ln(1/h)] \ , \tag{3.106}$$

where V_w is the molar volume (18 cc/mol), S is the surface tension, R the gas constant and T the Kelvin temperature. In cgs units r is given in centimeters. Table 3.8 lists some values of r for several values of h at 300 K (27 °C). The values tabulated below about h = 0.8 are suspect, since the calculated radii are less than 5 nm, the capillary diameter of the 10 nm being only 30 times that of a water molecule. This is because the concept of surface tension assumes that large

Table 3.8. Relationship of activity (h), capillary pressure ($P_0 - P$) and capillary meniscus radius at 27 °C

h	$P_0 - P$		Radius (r) (μm)
	(Atm.)	(MP_a)	
1.0000	0.0	0.0	–
0.9999	0.13	1.32	10.9
0.999	1.3	13.2	1.09
0.99	14.0	142	0.105
0.9	142	1440	0.0101
0.8	300	3040	0.0048[a]
0.6	680	6900	0.0021[a]
0.4	1210	12120	0.0012[a]

[a] These capillary radii values are questionable (see text).

numbers of molecules are involved which ceases to be true. Nevertheless, Eq. (3.106) can be used with caution at larger values of h.

An intuitive understanding of the Kelvin equation may be gained by considering the relative attractions of surface water molecules to neighboring liquid water molecules when the surface curvature changes. For example, when the air-water interface is flat, there is a greater tendency for water molecules near the surface to escape than from a concave curved surface. Molecules which leave a flat surface from almost any angle can escape, whereas those which leave from the concave surface at a low angle are likely to collide with the surface again and to be recaptured. Thus, fewer molecules will escape from a curved than from a flat surface, and the vapor pressure p over the curved surface will be lower than over a flat surface.

An alternative viewpoint is that there are more neighboring surface molecules to a given surface molecule on the concave curved surface and therefore a higher attractive force between adjoining molecules. The potential energy of a surface molecule is therefore at a lower level, and the kinetic energy which it must attain in order to evaporate is higher, resulting in a lower vapor pressure.

The Kelvin equation also applies to convex surfaces such as raindrops or fog droplets. In this case, however, the relative vapor pressure h is greater than unity, since the escaping tendency for molecules near the surface is greater for a convex than for a flat surface. Equation (3.106) can be used to predict the vapor pressure elevation over a convex curved surface such as a fog droplet by using a negative value for the droplet radius r.

In older literature, the Kelvin equation was sometimes used to explain the shape of the sorption isotherm. This was based on the assumption that pre-existing capillaries in the cell wall enlarged as moisture content increased, thus increasing their capillary radii and corresponding equilibrium humidities. However, as noted above, the Kelvin equation is not valid except at high humidities.

Sorption hysteresis has also been attributed, at least in part, to capillary condensation. The reasoning is that when a meniscus is receding in a tapering capillary, such as during desorption, the contact angle at the wall surface will be

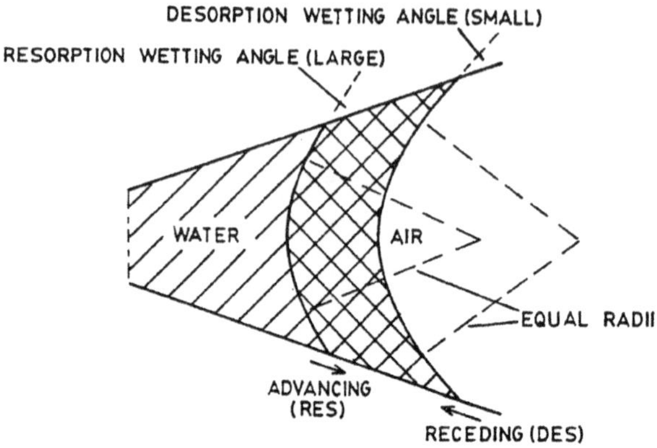

Fig. 3.19. Diagram illustrating the increased sorption in tapering capillaries when receding (desorption) than when advancing (resorption), for a given meniscus radius (after Marton and Hearle 1975)

smaller than when it is advancing as during resorption (Fig. 3.19). Therefore, more sorbed water will be present during desorption at a given meniscus radius and ambient humidity, than during resorption.

Simpson (1973) used the Kelvin equation to modify the BET sorption isotherm. He added the volume V_c of water attributed to capillary condensation to that predicted by the BET isotherm, assuming cylindrical capillaries of total length L and radius r per unit dry mass W_0 of the wood. Thus, the volume of capillary water V_c is given by

$$V_c = 2\pi r^2 L \ , \tag{3.107}$$

and the moisture content m_c due to capillary condensation is given by

$$m_c = \varrho V_c / W_0 \ . \tag{3.108}$$

Combining Eqs. (3.106) to (3.108) gives

$$m_c = (2\pi L/w_0)\ [2V_w S/(RT \ln(1/h))]^2 \ . \tag{3.109}$$

Simpson added Eq. (3.109) to the moisture content obtained from the BET equation [Eq. (3.36)], so the total sorption was the sum of the two. He then fitted the sorption data to this modified BET isotherm. Figure 3.20 shows his results for sorption data at 40 °C. The capillary water thus calculated becomes important at high humidities. It should be noted that four constants now appear in the total equation; one of which is the calculated ratio of L/w_0, the magnitude of which is approximately 10^{10} cm per gram of dry wood, based on Simpson's analyses.

In principle, Eq. (3.109) or some variation of it could be used to modify some of the other sorption models. In this case one would expect a smaller contribution of capillary moisture to the total isotherm, since most other models give a larger increase in secondary water at high humidities than does the BET model alone with reasonably small numbers of allowed layers n [Eq. (3.38)].

Kollmann (1962) separated capillary sorption into two parts, one designated as "submicroscopic" and the other "microscopic". Both increased exponentially with increasing humidity, the former beginning near $h = 0.7$ and the latter near $h = 0.9$ (Fig. 3.21). Both have a form similar to the capillary moisture shown in Fig. 3.20. Six constants are required to completely describe the total isotherm.

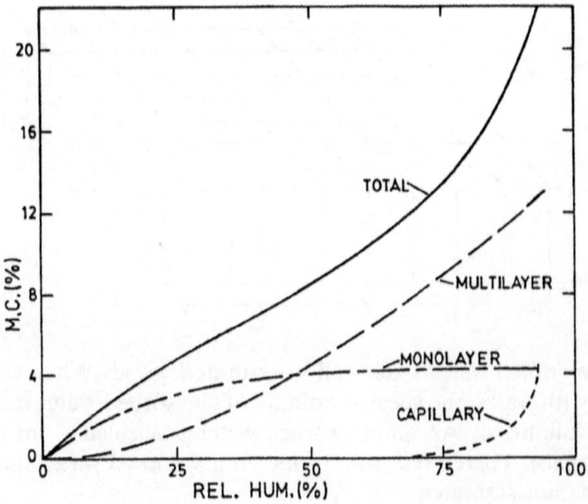

Fig. 3.20. Total sorption isotherm calculated by Simpson (1979) for wood at 40 °C using the BET model [Eq. (3.36)] modified to account for capillary sorption at high humidities. The BET monolayer components as well as the calculated capillary condensation, are also shown

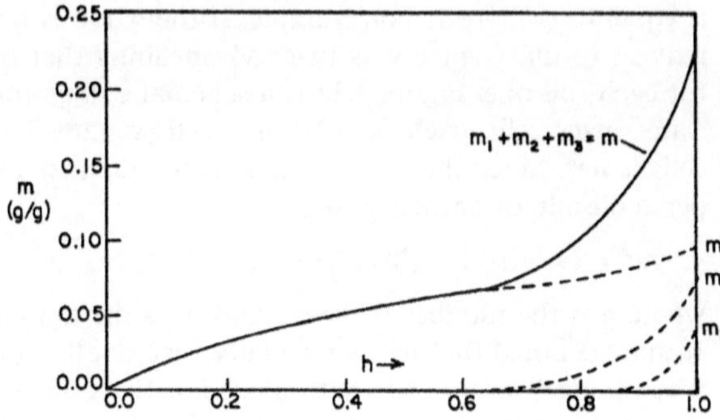

Fig. 3.21. Total sorption isotherm and its three components, m_1 the hygroscopic or molecular sorption, and m_2 and m_3 the submicroscopic and microscopic capillary condensations respectively, based on the Kollmann (1962) model

3.2.8 Malmquist's Sorption Model

The sorption theory of Malmquist (1958, 1959, 1967) treats the sorption of water by wood and other hygroscopic materials in terms of *space-dimensional* factors within the cell wall. It also considers the cohesive properties of the cell wall which might limit the swelling associated with water sorption and therefore the extent of water sorption by the wood. The general theory also takes into account whether sorption occurs with three, two, or one degree of freedom, corresponding to three-dimensional sorption volumes, two-dimensional surfaces or one-dimensional lines, respectively. Only the two-dimensional or surface sorption case is considered here, since this case seems to apply most closely to moisture sorption by wood according to Simpson's (1973) analysis.

This theory considers that there is a sorption space within the cell wall which sorbed water molecules can occupy. There is also the surrounding vapor space, defined as the space which the water-vapor molecules occupy.

When the cell wall is fully saturated with water, the specific volume of the sorbed water molecule is taken as v_f', which is then also equal to the volume of

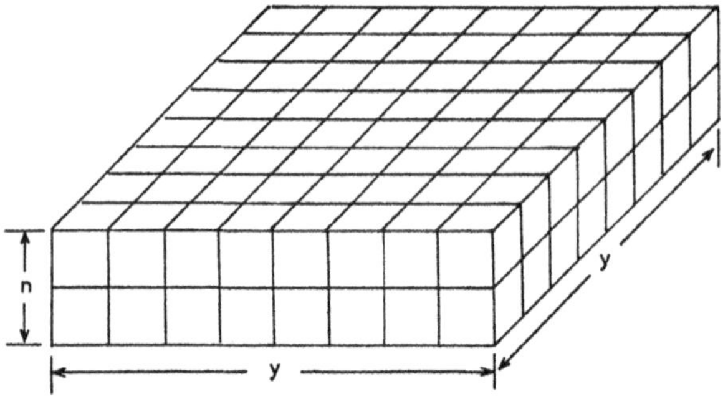

Fig. 3.22. Schematic diagram showing two layers of sorbed water space cells, each layer being y cells square, based on the sorption model of Malmquist (1959)

a sorbed-water space-cell in saturated wood. When the cell wall is not saturated with water, the specific volume of the sorbed water molecule is v', larger than the volume v'_f. An empty sorbed-water space-cell is an empty cell in the sorption space. There are always some empty sorbed-water space-cells when the cell wall is not saturated.

The number of empty sorbed-water space-cells per molecule of sorbed water is equal to $(v'/v'_f)-1$. For example, if the wood is half-saturated, this expression reduces to unity since v' is twice v'_f, meaning that there is one empty space-cell for each full one. Figure 3.22 is a schematic diagram showing n layers of sorbed water space cells, each layer being y cells square. Therefore, the total number of cells is ny^2. Since this corresponds to the number of empty sorption space cells per molecule of sorbed water,

$$ny^2 = (v'/v'_f)-1 = (M/M_f)-1 \;, \qquad (3.110)$$

where n is the number of layers and y^2 is the number of surface cells per layer. It may be noted that ny^2 is normally very small, equivalent to unity for example when the cell wall is half-saturated. For the general case treated by Malmquist, the exponent two in Eq. (3.110) is replaced by a variable term i, which may vary from zero to three.

Malmquist (1959) also considered the volume occupied by water vapor molecules in the adjacent space. The term v'' refers to the specific volume or space volume per vapor molecule. The value of v'' for completely saturated vapor space is designated as v''_f, the equilibrium value of v'' when $v' = v'_f$, that is, when the cell wall is completely saturated. It is assumed that

$$y^3 = (v''/v''_f)-1 = (p_f/p)-1 \;, \qquad (3.111)$$

since $v''/v''_f = p_f/p$ from the ideal gas law, where p and p_f are the equilibrium and saturated vapor pressures of the vapor space.

Eliminating y between Eqs. (3.110) and (3.111) gives

$$M_f/M = 1+n\,[(p_f/p)-1]^{2/3} \;. \qquad (3.112)$$

Up to this point the theory has not considered the "cohesion factor" which Malmquist (1959) used to account for the limits to sorption imposed by the resistance of the wood to swelling. He defined the cohesion factor K as

$$K = [1-(p_f/p_0)]/[1-(M/M_f)] \;, \qquad (3.113)$$

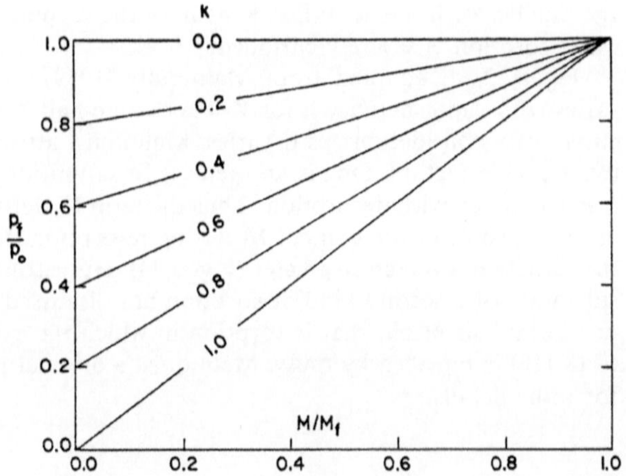

Fig. 3.23. Series of curves showing how the ratio p_f/p_0 increases with the ratio M/M_f for several values of the cohesion factor K, according to Malmquist (1959)

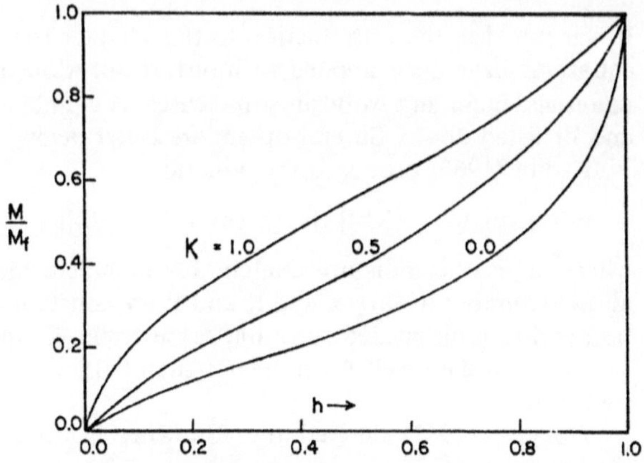

Fig. 3.24. Effect of the cohesion factor K on the sorption isotherms plotted as the ratio M/M_f against h, based on the Malmquist (1959) sorption model

where p_0 is the normal saturated vapor pressure of water in free space. According to the model, p_f may be lower than p_0 because of the cohesive forces in the cell wall which resist the swelling associated with the entry of water.

Equations (3.112) and (3.113) can be combined to give

$$M_f/M = 1 + n \, \langle\{[1 - K\,(1 - M/M_f)]/h\} - 1\rangle^{2/3} \,, \tag{3.114}$$

where the activity $h = p/p_0$. For $K = 0$, this reduces to

$$M_f/M = 1 + n \, [(1 - h)/h]^{2/3} \,. \tag{3.115}$$

For $K = 1$, Eq. (3.114) becomes

$$M_f/M = 1 + n \, [(M_f/M)/h - 1]^{2/3} \,. \tag{3.116}$$

The relationship of p_f/p_0 to M/M_f is shown in Fig. 3.23 for several values of the cohesion factor K. When K is zero, $p_f = p_0$, and $p/p_f = h$. In this case the sorption space surfaces are assumed to be spaced sufficiently far apart that there is no restraint to wood swelling other than that due to the limitation on the average number of layers n. When K is unity the sorption space is so restricted that water sorption is severely curtailed.

Figure 3.24, adapted from Malmquist (1959), shows how the ratio M/M_f varies with vapor activity h for K = zero, one-half and unity. The shapes of the curves vary considerably as K varies. Malmquist attributed sorption hysteresis to two possible factors. One is an increase in saturation pressure p_f during resorption compared with desorption. Thus the term K, defined by Eq. (3.113), is lower at any given moisture content M during resorption than during desorption, and the activity h is therefore greater (Fig. 3.24). His second explanation involves consideration of a second kind of sorption not discussed here. He called this "soft" or "linear" sorption, that is sorption in which the exponent two in Eqs. (3.115) to (3.116) is replaced by unity. Malmquist's original papers should be consulted for more details.

3.2.9 Additional Sorption Isotherm Equations

It was noted in the introduction to this chapter that several dozens of isotherm equations have been applied to moisture sorption data obtained for biological materials, including wood in some cases. A number of these have been derived and/or listed above. Several others are listed below.

Kadita (1960) proposed the equation

$$m = [Km_0 h/(1+Kh)]\ \{1+[B\ (h-h^n)/(1-h)]\}\ , \tag{3.117}$$

where m_0 is the moisture content for complete monolayer coverage, n is the allowed number of layers, and K and B are constants related to the heat of sorption and to multilayered sorption, respectively. Kadita indicates that Eq. (3.117) fits sorption data well for h greater than 0.05, the constants all having physical significance.

Doppert (1967) derived the following isotherm based on statistical thermodynamics

$$p = [k\theta/(1-\theta)]\ \exp\,(2\theta u/RT)\ , \tag{3.118}$$

where p is the vapor pressure, θ is the ratio of the existing moisture content m to the saturated value m_f, k is a constant, 2u is the interaction energy between one sorbed water molecule and its nearest neighbors, R is the gas constant, and T is the Kelvin temperature.

A number of empirical equations have been used to represent sorption isotherms in a convenient mathematical form. Iglesias and Chirife (1982) have applied many such equations to the sorption isotherms for various food products such as cereals, meats, and dehydrated foods. A number of these equations require only two constants, A and B, relating moisture content M and vapor activity h, including those shown in Table 3.9. Some of the equations shown have some theoretical basis, although most are essentially empirical. The constants A and B are generally different for each equation shown.

The equations listed in Table 3.9 were fitted to desorption data of Kelsey (1957) on klinki pine at 25 °C over the relative humidity range from ca. 15% to

Table 3.9. List of two-parameter isotherm equations and calculated values of constants using sorption data of Kelsey (1957) at 25 °C

Source[a]	Equation[b]	A	B	R^2
Igles.-Ch. 1981	$M = A + B[h/(1-h)]$	7.0040	3.1211	0.916
Labuza 1972	$M = A + B[1/\ln(h)]$	5.7824	−3.0566	0.924
Smith 1947	$M = A + B[\ln(1-h)]$	4.6507	−8.9553	0.990
Bradley 1936	$M = A + B\ln[\ln(1/h)]$	9.1975	−6.2903	0.998
Oswin 1946	$\ln(M) = A + B\ln[h/(1-h)]$	2.4023	0.3991	0.995
Halsey 1948	$\ln(M) = A + B\ln[\ln(1/h)]$	2.1461	−0.5621	0.974
Henderson 1952	$\ln(M) = A + B\ln\{\ln[1/(1-h)]\}$	2.6516	0.5484	0.997

[a] Based on equations listed by Iglesias and Chirife 1982.
[b] Reduced to linear form with M as a function of h.

80%. The calculated values of A and B and the variance R^2 are included in Table 3.9.

All of the equations shown in Table 3.9 have been written in forms such that moisture content M is a function of activity h. Several of them were written in other forms by the various authors, but have been converted to the form shown for convenience. It may be noted that the two-constant form of the Bradley (1936) equation [Eq. (3.99)] is included in the table. It incidentally gave the best fit to the sorption data, its R^2 value of 0.998 being higher than any of the others listed. These equations are convenient to apply to sorption data over the humidity range shown, since simple first-order regression equations may be used to fit the data once they have been converted into the forms shown. Many other equations not cited here are compiled in the review paper by Van den Berg and Bruin (1981), of the 77 isotherm equations which have been applied to biological materials.

Two additional references, both concerned with the moisture sorption isotherms of food and fiber materials are cited here. The first, an extensive bibliography by Wolf et al. (1985), lists more than 2000 references relating to sorption isotherms of food materials. The second, authored by Labusa (1984) contains a large bibliography as well as a discussion of the measurement and use of moisture sorption isotherms, including their application in evaluating the moisture condition of natural food and fiber products. Moisture in food, as in wood, is an important indicator of its utility and storage behavior.

Chapter 4

Hygroexpansion in Wood

4.1 Introduction

In the living tree, the cell walls of wood are always in the fully swollen condition, at fiber saturation. Therefore no skrinkage or swelling occurs in the living tree except from hydrostatic tensions in the cell cavity water, and normal thermal expansion and/or changes in the fiber-saturation point with temperature. However, when green wood dries, hygroscopic shrinkage takes place, the extent depending on a number of factors. These include: amount of moisture loss, structural direction (tangential, radial, or longitudinal), kind of wood, drying stresses caused by moisture gradients, etc.

Wood is normally dried to a moisture content approximating its anticipated mean equilibrium moisture content (EMC) in use. However, since environmental conditions are rarely constant, the EMC is continually changing and the moisture content of the wood under normal exposure conditions is seldom equal to its EMC. Concurrent dimensional changes are also taking place. These have been called *movement* by Stevens (1963), to distinguish them from the initial shrinkage associated with drying the green wood. In the treatment which follows, both initial shrinkage and subsequent movement are considered under the general subject of "hygroexpansion".

4.2 Definition of Terms

A number of specialized terms are used in this chapter in relation to the hygroexpansion properties of wood. It is appropriate to define these terms prior to discussing the specific factors which affect moisture-related dimensional changes. The terms *density* ϱ, *specific gravity* G, and *density index* d, as used here, are defined in addition to those terms specifically related to hygroexpansion in wood.

Density ϱ is defined as the mass w per unit volume v of the wood, both measured at the same moisture content m,

$$\varrho_m = w/v = w_m/v_m \ , \tag{4.1}$$

where the subscript m indicates measurement at moisture content m. Since w and v both increase with m, the value of ϱ_m depends on their relative rates of increase, which vary among woods.

Another wood-density related term is *specific gravity*, symbolized here by G. In North America it is defined as

$$G_m = (w_0/v_m)/\varrho_w \ , \tag{4.2}$$

where ϱ_w is the density of water in the same units in which *dry weight* w_0 and *moist volume* v_m are measured. This term does not conform to the true definition of specific gravity, since w_0 and v_m are not measured at the same moisture condition except when m is zero. However, it provides an useful measure of the dry mass of wood per unit volume of moist wood, particularly of green wood. It is therefore a convenient parameter for converting green volume of wood into dry mass. This is of economic importance since wood quantities in the forest are usually measured in volume but the dry mass is most important from the viewpoint of utilization. The term G relates the two parameters directly.

Panshin and de Zeeuw (1980) suggested that the term *density index* be used in lieu of *specific gravity*. The term will be used in this book in a slightly different manner, as follows

$$d_m = w_0/v_m = \varrho_w G_m \ , \tag{4.3}$$

where d symbolizes density index, having the same dimensions as density but equal in numerical value to G, using cgs units. Depending on context and convenience, either of these terms, or density itself, may be used in discussing hygroexpansion.

Hygroexpansion in wood is traditionally given in terms of *percent shrinkage* s or *percent swelling* S, the two terms differing with respect to the reference dimension used. The moist dimension is used as the reference base for shrinkage s while the dry dimension is used in the case of swelling S. In the third case, designated here as Ss, the *percent dimensional change*, the *mean* of the moist and dry dimension is used. Thus, three terms may be defined. For volume change, for example, where v_1 and v_2 are the volumes at moisture contents M_1 and M_2, respectively, M_2 being greater than M_1, they are defined as

$$s_v = 100\ (v_2 - v_1)/v_2 \tag{4.4}$$

$$S_v = 100\ (v_2 - v_1)/v_1 \tag{4.5}$$

$$Ss_v = 200\ (v_2 - v_1)/(v_2 + v_1)\ . \tag{4.6}$$

It is apparent that the percent shrinkage s is less than the percent swelling S because v_2 is larger than v_1. Likewise the percent change Ss is intermediate between s and S, essentially equal to their mean value.

Another useful index of hygroexpansion is the ratio of the relative dimensional changes S, s, or Ss to the moisture change $M_2 - M_1$. This ratio is a measure of the proportionate change in dimensions associated with a given moisture content change. Various expressions have been used for such changes, including swelling gradient (Kollmann and Côté 1968), differential swelling (Keylwerth 1962, 1964), hygroexpansion coefficient (Skaar 1972), coefficient of moisture expansion (Skaar 1972), coefficient of swelling (Skaar 1972, Rosen 1974), ratio of swelling (Noack et al. 1973), coefficient of shrinkage (Schroeder 1972), and dimensional change coefficient (USDA Wood Handbook 1974).

Three different indices of the volumetric hygroexpansion ratio are defined here, as follows

$$X_v = (1/v)\ (dv/dm) \tag{4.7}$$

$$X_{vs} = (1/v_f)\ (dv/dm) \tag{4.8}$$

$$X_{vS} = (1/v_0)\ (dv/dm)\ . \tag{4.9}$$

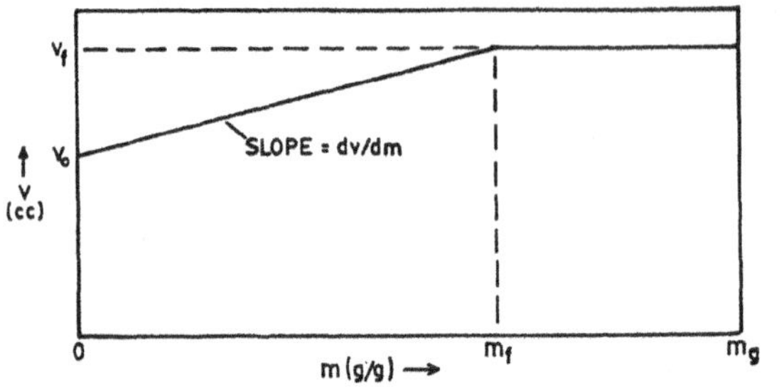

Fig. 4.1. Curve showing idealized volumetric hygroexpansion of wood volume V against moisture content m

Equation (4.7) defines the *volumetric moisture expansion coefficient* X_v in which the volume v is measured at moisture content m. Equation (4.8) defines the *volumetric shrinkage coefficient* X_{vs} with the volume v_f fixed at fiber saturation, and Eq. (4.9) defines the *volumetric swelling coefficient* X_{vs}, since the reference volume v_0 is measured in the dry condition.

Mean values of these coefficients over any given moisture range, m_1 to m_2, or M_1 to M_2, are obtained by dividing Ss, s, or S, as given by Eqs. (4.4) to (4.6), respectively, by $M_2 - M_1$. Thus

$$\bar{X}_v = Ss_v/(M_2 - M_1) = 200\ (v_2 - v_1)/[(v_2 + v_1)\ (M_2 - M_1)] \tag{4.10}$$

$$\bar{X}_{vs} = s_v/(M_2 - M_1) = 100\ (v_2 - v_1)/[v_2\ (M_2 - M_1)] \tag{4.11}$$

$$\bar{X}_{vs} = S_v/(M_2 - M_1) = 100\ (v_2 - v_1)/[v_1\ (M_2 - M_1)]\ , \tag{4.12}$$

where $\bar{X}_x$, $\bar{X}_{vs}$, and $\bar{X}_{vs}$ are the *mean* values of the volumetric moisture expansion coefficients over the moisture range given. They are equivalent to Eqs. (4.7) to (4.9) if M_1 and M_2 are equal to M_0 and M_f, respectively, and if $(v_2 - v_1)/(M_2 - M_1)$ is equal to dv/dM, that is, the volume v increases linearly with moisture content over the entire moisture range.

Figure 4.1 shows an idealized linear curve of volume v against moisture content, in which the slope dv/dm is constant from m = 0 to m_f. This approximation of linearity is convenient and is often used. However, detailed studies, described in Sect. 4.4.2, indicate departures from linearity at both low and high moisture contents in the hygroscopic range.

It is often important to know the hygroexpansion properties of wood with respect to humidity changes such as those to which wood is exposed during use. These properties are given here in terms of the *humidity expansion coefficient Y*. For *volumetric* hygroexpansion it is defined as

$$Y_v = (1/v)\ (dv/dh) = ZX_v\ , \tag{4.13}$$

where $Z = (dM/dH)^* = (dm/dh)^*$, is the effective slope of the sorption isotherm.

During the initial drying of lumber the term Z is equal to the slope dM/dH of the standard initial desorption isotherm. Likewise when wood resorbs water from the dry condition Z is equivalent to the slope of the resorption isotherm. However, under conditions of oscillating sorption, hysteresis reduces the effective slope $(dM/dH)^*$, as is illustrated in Fig. 4.2. Here, the mean value of Z over the humidity range H_1 to H_2 is $(M_{2r} - M_{1d})/(H_2 - H_1)$. This is appreciably lower than

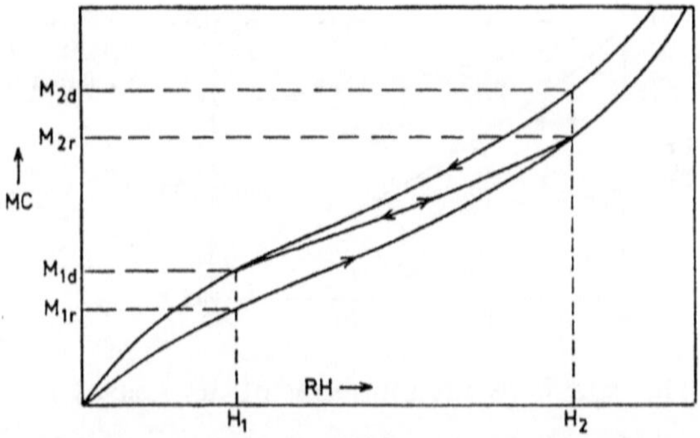

Fig. 4.2. Typical desorption and resorption isotherms showing the effect of hysteresis on the effective slope Z of the sorption isotherms

the mean slopes $(M_{2d}-M_{1d})/(H_2-H_1)$ and $(M_{2r}-M_{1r})/(H_2-H_1)$, respectively, of the desorption and resorption isotherms over the same humidity range.

From the wood utilization standpoint the humidity expansion coefficient Y is a more important wood property than the moisture expansion coefficient X. This is because Y determines the degree of dimensional change associated with atmospheric humidity changes. However, because of tradition and the fact that it is easier to measure M than H, more information is available for calculating X than Y. The term Z, which relates X and Y, is a complex function of species, temperature, moisture content, and specimen size as well as specimen history, particularly in relation to sorption hysteresis, as described in Section 4.5.2.

Hygroexpansion coefficients may also be defined for shrinking and swelling in one dimension, such as in one or more of the principal structural directions for wood. These are similar to those defined in Eqs. (4.4) to (4.13) except that the volume subscript v is replaced by the appropriate dimensional subscript t, r, or l for tangential, radial, and longitudinal hygroexpansion, respectively. These directional hygroexpansion coefficients and their relationships to the volumetric coefficients described above are discussed in Section 4.5.

4.3 Volumetric Hygroexpansion of the Wood Cell Wall

The cell wall of wood is anisotropic, and therefore hygroexpansion is also anisotropic. However, in the present discussion only volumetric changes in the cell wall are considered, and anisotropy is neglected. Furthermore, the notation convention followed is to designate a cell-wall parameter by means of a *prime*. For example, if G is the specific gravity of the gross wood, G′ is that of the cell wall.

The density ϱ'_0 of the dry cell wall is approximately 1.5 g/cc based on pycnometric measurements. Stamm (1964), Skaar (1972), and Siau (1984) discussed the factors which affect cell-wall density but for the discussion here it is convenient to use 1.5 g/cc. When the cell wall takes up water, both its mass w′ and volume v′ increase, the former proportionally less than the latter. Since $w'_m = w'_0(1+m)$, and $v'_m = v'_0 + v_w$, the density ϱ'_m of the moist cell wall is

$$\varrho'_m = w'_m/v'_m = w'_0(1+m)/(v'_0+v_w) \ . \qquad (4.14)$$

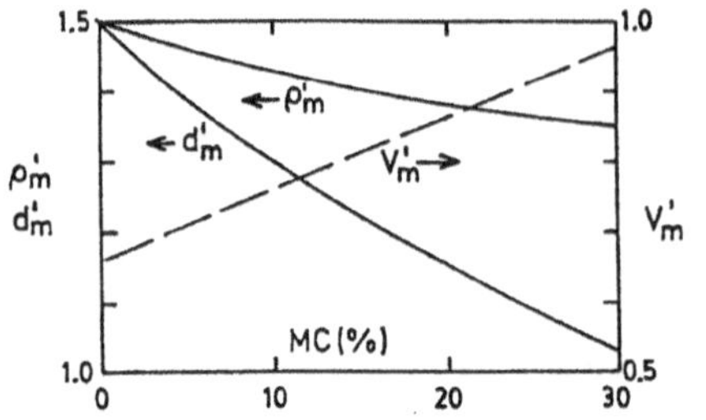

Fig. 4.3. Curves of density ϱ_m, density index d_m, and specific volume of the cell wall, in relation to its moisture content M

This may be written in terms of dry density ϱ_0', as follows

$$\varrho_m' = \varrho_0' \varrho_w (1+m)/(\varrho_w + \varrho_0' m) \ . \tag{4.15}$$

Likewise the density index d_m' and specific gravity G_m' of the moist cell wall may be written

$$d_m' = d_0' \varrho_w/(\varrho_w + d_0' m) = \varrho_m'/(1+m) \tag{4.16}$$

$$G_m' = G_0'/(1+G_0' m) = d_m'/\varrho_w \ . \tag{4.17}$$

The specific volume V_m', defined as the moist volume per unit mass of dry cell wall, is equal to the reciprocal of d_m', or

$$V_m' = 1/d_m' = (\varrho_w + d_0' m)/d_0' \varrho_w \ . \tag{4.18}$$

Figure 4.3 shows plots of density, density index, and specific volume as functions of cell-wall moisture content, based on Eqs. (4.16) to (4.18). A simple linear relationship between volume and moisture content has been assumed. For a more detailed discussion of this relationship the reader is referred to Stamm (1964), Skaar (1972), Kellogg et al. (1975) and Siau (1984). It should be mentioned, however, that the linear swelling predicted by Eq. (4.18) is an over-simplification of the actual situation (Sect. 4.4.2). Nevertheless, it is a convenient approximation for the present discussion.

The maximum volumetric shrinkage s_{vf}' and swelling S_{vf}' of the cell wall may be estimated from Eqs. (4.4) and (4.5), if the fiber saturation point m_f is known. In this case v_2 and v_1 are equal to V_f' and V_0', the specific volumes of the cell wall at saturation and in the dry condition, respectively. It can be shown that Eqs. (4.4) and (4.5) reduce to

$$s_{vf}' = M_f G_f' \tag{4.19}$$

$$S_{vf}' = M_f G_0' \ . \tag{4.20}$$

According to Eqs. (4.19) and (4.20), the maximum possible percent shrinkage and swelling of the cell wall are linear functions of the fiber saturation point and specific gravity. The dry specific gravity G_0' is taken to be constant at 1.5 but the saturated value G_f' decreases with increase in M_f (Fig. 4.3). The value of G_f' may be calculated from Eq. (4.17) written in the form

$$G_f' = G_0'/(1+G_0' m_f) \ . \tag{4.21}$$

Figure 4.4 shows how s_{vf}' and S_{vf}', calculated from Eqs. (4.19) to (4.21), increase with increasing M_f from $M_f = 15$ to 40 percent.

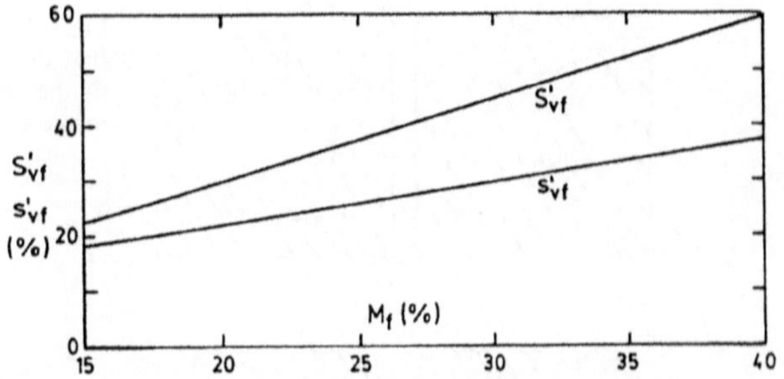

Fig. 4.4. Curves of the maximum volumetric swelling $(S_v)_{max}$ and shrinkage $(s_v)_{max}$ of the cell wall in relation to the moisture content M_f at fiber saturation

4.4 Volumetric Hygroexpansion of Wood

In the preceding section volume changes in the cell wall of wood were discussed. In this section we will consider volumetric hygroexpansion in the wood itself, which contains air spaces in addition to cell-wall material. It is anticipated therefore that volumetric hygroexpansion of the wood should be different from that of the cell wall. Directional hygroexpansion in the three primary structural directions is neglected here but is treated in Section 4.5.

The total volumetric shrinkage and/or expansion of the gross wood between the dry and water-saturated condition will be discussed first, followed by consideration of the volumetric hygroexpansion coefficient over smaller moisture ranges.

4.4.1 Maximum Volumetric Shrinkage and Swelling

If the cell-wall structure is homogeneous, volumetric shrinkage or swelling should be independent of the fraction of void volume in the wood, as Stamm and Loughborough (1942) noted. For example, suppose that a single large hole (Fig. 4.5), or many tiny holes, are bored into a dry solid wood wafer, such that much of the wood volume is removed. If it is soaked in water, and swelling stresses due to moisture gradients are neglected, the wafer will swell to the same extent as a wafer without bored holes. In other words, the amount of swelling is independent of the overall "density" of the wood, as is shown in Fig. 4.5. The percent of volumetric swelling S_v in this case is equal to S'_v, that of the cell wall of wood, as given by Eq. (4.20). The situation would be the same if the cell cavities in wood consisted simply of air spaces dispersed in a homogeneous cell-wall material. In this case all woods would shrink or swell to the same extent, the maximum value determined only by M_f. This would be true even if the cell wall itself was anisotropic with respect to degree of swelling. For example, if it consisted entirely of cellulose molecules oriented in the same direction such that all hygroexpansion occurred at right angles to the direction of orientation, the same conditions would apply. The extent of swelling would still be independent of wood density or proportion of solid wood volume.

Fortunately, the volumetric hygroexpansion of wood is considerably less than that of the cell-wall material with or without the simple air spaces as described

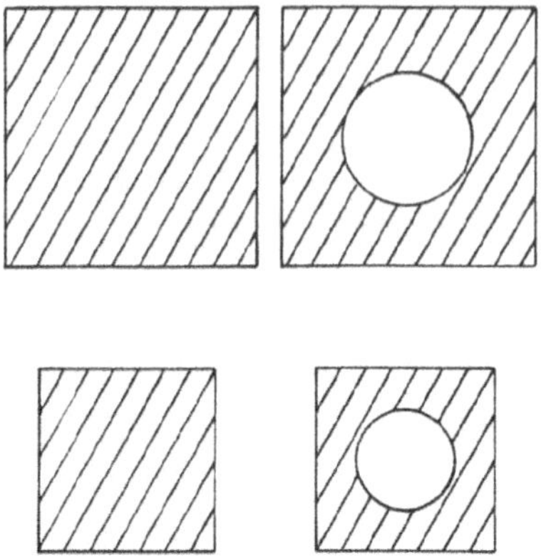

Fig. 4.5. Schematic diagrams showing that the swelling of wood is not affected by the presence of a simple hole drilled in the wood

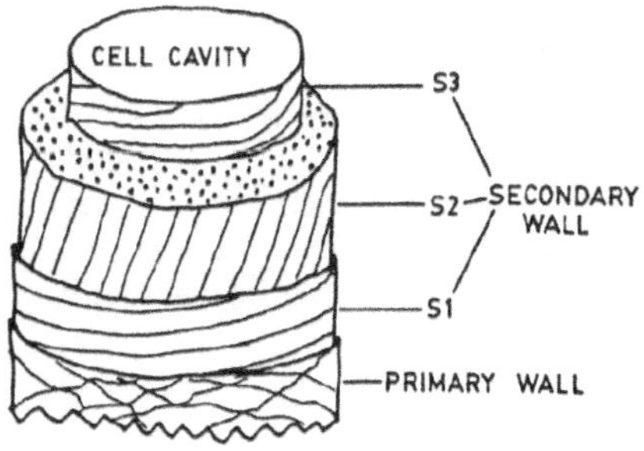

Fig. 4.6. Diagram showing the fibril orientations and relative thicknesses of the S1, S2, and S3 layers of the secondary wall

above. This is because of the unique structure of typical wood cells, shown schematically in Fig. 4.6. This depicts the orientation of the fibrils in the three layers of the secondary wall of a normal conifer tracheid. It should be recalled that hygroexpansion normally occurs at right angles to the fibrils and that the fibrils strongly resist expansion along their lengths.

The fibril orientation of the central or S2 layer, the thickest layer, is nearly parallel to the long axis of the cell. It expands transversely in proportion to the moisture change. The fibril orientations in the thinner outer S1 and inner S3 layers, however, are more nearly perpendicular to the cell axis and tend to restrain the external and internal dimensional changes in the secondary wall. In order for the dominant S2 layer to swell transversely, it must stretch the other two layers in the fibril direction. They resist such stretching because they are strong in this direction. The net result is reduced swelling compared with that in a homogeneous wall. In some cases, partial swelling may take place into the cell cavities. In others, the cell cavity may remain constant or even expand somewhat.

Stamm and Loughborough (1942) first suggested that the cell cavity remains essentially constant when wood shrinks or swells. If this is the case Eqs. (4.19) and (4.20) may be modified to apply to wood of any specific gravity. Thus

$$s_{vf} = M_f G_f : S_{vf} = M_f G_0 \ , \tag{4.22}$$

where s_{vf} and S_{vf} are the predicted *maximum* percent shrinkage and swelling, respectively, between fiber saturation M_f and the dry condition M_0, and G_f and G_0 are the corresponding specific gravities.

Equation (4.22) written as

$$M_f = s_{vf}/G_f = S_{vf}/G_0 \tag{4.23}$$

predicts a constant ratio between the percent volumetric *shrinkage* s_{vf} and the swollen volume specific gravity G_f, and therefore between the percent volumetric *swelling* $(S_v)_{max}$ and the dry-volume specific gravity G_0, based on total volume change between M_f and zero moisture content. Furthermore, the ratio in each case is equal to M_f. The validity of the assumptions made in deriving Eq. (4.23) can be determined if data on s_{vf} or S_{vf} and G_f or G_0 are available for a large number of specimens and/or species of wood representing a wide range of densities.

There are two methods for testing Eq. (4.23). One is to calculate the mean value of the ratio s_{vf}/G_f or S_{vf}/G_0 and the distribution of the individual values about the mean. The second is to calculate the regression coefficients A and B, or C and D, in the following equations

$$s_{vf} = A + B(G_f) \tag{4.24}$$

$$S_{vf} = C + D(G_0) \; . \tag{4.25}$$

Newlin and Wilson (1919) reported that the volumetric shrinkages of green woods appeared to be a linear function of green volume specific gravity G_g such that

$$s_{vf} = 28\, G_g \; , \tag{4.26}$$

based on shrinkage data on more than 100 different wood species grown in the United States, including both softwoods and hardwoods. They did not apply linear regression techniques to the data but calculated Eq. (4.26) by simply plotting s_{vf} against G_g and drawing a line through the origin and the mean values of s_{vf} and G_g.

Stamm and Loughborough (1942) reported a mean ratio for s_{vf}/G_f of 27 for 107 North American hardwood species, approximately equal to M_f as predicted by Eq. (4.23). There was considerable variation among the woods, half of them falling between approximately 24% to 30%, and three-fourths between 21% to 33%. This wide variation suggests that either the values of M_f vary between those limits, or else that the cell cavities shrink or swell to some extent, or both. They found a similar relationship for 52 different softwood species, for which the mean ratio was 26, with slightly less variation among the different species than was the case for the hardwoods.

Greenhill (1936), according to Stamm and Loughborough (1942), reported a mean ratio for s_{vf}/G_g of 27 for 170 species of Australian woods. There was more scatter in this case, only half of the species' ratios falling between 21% to 33% and three-fourths between 18% and 36%.

Several other sets of volumetric shrinkage data have been analyzed by means of Eqs. (4.23) to (4.25), the results of which are summarized in Table 4.1. From

Table 4.1. Summary of volumetric shrinkage data on woods grown in temperate and tropical climates, based on use of Eqs. (4.23) and (4.24)

	Temperate	Tropical
source[a]	1, 2, 3	4, 5, 6
Number of woods	198	209
$M_f \pm SD$ (Eq. 4.23)	27.50 ± 4.94	20.32 ± 5.48
A (Eq. 4.24)	3.94	4.41
B (Eq. 4.24)	18.51	12.22
R^2 (Eq. 4.24)	0.468	0.294
$G_g \pm SD$	0.459 ± 0.0999	0.581 ± 0.149

[a] 1. US For Prod Lab 1955. 2. Can For Prod Lab 1951. 3. Chen 1962. 4. Vink 1965. 5. Ma 1960. 6. Limaye and Seaman 1933.

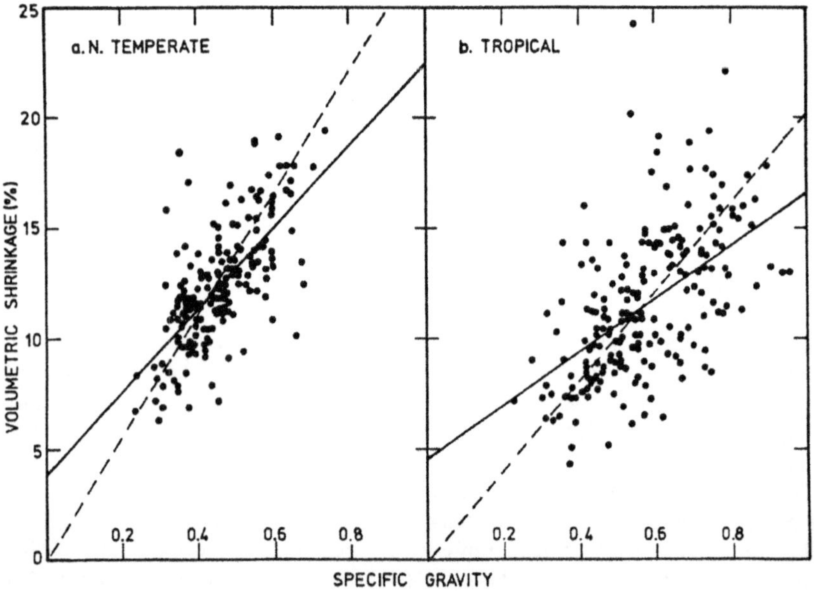

Fig. 4.7. Plotted points and regression curves (*solid lines*) of volumetric shrinkage against specific gravity (G_g) for woods grown in: **a** North temperature forests (198 woods), and **b** Tropical forests (209 woods). Data are summarized in Table 4.1. *Broken lines* are drawn through the origin and the mean values in each case

these results it appears that, as a general rule, volumetric shrinkage is proportional to wood density. For woods grown in temperate climates the relationship given by Eq. (4.23) yields mean values for M_f close to those usually associated with fiber saturation, in the range from 25% to 30% (Fig. 4.7 a). For those grown in tropical or semi-tropical areas, the mean values obtained for M_f using Eq.

(4.23) are generally considerably smaller (Fig. 4.7b). Wangaard (1951) first reported this difference in his studies of the shrinkages of woods grown in the tropical regions of Central and South America. He suggested that one reason for the difference is the lower fiber saturation points of tropical woods.

Chafe (1986) reported on a detailed study of the shrinkage-specific gravity relationships of 218 woods of Australia and the southwest Pacific region. He noted in his study, which included 69 different eucalypts, that there was a general correlation between shrinkage and specific gravity, based on all 218 species. However, two factors reduced this correlation, the excessive initial shrinkage due to collapse, and the anomalous behavior of the eucalypts. Use of shrinkage data based on reconditioned samples from which collapse had been removed (Sect. 4.4.2) greatly improved the correlation. Further improvement resulted from removing shrinkage data obtained on the 69 eucalypt species, leaving 149 non-eucalypt woods.

Chafe further demonstrated that the 69 eucalypt species showed no positive relationship between volumetric shrinkage and specific gravity. In fact, the initial shrinkage of green eucalypt wood was negatively correlated with specific gravity. Presumably this is because the excessive initial shrinkage or collapse (Sect. 4.4.2) so common in eucalypt wood is greater in low than in high density wood. However, even in reconditioned wood in which collapse had been presumably eliminated, there was no correlation between volumetric shrinkage and specific gravity in eucalypts, in contrast with the general trend. Kanagawa and Hattori (1984), in their study of the shrinkage of 41 woods from Sulawesi, Indonesia, noted that collapse occurred primarily in ray tissue.

Quirk (1984), using optometric measuring techniques, found that the cell lumens of the earlywood cells of Douglas-fir expanded somewhat during shrinkage of the wood while those of the latewood contracted. The reason for the difference in lumen behavior is not known, but is probably related to the relative thicknesses, swelling, characteristics, fibril orientation and stiffness of the cell wall layers (Sect. 4.5). Quirk interprets his findings on Douglas-fir to indicate that the lignin-rich middle lamella is more effective in reducing shrinkage in the thin-walled earlywood cells than in the thick-walled latewood cells.

The validity of Eqs. (4.22) to (4.23) within a given species can best be determined for those woods, such as the southern yellow pines, and Douglas-fir, which show a large variation in density. Stamm and Loughborough (1942), for example, reported s_{vf}/G_g ratios of 25.4, 26.2, 24.0, and 24.8 for loblolly pine boards whose green-volume specific gravities were 0.63, 0.54, 0.47, and 0.36, respectively. They calculated a mean value of 28 for the ratio s_{vg}/G_g from measurements on thin tangential sections of Douglas-fir containing varying proportions of earlywood and latewood.

Choong and Barnes (1968) reported that the volumetric swelling S_{vf}, between the dry and water-soaked condition, of four southern hard pines (*P. echinata*, *P. palustris*, *P. taeda*, *P. eliottii*), increased essentially linearly with dry specific gravity G_0 over the range from ca. 0.4 to 0.8. They measured core wood (juvenile wood) and mature wood separately and noted that more consistency was obtained with mature wood than with core wood. The combined regression equations for the four species are: $S_{vf} = 4.7 + 16.5\,G_0$, $2.0 + 24.6\,G_0$, and $2.8 + 21.6\,G_0$ for the

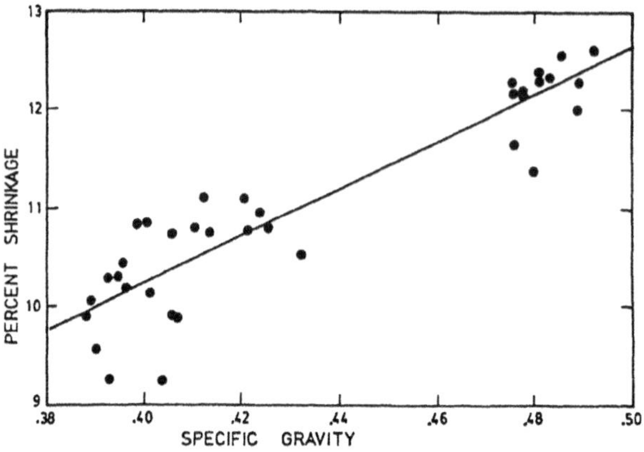

Fig. 4.8. Plotted points and regression curve of volumetric shrinkage s_{vf} against green-volume specific gravity G_g for *Pinus taeda*. (Yao 1969)

core wood, mature wood, and combined core wood and mature wood, respectively. The corresponding values of R^2 (coefficient of determination) were 0.20, 0.72, and 0.37. Therefore, the best regression equations were obtained from mature wood data, the poorest from core wood. The corresponding values of M_f, obtained from the mean value of S_{vf}/G_0, based on Eq. (4.23), were 24.2, 27.7, and 26.0, in reasonable agreement with other values reported for softwoods. In a subsequent paper, Choong (1969) reported that extraction of the core wood with hot water and organic solvents greatly improved the regression of S_{vf} against G_0, such that R^2 was raised from 0.20 for unextracted wood to 0.59 for the extracted wood. The percent swelling S_{vf} also became more sensitive to G_0 such that S_{vf} changed from $4.7 + 16.3\ G_0$ to $1.6 + 26.3\ G_0$, the mean value of the ratio S_{vf}/G_0 increasing from 23 to 30 after extraction.

Measurements by Yao (1969) on shrinkage of over 1000 wood samples from 28 trees of *Pinus taeda* are graphed in Fig. 4.8, in terms of percent shrinkage against specific gravity. The regression equation of s_{vf} against G_g is given by $0.48 + 23.75\ G_g$ with R^2 equal to 0.83. Each of the 36 points plotted represents the mean value of shrinkage and specific gravity for all 28 trees at the same relative location in each tree.

4.4.2 Volumetric Hygroexpansion Coefficients

It has been assumed above that volumetric hygroexpansion is linear with moisture changes in the hygroscopic range as depicted in Fig. 4.1. This is a convenient approximation but is not usually observed when careful measurements are made. These curves are linear only over a portion of the hygroscopic moisture range.

Figure 4.9, taken from Keylwerth (1964), shows the volumetric v, tangential t, and radial r swelling of European birch from the dry condition during resorption to cell-wall saturation. All three curves are slightly sigmoid. During the initial sorption from the dry condition, the rate of swelling dv/dm is relatively small, increasing with M up to about 4% or 5%. This may be due to the fact that the first water molecules sorbed find their way into interstices in the cell wall matrix and

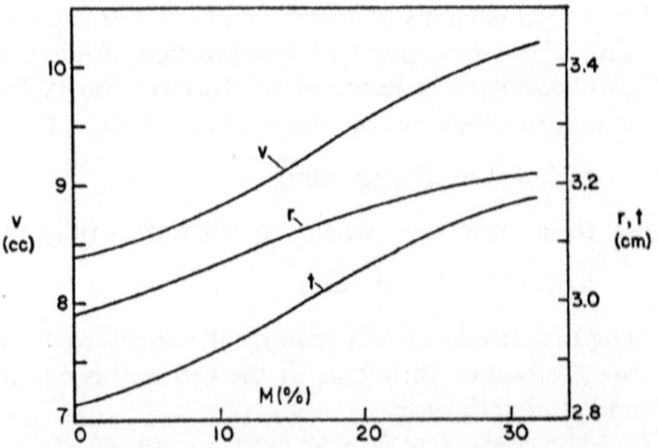

Fig. 4.9. Curves of volume v and of radial r and tangential t dimensions as functions of wood moisture content M for birch, as given by Keylwerth (1964)

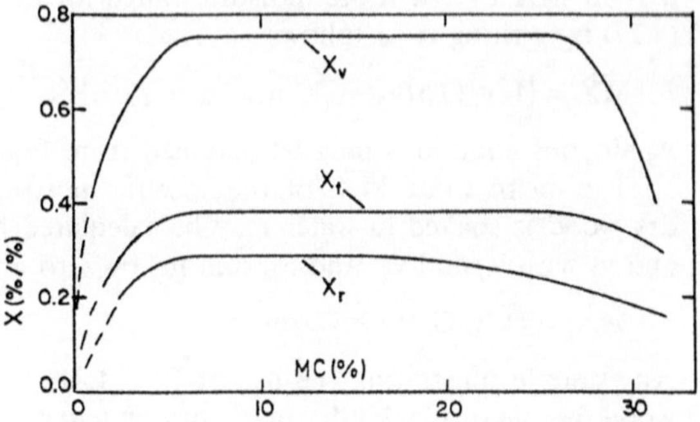

Fig. 4.10. Curves showing the volumetric X_{vs}, radial X_{rs} and tangential X_{ts} swelling coefficients as functions of wood moisture content. (After Keylwerth 1964)

do not contribute their full volumes to the cell wall (Chap. 3.2). Above about 5% moisture content, the volume increase is essentially linear to about 25%, above which dv/dm decreases again. This decrease may be caused by capillary condensation in the larger microvoids as predicted by the Kelvin equation [Eq. (3.106)]. It may also be caused by swelling stresses which tend to reduce the total swelling.

Keylwerth (1964) replotted the volume curve shown in Fig. 4.9 in terms of X_{vs}, the *volumetric swelling coefficient*, as defined by Eq. (4.9). He took the slope dv/dm of the curve as a function of moisture content and divided it by v_0, the dry volume. The resulting curve is shown in Fig. 4.10, together with those for the tangential and radial swelling coefficients, X_{ts} and X_{rs}. It is apparent from Fig. 4.10 that X_{vs} is constant over most of the moisture range except for extremes.

The ratio X_{vs}/G_0, where G_0 is the dry-volume specific gravity, gives the increase in wood volume per unit volume of liquid water added. Keylwerth (1964) defined this ratio as the *specific volume swelling*. Actually his definition was equivalent to $(0.01)(X_{vs}/G_0)$, since he defined X_{vs} as $(1/v_0)(dv/dm)$ where M is in percent, rather than as $(1/v_0)(dv/dM)$. In this book we will use the term *specific moisture expansion coefficient*, symbolized by MX, for this ratio. Thus

$$MX = X_{vs}G_0 = (1/v_0G_0)(dv/dm) \quad , \tag{4.27}$$

having dimensions of wood volume per unit liquid water volume, since G_0 is the ratio of the dry density of wood to that of liquid water. The coefficient MX may also be written in terms of the specific gravity G_m based on swollen volume v_m at moisture content m, since $v_0 G_0 = v_m G_m$. Thus,

$$MX = (1/v_m G_m)(dv/dm) \ . \tag{4.28}$$

At fiber saturation, when the cell wall is fully saturated,

$$MX = (1/v_f G_f)(dv/dm) \ . \tag{4.29}$$

The magnitude of MX is *unity* if the cell cavity remains constant in size and if the swelling or shrinkage of the cell wall is equal to the volume of liquid water added or removed.

The coefficient MX as given in Eqs. (4.27) to (4.29) may vary with moisture content since the slope dv/dm may not be constant (Fig. 4.9). An *average* value $\overline{MX}$ of MX over a finite moisture range m_1 to m_2 may be obtained from Eq. (4.27) by writing it as follows

$$\overline{MX} = (1/v_0 G_0)(v_2 - v_1)/(m_2 - m_1) \ . \tag{4.30}$$

Analogous equations may be obtained from Eqs. (4.28) and (4.29).

The mean value $\overline{MX}_f$ of the specific moisture expansion coefficient, when dry wood is soaked in water may be calculated from Eq. (4.30) by replacing v_1 and v_2 with v_0 and v_f, and m_1 and m_2 by zero and m_f, respectively. In this case

$$\overline{MX}_f = (1/v_0 G_0)(v_f - v_0)/m_f \ . \tag{4.31}$$

An example illustrating the use of Eq. (4.31) is given here. Assume that a dry wood sample of two kg dry mass, having a dry volume specific gravity G_0 of 0.5, and fiber saturation point M_f of 30%, is soaked in water. Its dry volume v_0 is 4 l. If the cell cavity size remains constant and the wood swelling is equal to the volume of water taken up, the wood volume at 30% moisture is 4 l, plus 0.6 l of water, or 4.6 l (Fig. 4.11 a). The *mean* hygroexpansion coefficient X_{vs} (Eq. 4.9) is

$$\bar{X}_{vs} = (1/v_0)(\Delta v/\Delta m) = (1/4)(0.6/0.3) = 0.5(\%/\%) \ .$$

The mean value $\overline{MX}_f$ of the specific expansion coefficient (equal to $\bar{X}_{vs}/G_0$) therefore is 0.5/0.5 or 1.0 in this case, since the volume change, $v_f - v_0$, of 0.6 l is equal to the volume of liquid water taken up by the wood.

If half of the swelling (0.3 l) of the cell wall takes place into the cell cavity and half externally, then $\bar{X}_{vs}$ is only half of 0.5, or 0.25 (Fig. 4.11 b). In this case $\overline{MX}_f$ is 0.25/0.5 or 0.5, rather than 1.0.

In contrast, if the cell cavity *expands* by 0.3 l rather than contracts, the external swelling of the wood is 0.9 l (Fig. 4.11 c). The corresponding values of $\bar{X}_{vs}$ and $\overline{MX}_f$ are 0.75 and 1.5, respectively.

A fourth scenario is that the percent expansion in the cell cavity is the same as that of the cell wall itself. In this case the cavity expands by 1.2 l, twice that of the cell wall, since it occupies twice as much volume (Fig. 4.11 d). The external swelling of the wood is then 0.6 plus 1.2, or 1.8 l, and $\bar{X}_{vs}$ and $\overline{MX}_f$ are 1.5 and 3.0, respectively. Such would be the case if the cell-wall structure was homogeneous.

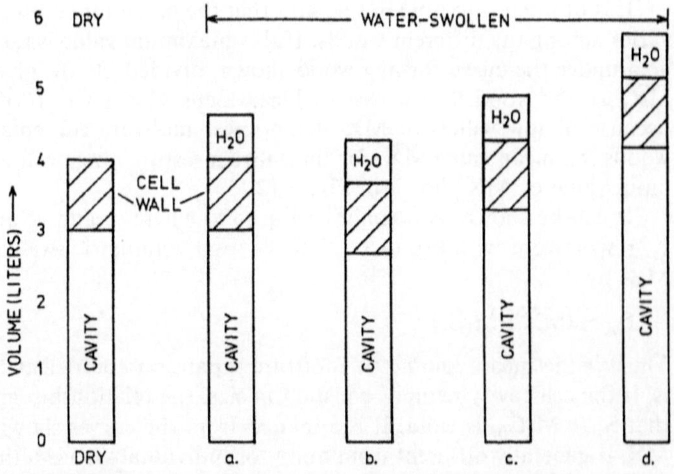

Fig. 4.11. Schematic diagrams showing the changes in wood volume due to moisture sorption for four cases with respect to cell cavity changes: **a** Cavity remains constant. **b** Cavity shrinks. **c** Cavity expands slightly. **d** Cavity expands by the same percentage as does the moist cell wall (see text)

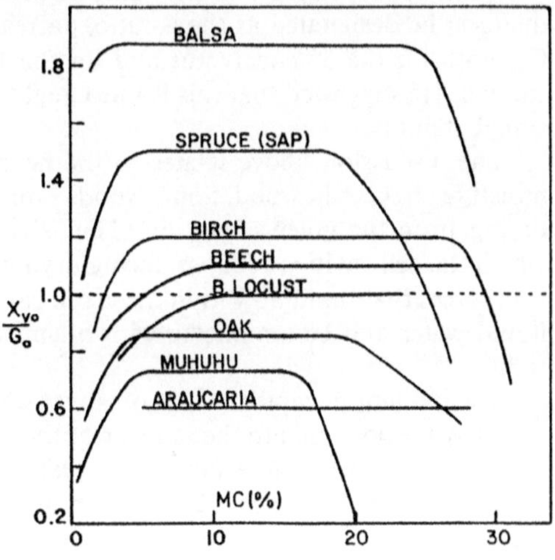

Fig. 4.12. Curves showing the specific volumetric hygroexpansion coefficients MX, or X_{vs}/G_0, as functions of wood moisture content for several different woods (After Keylwerth 1964)

In the hypothetical examples just given, the mean values $\bar{X}_{vs}$ and $\overline{MX}_f$ over the total moisture range zero to fiber saturation were used. Keylwerth (1964) showed curves of MX, or X_{vs}/G_0, as functions of wood moisture content for several kinds of wood, including birch (Fig. 4.12). For some of the woods, those with MX less than 1.0, there was some contraction in the cell cavity (Fig. 4.11 b). In those for which MX is greater than 1.0, the cavity apparently expanded somewhat during water resorption (Fig. 4.11 c). In the case of *B. locust* (Fig. 4.12) for which MX is unity over much of the moisture range, the cell cavity is presumed to remain essentially constant (Fig. 4.11 a).

It is of interest to note in Fig. 4.12 that the maximum moisture content shown varies among the different woods. If this maximum value is taken as M_f then the area under the curve for any wood shown, divided by M_f, gives the mean value $\overline{MX}_f$ for the wood. In the case of black locust the value of $\overline{MX}_f$ is less than 1.0 because of low values of MX at the lower moisture contents. For each of the woods the mean value $\overline{MX}_f$ for the entire moisture range is lower than the maximum value of MX shown in Fig. 4.12.

It can be shown by combining Eq. (4.5) where v_1 and v_2 are taken as v_0 and v_f, respectively, with Eq. (4.31) that the total volumetric swelling S_{vf} is related to $\overline{MX}_f$ by

$$S_{vf} = (\overline{MX}_f)(M_f G_0) \ . \tag{4.32}$$

Thus, if the mean volumetric moisture expansion coefficient $\overline{MX}_f$ is unity, that is, if the cell cavity remains constant in size, the relationship given by Eq. (4.22), that $S_{vf} = M_f G_0$, is valid. It is apparent from the curves shown in Fig. 4.12 that $\overline{MX}_f$ is generally different from unity for individual woods although on the average it may be close to unity. Also it is noteworthy that M_f itself varies among different woods, so that the product of $\overline{MX}_f$ and M_f which determines the ratio S_{vf}/G_0, or s_{vf}/G_f also varies among woods, and even among wood samples from a given tree.

Chafe (1986) noted that the value of $\overline{MX}_f$, which is essentially identical with the term he designated as the R-ratio, decreased with increase of specific gravity G_g, both for the 69 eucalyptus and for the 149 non-eucalyptus woods which he studied. He suggested that this R-ratio might be used as a general index of dimensional stability.

The discussion above relates to the normal hygroexpansion associated with moisture changes in wood. Some woods exhibit abnormal shrinkage during initial drying from the green condition. This is known as *collapse*, a phenomenon in which the cell cavities collapse during drying. It is due primarily to capillary tension forces that build up within those cell cavities which are completely filled with liquid water and communicate with other cells only through pit pores of small radii.

The maximum capillary tension stress which occurs under these conditions is inversely proportional to the radius r of the largest pore connecting the particular cell with an air-water interface. It is equal to $2\sigma/r$, where σ is the *surface tension* of the air-water interface, 72 dynes/cm at room temperature. The magnitude of the capillary tension is approximately 0.14, 1.4, 14, 140, and 1400 atm for capillary radii of 10, 1, 0.1, 0.01, and 0.001 μm, respectively, at room temperatures. These decrease slightly with increase in temperature, since σ decreases somewhat. However, the strength of wood decreases much more rapidly than does capillary tension, so that collapse is more likely to occur at elevated temperatures, based on this mechanism.

Another mechanism may also contribute to collapse or abnormally high shrinkage during drying of lumber. This is due to mechanical stresses induced by moisture gradients. During the early stages of drying the dry surface layers shrink and exert a compressive stress on the moist interior of lumber. These compressive stresses also tend to collapse the cells in the moist interior. The combination of

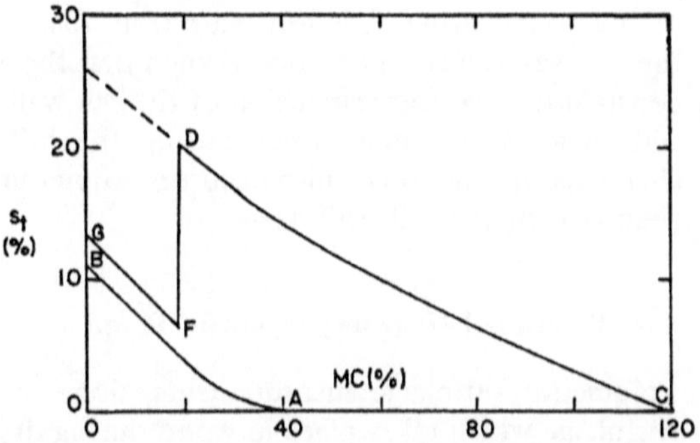

Fig. 4.13. Tangential shrinkage s_t in relation to wood moisture content M for a collapse-prone wood for different moisture histories. (After Kaumann 1964) (see text)

this compressive stress with the capillary tension forces in the cell cavity causes a greater tendency for collapse of the wood than either mechanism separately.

A more detailed discussion of the mechanisms of collapse due to capillary tension forces is given by Stamm (1964) and by Siau (1971, 1984). Additional consideration of the association of wood shrinkage and drying stresses is given in Section 4.6.

The effect of collapse on the initial shrinkage of wood during drying from the green condition is shown in Fig. 4.13. This illustration, adapted from Kauman (1964) shows the shrinkage-moisture content relationships for a collapse-prone wood dried under three conditions. The curve AB represents the normal shrinkage based on green dimensions which would result in wood which dries from the green condition without collapse. The curvilinearity between 20% and 40% moisture content may represent the effect of stresses caused by moisture gradients. For example, the outer surfaces of the wood dry first and tend to shrink, forcing the more moist but weaker wood in the interior of the piece to shrink, even though the mean moisture content of the entire sample is somewhat above fiber saturation.

The curve CDF shows the shrinkage for a sample which collapses severely as it dries. Note that it begins to shrink long before it reaches the fiber saturation point of approximately 30%. A conditioning treatment originally ascribed to Tiemann (1944), which restores the wood almost to its collapse-free condition, may be applied when the moisture content reaches 15% to 20%, represented by point D. This treatment consists of raising the temperature and humidity of the dry kiln for a brief time. This results in raising the water vapor pressure of the wood and also its temperature, as well as adding a slight amount of water to the wood. The increased vapor pressure in the cell cavities, coupled with the softening of the wood by heat and moisture, effectively restores the collapsed cells almost to their original dimensions, shown by point F. Subsequent drying follows the curve FG which shows somewhat higher shrinkage than the collapse-free curve AB.

Kübler (1970) has shown that heat recovery of dried red oak samples with excessive initial shrinkage or collapse can be accomplished by heating them in sealed vapor-tight compartments at moisture contents as low as 11%. Heating sealed samples in the moisture range between 11% and 15% gave the most favorable recovery of this excessive shrinkage.

As noted earlier, it is fortunate from the viewpoint of wood utilization that the cell-wall structure of wood is such that the cell cavity volume remains nearly constant during hygroexpansion of the cell wall. Thus, except for collapse of the cell cavity which may occur during the initial drying of some woods, the shrinkage and swelling which occurs in wood during normal exposure is much less than that of the cell wall itself.

4.5 Directional Hygroexpansion in Wood

Volumetric shrinkage measurements alone are not sufficient to describe the shrinkage which takes place in wood during drying because wood is orthotropic with respect to shrinkage as well as in most of its other physical properties. The reasons for the differential shrinkages in the three primary directions, longitudinal, radial, and tangential, are discussed below. At this point we will only state that longitudinal shrinkage s_l is least, usually in the order of 0.1% to 0.3 % from fiber saturation to the completely dry condition in *normal* wood. Total radial shrinkage s_r ranges from about 3% to 6% and tangential shrinkage s_t from about 6% to 12% for most woods over the same moisture range.

Volumetric shrinkage s_v is slightly less than the sums of the three directional components. The exact expression (Greenhill 1936) for a prismatic sample is

$$\begin{aligned} s_v &= 100\,[1-(1-0.01\,s_l)(1-0.01\,s_r)(1-0.01\,s_t)] \\ &= s_l+s_r+s_t-(0.01)(s_l s_r+s_r s_t+s_t s_l)+(0.0001)(s_l s_r s_t) \ , \end{aligned} \tag{4.33}$$

where the shrinkages are given in percent. A slightly less exact but practically identical equation can be obtained from Eq. (4.33) by eliminating the terms in which the longitudinal shrinkage s_l occurs as a multiple with s_r or s_t or both. Since s_l is so small for normal mature wood these terms become negligibly small even when s_r and s_t are large. This simplification reduces Eq. (4.33) to

$$s_v \simeq s_l+s_r+s_t-0.01\,s_r s_t \ , \tag{4.34}$$

which is sufficiently accurate within the range of s_r and s_t values normally found for wood. If the last term involving the product $s_r s_t$ is neglected, the equation reduces further to the simple form

$$s_v \simeq s_l+s_r+s_t \ , \tag{4.35}$$

or if s_l is completely negligible compared with s_r and s_t

$$s_v \simeq s_r+s_t \ . \tag{4.36}$$

The magnitude of the errors involved in using Eq. (4.35) rather than the exact Eq. (4.33) or the nearly exact Eq. (4.34) depends on the magnitudes of s_r and s_t. Ordinarily s_t is approximately twice s_r, the ratio of s_t/s_r commonly being called the T/R ratio. Table 4.2 shows the values of s_v calculated from Eqs. (4.33) and (4.34) over the range of s_v normally found for wood, assuming T/R ratios of 1, 2, and 3, and that the longitudinal shrinkage s_l is small compared with s_r and s_t. It is clear that the difference between the calculated volumetric shrinkages based on the exact and approximate equations increases as the shrinkage increases and as the T/R ratio decreases toward unity.

Table 4.2. Comparison of percent volumetric shrinkage s_v obtained using the approximate Eq. (4.34) and the exact Eq. (4.33) for T/R ratios of 1, 2 and 3

Calculated value of s_v (%) Approximate Eq. (4.34)	Exact Eq. (4.33) T/R = 1.0	T/R = 2.0	T/R = 3.0
0.00	0.00	0.00	0.00
3.00	2.98	2.98	2.98
6.00	5.91	5.92	5.93
9.00	8.80	8.82	8.85
12.00	11.64	11.68	11.73
15.00	14.50	14.50	14.58

Having discussed the relationships between volumetric and directional shrinkages, we return now to a more general treatment of dimensional changes associated with moisture changes after initial drying. As is the case with shrinkage, it is necessary to distinguish between volumetric and directional dimensional changes. The symbol X_v has been used to represent the coefficient of volumetric moisture expansion in Eq. (4.7). Similar expressions for directional moisture expansion coefficients are

$$X_l = (1/l)(dl/dm) \quad \text{(longitudinal)} \tag{4.37}$$

$$X_r = (1/r)(dr/dm) \quad \text{(radial)} \tag{4.38}$$

$$X_t = (1/t)(dt/dm) \quad \text{(tangential)} . \tag{4.39}$$

The corresponding expressions for the directional *humidity expansion coefficients* Y_l, Y_r, and Y_t are

$$Y_l = (1/l)(dl/dh) = ZX_l \tag{4.40}$$

$$Y_r = (1/r)(dr/dh) = ZX_r \tag{4.41}$$

$$Y_t = (1/t)(dt/dh) = ZX_t \tag{4.42}$$

where $Z = dM/dH = dm/dh$, the effective slope of the sorption isotherm, as in Eq. (4.13).

It has already been pointed out that wood is anisotropic with respect to its dimensional changes with changing moisture content. Normal wood is most stable dimensionally in the parallel-to-grain direction and least so in the tangential direction. Fortunately, wood also possesses its highest mechanical strength when external forces are applied so that the stresses in the wood are parallel to the grain. For example, wooden beams in the floors of buildings are loaded in such a way that the internal compression and tensile stresses are maximized along the grain. Likewise hygroexpansion is smallest in the long direction of the beams, thus reducing length changes. The same factor is also important when wood is used for sheathing or flooring, since the larger dimension is always parallel to the grain. We will therefore first consider the longitudinal shrinkage of wood and the factors which affect it before discussing transverse shrinkage in the radial and tangential directions.

4.5.1 Longitudinal Hygroexpansion

Longitudinal or axial hygroexpansion in normal mature wood is generally one or two orders of magnitude less than in the transverse directions. For example, initial shrinkage ranges from 0.1% to 0.3% when normal mature wood dries from the green to ovendry condition. Sometimes negative shrinkage occurs along the grain; that is, the wood may be slightly longer in the air-dry condition than in the green condition (Hann 1969). Therefore, normal mature wood from the outer portions of straight older trees shrinks so little along the grain that there is no shrinkage problem in use. However, the wood known as juvenile wood, which comes from young trees or from near the pith of older trees, may have appreciably higher longitudinal shrinkage than those given above. This is believed to be the result of the difference between the fibril angles in the S2 layer in the cell walls of juvenile wood and those of mature wood. The increased use in recent years of young trees which contain a substantial amount of juvenile wood has increased interest in the longitudinal shrinkage of wood. Also, the increased use of reaction wood, designated as compression wood in softwoods and tension wood in hardwoods, both of which also show excessive longitudinal shrinkage (Panshin and de Zeeuw 1980), focuses further attention on this factor.

Koehler (1960) listed three conditions of use under which longitudinal shrinkage may be considered excessive. These are:

1. When it causes undue shortening in length so as to throw adjoining members into serious disalignment, as in posts and columns, ... and opening of butt joints, as in flooring and siding.

2. When it causes crook or crossbreaks in lumber or waviness in veneer due to unequal shrinkage.

3. When it causes stresses in lumber that result either in pinching of the saw in ripping or crosscutting, or in longitudinal cracking of boards in ripping or machine planing.

There are other instances in which excessive shrinkage causes utilization problems in wood. For example, longitudinal crooking and/or bowing often occur in wood members which have reaction wood or juvenile wood on one edge or face, both of which have excessive longitudinal shrinkage. The use of juvenile wood in the lower chord members of wooden trusses has been shown (Gorman 1985) to produce arching due to excessive hygroexpansion of these members with seasonal changes in moisture content.

Several theories have been proposed to explain the factors which affect longitudinal shrinkage of wood. These are usually based on the fibril angles in the S2 layers of the cell walls of the longitudinal cells. Cockrell (1946) proposed a simplified lattice structure to explain longitudinal shrinkage, with some success. A more quantitative theory is that of Barber (1968), which is a refinement of the earlier theory of Barber and Meylan (1964). These theories are based on models which consider the cell wall of wood longitudinal cells to consist of an amorphous hygroscopic matrix in which are embedded parallel crystalline microfibrils which act to restrain hygroexpansion in the direction parallel to their axes. Later theories by Barrett et al. (1972), Cave (1972) and by Cousins (1976) accounted more fully for the behavior of the constituent components of the cell wall including cellu-

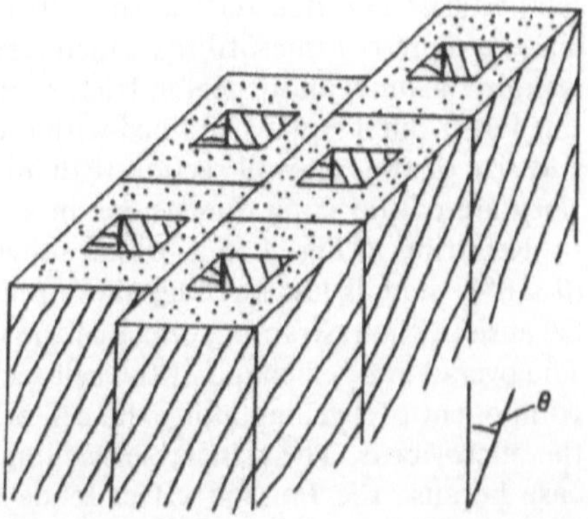

Fig. 4.14. Simplified cell-wall model showing mean fibril angle. (After Barber and Meylan 1964)

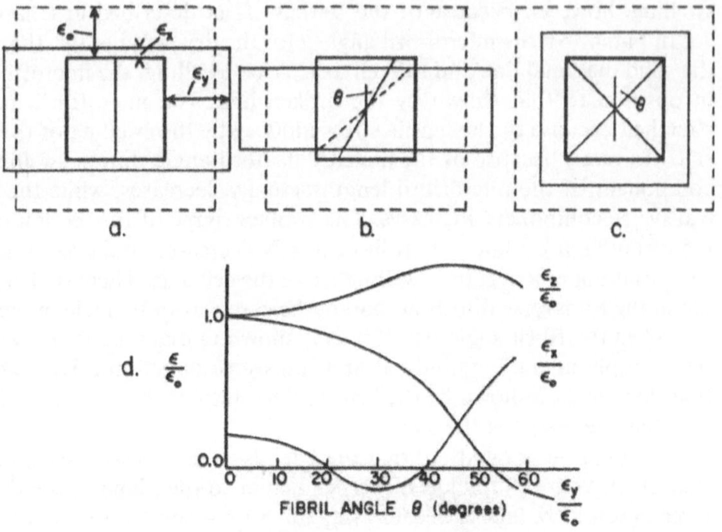

Fig. 4.15. Diagrams showing anisotropic swelling in the X (ε_x) and Y (ε_y) directions for fibril angles of: **a** Zero degrees. **b** 30°. **c** 45°. **d** Variable from 0° to 60°. (After Barber and Meylan 1964) (see text)

lose, hemicellulose, and lignin. The theories of Barber and Meylan (1964) and of Barber (1968) are discussed first, followed by consideration of Cave's theory.

In the early model of Barber and Meylan (1964), all of the fibrils are assumed to be oriented at the same angle θ to the long axis of the cell and the cells are assumed to be square in cross-section (Fig. 4.14). There is no tendency for the individual cells to twist as they swell because the fibrils in adjacent walls have the same angle θ but are of opposite sense since the walls are cemented together.

The principles involved in anisotropic swelling can best be visualized by reference to Fig. 4.15, adapted from Barber and Meylan (1964). In each of the diagrams, a, b, and c, an inner square represents the original dry dimensions of a section of the cell wall as seen from the cell cavity, with the long axis of the cell vertical. The fine vertical solid line in a and the diagonal fine solid lines in b and c represent microfibrils which resist length changes and therefore partially restrain the amorphous region from swelling in the direction parallel to their axes. The larger outer square outlined with the broken line represents the swollen shape and size of the cell-wall section if there were no microfibrils to restrain swelling. They are of the same dimensions in a, b, and c and indicate isotropic swelling.

Referring to Fig. 4.15 a, which shows the microfibrils parallel to the cell axis ($\theta = 0°$), there is less swelling along the cell axis than in the transverse direction because of the restraining effect of the microfibrils in the longitudinal direction. The overall swollen shape of the cell section is now rectangular with only a small component of swelling along the cell axis because of the high tensile strength of the microfibrils. The transverse swelling, however, is larger than for the isotropic case because the Poisson effect tends to keep the total swollen volume nearly equal to that for the isotropic case. One would expect some reduction in total swelling, however, because of the Barkas effect described in Chapter 2.

In Fig. 4.15 b the microfibril angle θ for the dry wood is 30°, the angle between the solid diagonal line and the cell axis. After swelling, the microfibril has shifted in position to that shown by the broken line. The microfibril may be slightly stretched because of the tensile stress induced by the swelling of the matrix material. However, because of the increase in fibril angle during swelling the vertical component of the microfibril length actually decreases, while the horizontal or transverse component increases. The swollen shape of the section is as shown by the rectangular outline in b. It has actually decreased in the axial direction of the cell, resulting in a negative swelling along the cell axis. There is also a larger swelling in the transverse direction than in the case shown in a where the angle θ is 0°.

When the fibril angle θ is 45°, as is shown in diagram c, the swelling tends to be isotropic in the longitudinal and transverse directions. The swelling is lower than for the case shown by the broken line outline where no microfibrils are present but the shape is the same.

It should be pointed out that there tends to be excessive swelling in the direction of the cell-wall thickness (perpendicular to the plane of the diagrams) in all three cases, a, b, and c, because there is no restraint in this direction. However, when the cells are square, the increase in thickness in one wall is equalled by the increase in thickness of the adjoining wall at right angles to it. Therefore, the overall cross-sectional shape of the cell remains square after swelling for all fibril angles, based on this model.

Figure 4.15 d shows three curves which illustrate qualitatively how the ratio $\varepsilon/\varepsilon_0$ of the expected swelling ε in each of the three principle directions when microfibrils are present to the isotropic swelling ε_0 which would occur in the matrix if the restraining effect of the microfibrils was absent. Looking first at $\varepsilon_x/\varepsilon_0$ where ε_x is the swelling in the vertical direction, parallel to the cell axis, it appears that ε_x decreases with increasing fibril angle θ until 30°, above which it increases. From $\theta = 20°$ to 40° the swelling is somewhat negative, indicating that longitudinal shrinkage occurs as the wood gains moisture.

The transverse swelling ratio $\varepsilon_y/\varepsilon_0$ begins at a maximum value when the fibril angle is zero and decreases slowly up to about 30° at which point it decreases rapidly, becoming equal to $\varepsilon_x/\varepsilon_0$ at $\theta = 45°$, as is indicated in Fig. 4.15 c. The

Fig. 4.16. Cylindrical cell model showing inner r_1 and outer r_2 radii and fibril angle. (Barber 1968)

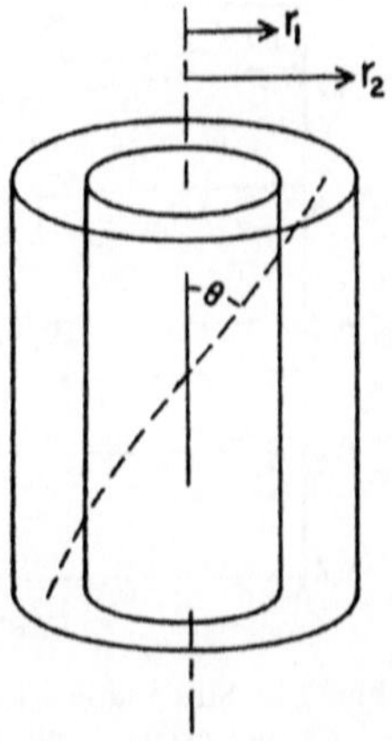

thickness swelling ratio $\varepsilon_z/\varepsilon_0$ is equal to the transverse swelling ratio $\varepsilon_y/\varepsilon_0$ at $\theta = 0°$, but increases with increase in θ while $\varepsilon_y/\varepsilon_0$ decreases.

The improved cell model of Barber (1968) shown in Fig. 4.16 differs from the earlier model in two respects; it is circular in cross-section and also provides for a thin constraining sheath outside the cylinder which acts to reduce transverse swelling. The cell wall therefore is considered to consist of two layers – the thin constraining sheath representing the S1 layer and the thick dominant layer of thickness $(r_2 - r_1)$, representing the S2 layer (Fig. 4.6), where r_1 and r_2 are the inner and outer radii of the cell wall, essentially equivalent to the same radii for the S2 layer. For convenience in deriving his general equation, Barber also mades the simplifying assumption in his cylindrical model that there are two sets of microfibrils in the S2 layer winding in opposite directions but at the same angle θ from the cell axis. This overcomes the twisting tendency which exists in a single isolated cell and has essentially the same effect as having two adjacent walls with microfibrils oriented in opposite senses (Fig. 4.14) and connected together by the middle lamella as they are in wood.

The equations used by Barber (1968) and by Barber and Meylan (1964) are derived in their papers and will not be given here. However, a graphical presentation of some of Barber's calculated results is shown in Fig. 4.17, adapted from Barber, in the form of six curves. All of the curves are calculated on the assumption that the rigidity of the restraining sheath is 1% of that of the S2 layer in the direction parallel to the microfibrils. Essentially this means that the sheath contains microfibrils oriented perpendicularly to the cell axis and that these have a total stiffness in restraining extension of only 1% compared with the total stiffness of those in the S2 layer. Presumably this small percent is based on the small volume of microfibrils in the thin sheath.

Two of the curves in Fig. 4.17 are qualitatively similar to the $\varepsilon_x/\varepsilon_0$ curve for longitudinal swelling shown in Fig. 4.15d, and are labeled in the same way. The two curves differ from each other in the ratio E/S, where E is the effective stiffness of the microfibrils in resisting elongation by the swelling matrix, and S is the rigidity or shear modulus of the swelling matrix, proportional to its resistance to

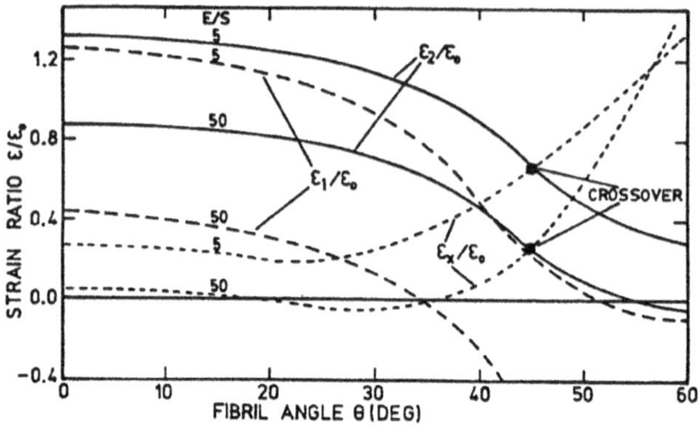

Fig. 4.17. Strain ratio $\varepsilon/\varepsilon_0$ curves of Barber for longitudinal $\varepsilon_x/\varepsilon_0$, transverse internal cell dimensions $\varepsilon_1/\varepsilon_0$, and external cell dimensions $\varepsilon_2/\varepsilon_0$, as functions of fibril angle, for two ratios E/S of the relative stiffness of the restraining sheath to that of the swelling cell-wall material. (Barber 1968) (see text)

shear deformation. It is clear that the relative longitudinal swelling at low fibril angles is smaller when the relative stiffness ratio E/S is larger. This is as expected because the stiffer the microfibrils are compared with the swelling matrix, the greater effect they will have in reducing longitudinal swelling at low angles of θ. It is also clear that the calculated longitudinal swelling ε_x is only a small fraction of the swelling ε_0 expected for the isotropic matrix at small values of θ. As the angle θ increases from zero the longitudinal swelling approaches zero and actually becomes negative, reaching a minimum near 30° and then increasing sharply above this point.

A second pair of curves in Fig. 4.17 shows the ratio $\varepsilon_2/\varepsilon_0$, where ε_2 is the calculated external transverse swelling (increase in radius r_2) when microfibril restraint is considered and ε_0 is the external transverse swelling expected for the isotropic matrix. Each of these curves is also for a different E/S ratio. It is clear from the figure that for low fibril angles the calculated value of swelling ε_2 is appreciably larger than the isotropic swelling ε_0. It remains essentially constant up to a fibril angle near 25° after which it decreases rapidly, crossing over the rapidly rising curves of $\varepsilon_x/\varepsilon_0$ at a fibril angle θ near 45°.

The third pair of curves shows the ratio $\varepsilon_1/\varepsilon_0$, where ε_1 is the swelling of the cell cavity or the increase in radius r_1 (Fig. 4.16). In most cases the relative swelling ε_1 is smaller than that of ε_2, the external cell swelling. It may approach zero for high ratios of E/S when the restraining sheath is sufficiently rigid. This would be the case found for those woods whose cell cavity appears to remain nearly constant in size during moisture changes, as discussed above (Sect. 4.4.1).

Experimental confirmation of the effect of fibril angle on the longitudinal and transverse shrinkage, including „crossover“ of the two shrinkage components at large fibril angles, is shown in Fig. 4.18, taken from Meylan (1968). Here are shown the relationship between longitudinal and tangential shrinkage and microfibril angle for *Pinus jeffreyi*. The crossover point occurs at a microfibril angle between 45° and 50° at which the two shrinkage components are equal. The

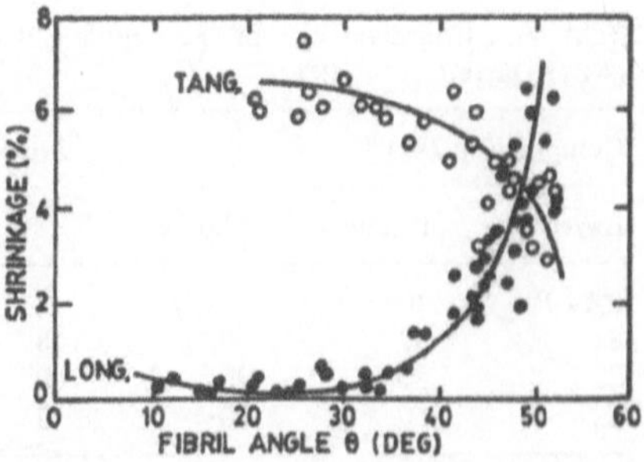

Fig. 4.18. Experimentally measured points and fitted curves of longitudinal and tangential shrinkages of *Pinus jeffreyi* in relation to mean fibril angle. (Meylan 1968)

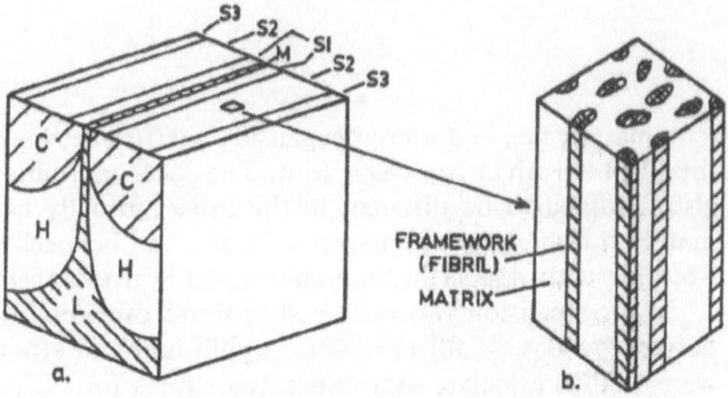

Fig. 4.19. Schematic diagram showing: **a** The seven layers in two contiguous cell walls, including the S1, S2, and S3 layers of each wall and compound middle lamella joining the two cell walls. The relative vertical magnitudes, designated by C, H, and L, show the relative distributions of cellulose, hemicellulose and lignin at any location in the cell wall. **b** The two components, framework and matrix, into which each layer is subdivided. (After Barrett et al. 1972)

shapes of the curves are similar to the theoretical curves derived by Barber (1968) (Fig. 4.17). The exact shape depends on factors which are difficult to evaluate quantitatively, but the general agreement is good.

Barrett et al. (1972) derived an equation for predicting the longitudinal hygroexpansion of softwoods, based on an idealized model of the conifer tracheid. They formulated the equation using a "mechanics of materials" approach, treating the wood structure as a composite material. Their model consisted of seven layers, including the middle lamella (M) and two primary walls (P) combined into a single layer (MP). The remaining six layers consisted of the S1, S2, and S3 layers of the secondary wall of each of two contiguous tracheids (Fig. 4.19a). They used this double cell-wall model to determine the effects on hygroexpansion of the thicknesses, chemical compositions, and fibril angles of the seven layers. The use of two contiguous walls as a double cell wall simplified the model by providing complete shear restraint and symmetry, as in the simpler model (Fig. 4.14) used by Barber and Meylan (1964).

Each of the seven layers was divided into two components, a framework and a matrix (Fig. 4.19b). The framework was presumed to consist of cellulose microfibrils combined with hemicellulose. It is therefore anisotropic with respect to both elastic and hygroexpansion properties, having a Young's modulus about

Table 4.3. Characteristics of the double cell wall model used to calculate the curves of Fig. 4.20. (After Barrett et al. 1972)

Composition (%)[a]			Latewood		Earlywood		Compr. wood	
Layer	Frame	Matrix	%	Angle	%	Angle	%	Angle
(M+P)	10	90	7	90°	15	90°	7	90°
S1	50	50	8	80°	30	80°	28	80°
S2	50	50	80	θ	40	θ	65	θ
S3	50	50	5	70°	15	70°	–	–

[a] Percent composition of framework and matrix.

ten times greater, and a hygroexpansion coefficient about 40 times smaller, along the fibril axis than transverse to it. The shear moduli and Poisson's ratios were also assumed to be different in the three mutually perpendicular planes. The matrix, a homogeneous mixture of lignin and hemicellulose, was assumed to be isotropic with respect to both elastic and hygroexpansion properties.

Hygroexpansion was evaluated in terms of *humidity expansion coefficients*, defined by Eqs. (4.40) to (4.42). Equilibrium and strain-compatibility relations were used to calculate hygroexpansion strains for each of the seven layers for a given humidity change. The total external hygroexpansion for the total wall was then calculated from those of the individual layers, based on their individual stiffnesses, expansion coefficients, and relative thicknesses. The stiffnesses and expansion coefficients of each layer were functions of their component compositions and the microfibril angle with respect to the cell axis.

Barrett et al. (1972) calculated anticipated longitudinal (Y_l) and transverse (Y_x) humidity expansion coefficients for typical earlywood and latewood conifer tracheids, and also for compression wood, using the data shown in Table 4.3. In order to calculate these coefficients, they used published values of the Young's moduli, shear moduli, and Poisson's ratios, as well as the humidity expansion coefficients (Y) of the framework and matrix components. For the framework, these coefficients were different in the three orthotropic axes but for the isotropic matrix they were identical in all directions.

Figure 4.20 shows plotted curves of Y_l and Y_x as functions of the fibril angle θ in the S2 layer, for each kind of tracheid shown in Table 4.3. The compression wood is presumed to be lacking an S3 layer (Panshin and de Zeeuw 1980). The broken vertical lines shown in each diagram represent the mean fibril angles for the S2 layer in each case. These are approximately 10°, 40°, and 40° for typical latewood, earlywood, and compression wood, respectively. Several conclusions may be reached from the curves:

1. As the mean fibril angle in the S2 layer increases, longitudinal hygroexpansion increases and transverse expansion generally decreases.

2. Longitudinal hygroexpansion of typical earlywood (40° fibril angle) is much greater than that of typical latewood (10° fibril angle), the reverse being true for transverse expansion.

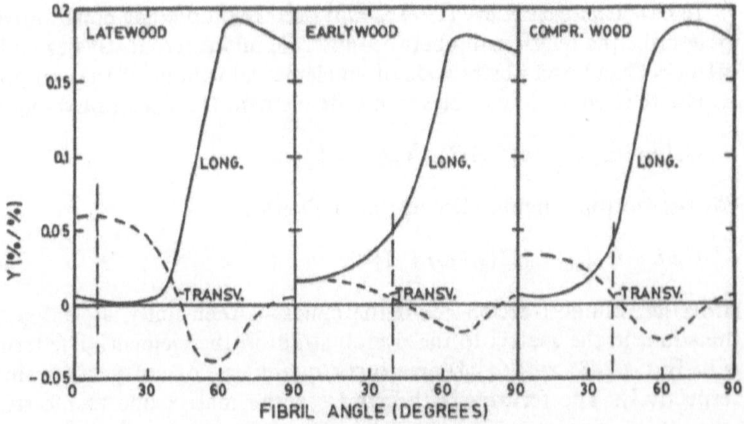

Fig. 4.20. Predicted curves of the longitudinal and transverse humidity expansion coefficients Y as functions of fibril angle, for latewood, earlywood and compression wood as calculated by Barrett et al. (1972)

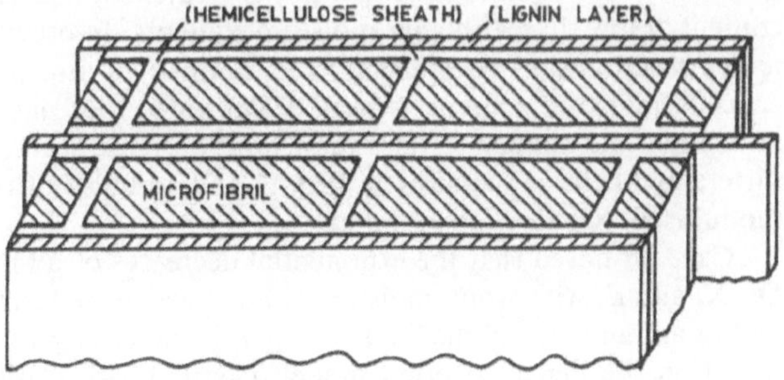

Fig. 4.21. Schematic diagram of the framework-matrix model used by Cave (1978a, b) consisting of parallel cellulose fibrils embedded in a hemicellulose matrix

3. Hygroexpansion of typical compression wood is similar to that of earlywood, although somewhat reduced (at 40° fibril angle) because of the greater lignin and reduced cellulose content of compression wood.

Cave (1978a, b) used a more complex model for the cell structure than that of Barrett et al. (1972). He replaced their framework matrix model (Fig. 4.19b) with an "array of parallel cellulose microfibrils embedded in a hemicellulose matrix", each of which is bonded to adjacent similar elements by a lignin microlayer (Fig. 4.21). Each of the cell wall layers, middle lamella (M), primary wall (PP), and the S1, S2, and S3 layers of the secondary wall were composed of these arrays of "basic cell-wall elements" bonded by lignin microlayers. Cave also used a double cell-wall model similar to that of Barrett et al. (Fig. 4.19a). Thus, the two physical models differ primarily in the organization of the constituents within the individual layers of the cell wall, as noted above.

In earlier papers, Cave (1972, 1975) used the following constitutive equations to describe the relationship between internal and external stresses and the overall stiffness C and overall strain $\Delta\varepsilon$ of an elemental volume of the composite material. For the cellulosic or polysaccharide element the constitutive equation is

$$C\Delta\varepsilon = F_m[C_m\Delta\varepsilon_0 - \Delta C_m(A_m\varepsilon - \varepsilon_0)] + \Delta\sigma \ . \tag{4.43}$$

For the isotropic lignin element this reduces to

$$C\Delta\varepsilon = C\Delta\varepsilon_0 - \Delta C(\varepsilon - \varepsilon_0) + \Delta\sigma \ , \tag{4.44}$$

since the volume fraction F_m of the matrix is then unity, as is A_m, the ratio of the strain in the matrix to the overall strain in the element. The terms C, ε and σ in Eqs. (4.43) and (4.44) are tensor quantities, of stiffness, strain and stress, respectively. The term ε_0 is the strain in the matrix due to moisture sorption when there is no restraint to swelling. The stiffness term C_m is that of the matrix in Eq. (4.43).

Several workers, including Cockrell (1947), Kelsey (1963), Hann (1969), Meylan (1972), Espenas (1974), and others, have reported that longitudinal hygroexpansion of wood is often a nonlinear function of moisture content. For *Pinus radiata*, Meylan found that it was greater for a given increment of moisture content at low than at higher moisture contents. In other words, the longitudinal moisture expansion coefficient X_l as defined by Eq. (4.37) decreased with increasing moisture content. Cave (1978a) noted that this decrease was nearly exponential, and similar to the exponential decrease with moisture content, of the differential heat of sorption Q_s [Eq. (2.41)] and also of the longitudinal Young's modulus E_l (Chap. 1.5) of wood.

Cave proposed that the exponential decreases of all three of these properties, Q_s, X, and E, with wood moisture content are related to the exponential increase in the amount of "alpha" water m_a with increasing total moisture content m, based on the Peirce sorption model (Chap. 3.2.4). Cave (1978a) used the term "high energy" water for the alpha water since it is more strongly bonded to the wood than is the "beta" or weakly bound water. As noted in Chapter 3.2.4, the alpha water is similar to the water of hydration m_h of the Hailwood-Horrobin sorption model and to the primary water m_1 of the Dent model.

Cave (1978a, b) modified his longitudinal hygroexpansion model by assuming that longitudinal expansion is proportional to the amount of high energy or alpha water taken up rather than to the total amount of water sorbed. With this assumption he calculated moisture expansion curves similar to those obtained experimentally by Meylan (1972) for *Pinus radiata* over small discrete moisture ranges (Fig. 4.22).

In order to make these calculations it was necessary to make assumptions as to the stiffness and hygroexpansion properties of the model components such as the lignin and hemicelluloses, as well as the cellulosic fibrils. Cave also combined the S1 and S3 layers of the secondary wall with the middle lamella (M) and primary wall (P) layers to form a "binding" or "hoop" layer, having a mean fibril angle of 70°. As was the case with the models of Barber (1968) and of Barrett et al. (1972), the S2 layer was taken to be the dominant layer, with a variable fibril angle

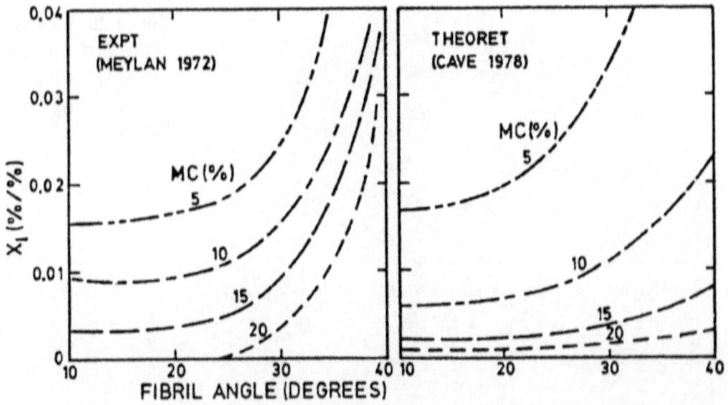

Fig. 4.22. Comparison of the curves of the longitudinal moisture expansion coefficients X_l at several moisture contents as functions of fibril angle, as measured by Meylan (1972), with those calculated by Cave (1978a, b) from his theoretical model

Table 4.4. Model for the average cell. (Cave 1976)

		S2 Layer	Binding Layer	Middle Layer
Cellulose	Vol. %	50.4	50.4	0
Hemicellulose	Vol. %	25.2	25.2	0
Pectic polysaccharide	Vol. %	–	–	15.8
Lignin	Vol. %	24.4	24.4	84.2
Percent layer volume	Vol. %	76.2	12.4	11.4
Mean fibril angle	–	variable	70°	–

θ the independent variable plotted in Fig. 4.22. Table 4.4 gives some of the parameters used by Cave (1978b) for characterizing his shrinkage model.

The large longitudinal hygroexpansion of reaction wood, compression wood in conifers, and tension wood in hardwoods is partly related to the large fibril angle in their cell walls. Several investigators (Harris and Meylan 1965, Boyd 1977, Harris 1977) have concluded that both the large fibril angle in the S2 layer and the great thickness of the tracheid wall contribute to the excessive axial hygroexpansion of compression wood. In tension wood, the S2 layer is often reduced in thickness, and the large fibril angle in the S1 layer is probably responsible for the large longitudinal dimensional changes. The gelatinous layer in tension wood fibers does not affect these changes according to Norberg and Meier (1966), as could be expected from the axial orientation of its cellulose microfibrils.

4.5.2 Radial and Tangential Hygroexpansion

It has already been noted that hygroexpansion along the grain of wood is one or two orders of magnitude smaller than across the grain except when the fibril angle θ is excessively large. Furthermore, there is considerable anisotropy in transverse

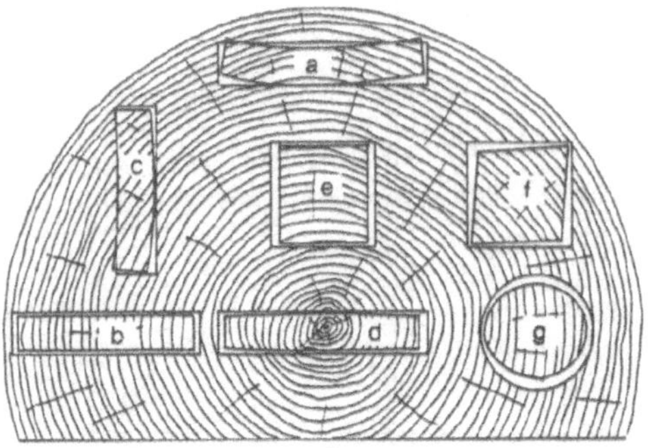

Fig. 4.23. Changes in cross-sectional shapes of boards, squares, and round members after drying in relation to location in a log. (After US For Prod Lab 1974)

hygroexpansion, it being approximately twice as great in the tangential as in the radial direction. This anisotropy is responsible for much of the warping of lumber associated with initial drying and during subsequent use.

Before discussing the mechanisms believed to cause transverse anisotropy in hygroexpansion we will first consider some of its practical effects. Figure 4.23 illustrates some of the more common kinds of transverse warping caused by tangential/radial anisotropy in shrinkage when lumber cut from various locations in the cross section of a typical log is dried.

The most striking change is in the flat-sawn board a which cups into a trough-like shape after drying. This effect can be reduced if the lumber is restrained from cupping during drying by the application of weights or other restraints to force the lumber to remain more nearly flat. However, cupping cannot be eliminated completely since some residual stress which tends to cup the board is inevitably present. If a plane board is required for use, considerable surfacing of the cupped surfaces is required, resulting in a thinner board and much waste. Furthermore, after exposure to changing environmental humidity, the board will tend to cup cyclically with cyclic humidity changes.

The quarter-sawn board b does not cup, although its shape changes somewhat as shown. This is the best way of cutting lumber from the viewpoint of dimensional stability but in the normal practice of sawing a log it is only possible to obtain one or two truly quarter-sawn boards, the others having a ring orientation somewhere between the two cases shown in a and b, such as c. A board cut through the pith such as d shrinks less at the center than at the two edges, becoming somewhat convex. The square timber cut at e becomes rectangular after drying, while that cut at f becomes diamond-shaped. The round member cut at g distorts into an oval shape.

Hsu and Tang (1975) have derived equations for calculating the extent of distortion of cross-sections of lumber of shapes similar to those shown in Fig. 4.23 under stress-free conditions, that is those in which the lumber is unrestrained and there are no moisture gradients during drying. The latter assumption was necessary in order to obtain an explicit solution of the problem for the stress-free condition. For the case of a flat-sawn board such as is shown in Fig. 4.23 a, whose outer face is located at a distance r from the pith (Fig. 4.24), the amount of cupping, expressed as the ratio of the depth of cup d to the original width w of the board is given by them as

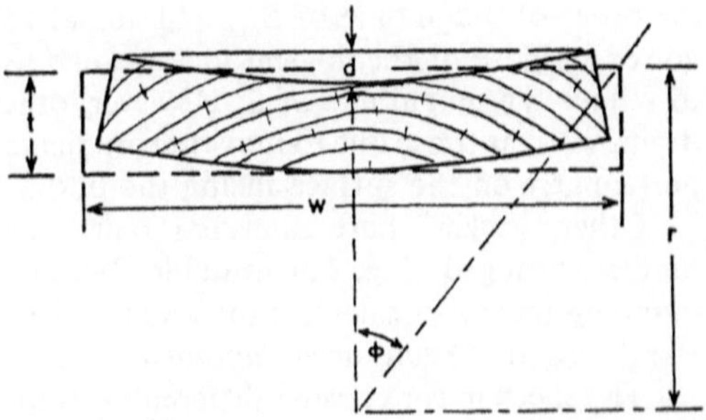

Fig. 4.24. Diagram showing the magnitude of cupping, defined by Hsu and Tang (1975) as the ratio of depth of cup d to the original w of the board

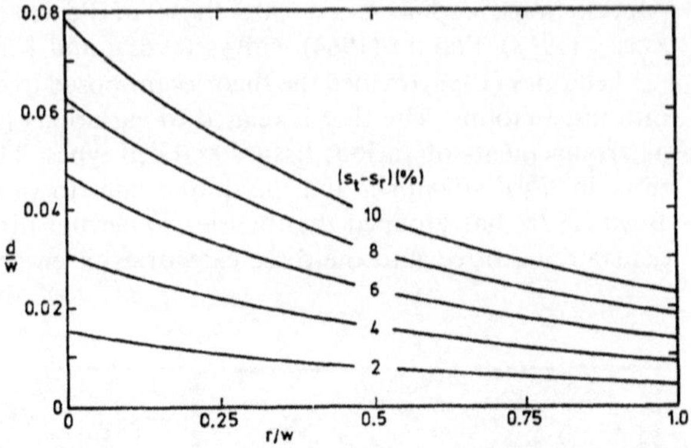

Fig. 4.25. Theoretical curves relating the cupping ratio d/w to the ratio r/w (Fig. 4.24) for several values of the difference between tangential s_t and radial s_r shrinkages as given by Hsu and Tang (1975)

$$d/w = 0.005\,(s_t - s_r)\arctan(w/2r)\ , \tag{4.45}$$

where s_t and s_r are the tangential and radial shrinkages, respectively, and arctan (w/2r) is the angle ϕ (Fig. 4.24).

According to Eq. (4.45) the extent of cupping d/w is directly proportional to the difference in tangential and radial shrinkages and also to the angle ϕ whose tangent is w/2r. For a board of a given width, w, the extent of cupping decreases with distance r of the outer surface from the pith. Figure 4.25 illustrates how d/w varies with the ratio r/w for several values of the difference $(s_t - s_r)$. As Hsu and Tang (1975) point out, Eq. (4.45) is valid only for flat-sawn boards which do not include the pith. If the pith is included, internal stresses arise which negate the equation. It is evident from Fig. 4.25 that cupping increases with the difference $(s_t - s_r)$, as well as with proximity to the pith.

Hsu and Tang (1975) also calculated the magnitude of the restraining force required to prevent cupping, treating the cross-section (Fig. 4.24) as a simply supported beam of span w, with the restraining force P applied at the center of the span. Using simple beam theory they calculated the maximum stress S_{max}, occurring at the center of the upper and lower surfaces to be

$$S_{max} = [0.03\,Eh\,(s_t - s_r)/w]\,[\arctan(w/2r)]\ , \tag{4.46}$$

where E is the Young's modulus for the wood in the tangential direction and h is the depth of the beam, that is the board thickness. Equation (4.46) gives only the order of magnitude of S_{max}. In actual fact, somewhat lower stresses are involved because of rheological factors such as relaxation of stress which operate to reduce the magnitude of S. However, other factors normally associated with drying such as those due to moisture gradients (Sect. 4.6) exacerbate the situation, particularly on the surface facing the bark.

Other workers have analyzed mathematically the shrinkage behavior of lumber during drying. For example, Imata (1975) has applied the concepts of mapping theory to calculate the magnitudes and distributions of stresses during the drying of *Cryptomeria japonica.*

The mechanism causing differential transverse hygroexpansion has been the subject of considerable controversy in the literature on wood science. One of the most thorough reviews of this subject is that of Boyd (1974), who cites 76 relevant papers. Some earlier reviews are those of Pentoney (1953), Bosshard (1956), Kelsey (1963), Stamm (1964), Crews (1965), and Kollmann and Côté (1968).

Pentoney (1953) divided the theories proposed to explain transverse anisotropy into three groups. The first is related to variations in *gross wood structure* such as arrangements of various tissues and cell types. The second is based on variations in *fibril alignment* and the third relates to variations in *cell-wall layering.* Boyd (1974) has grouped the theories somewhat differently, but his groups can generally be fitted into the three categories given above.

Fig. 4.26. Schematic drawings of a typical softwood or coniferous wood (*Pinus strobus* L.) on the left (200×), and a typical hardwood or porous wood (*Liquidambar styraciflua* L.) on the right (160×). The three surfaces shown in each case are the transverse or cross-section *A*, the radial surface *B*, and the tangential surface *C*. A portion of a wood ray is shown at the bottom edge of the radial face (*B*) in each diagram. A part of a second ray is also shown on both the *A* and *C* faces in each drawing. (Reproduced from Panshin and de Zeeuw, 1980, from original drawings by C. H. de Zeeuw)

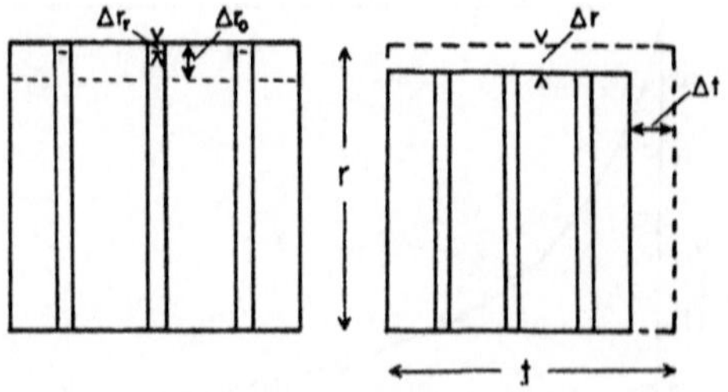

Fig. 4.27. Schematic diagram of model used to support the ray-restraint theory of anisotropic shrinkage, showing: **a** Radial shrinkages of isolated ray tissue Δr_r and ray-free tissue Δr_0. **b** Radial Δr and tangential Δt shrinkages of whole wood

Two explanations for transverse hygroexpansion anisotropy based on gross wood structure have been proposed. The oldest and most popular may be termed the *ray restraint theory*. It is based on the assumption that the radially oriented ray tissue (Fig. 4.26) shrinks and swells less radially than does the surrounding longitudinal tissue. Furthermore, because of its radially elongated cells it is presumed to be stronger than adjoining tissue, thus restraining radial expansion. The second is the *earlywood-latewood* theory which postulates that latewood shrinks more tangentially and is also stronger than earlywood. Thus it forces the earlywood to shrink more tangentially and less radially, because of the Poisson ratio effects, resulting in higher tangential than radial shrinkage. The ray restraint theory will be discussed first.

The ray restraint theory can best be understood by reference to Fig. 4.27, which shows a hypothetical cross-section for wood consisting of longitudinal and ray tissues. The assumption is made that the radial hygroexpansion coefficient X_{rr} of isolated ray tissue is lower than the radial coefficient X_{r0} of ray-free longitudinal tissue. The radial hygroexpansion coefficient X_r of the gross wood including rays should be intermediate between X_{rr} and X_{r0}, the actual value depending on the relative stiffness of the ray and ray-free tissues. It may be calculated approximately from the equation

$$X_r = (X_{rr}V_r + X_{r0}V_0K_{0r})/(V_r + V_0K_{0r}) \ , \tag{4.47}$$

where K_{0r} is the ratio of the Young's moduli E_0/E_r of the ray-free and ray tissues, respectively, in the radial direction, and V_0 and V_r are the volume fractions of ray-free and ray tissues, respectively.

Equation (4.47) is an approximation to a more exact equation given by McIntosh (1955). It differs also in that his equation is given in terms of percent radial shrinkage s_r of the wood in terms of the percent radial shrinkage s_{rr} of the ray tissue and s_{r0} of the ray-free tissue. McIntosh measured the isolated shrinkages s_r of ray tissue from red oak and American beech wood, and of wood containing various volume fractions V_r of ray tissue, using microtomed sections. He then measured the radial shrinkage s_{rr} of isolated rays and the radial shrinkage s_r of wood containing two different fractions V_r of ray tissue. From these two sets of measurements he was able to calculate both s_{r0} and the ratio E_0/E_r by solving his equation for two different sets of measured data on the other variables s_r, V_r, and s_{r0}, the latter value being taken as constant. The results he obtained for beech are shown in Fig. 4.28.

It is informative to note how the calculated values of radial wood shrinkage without rays compares with the observed radial shrinkage of the gross wood and

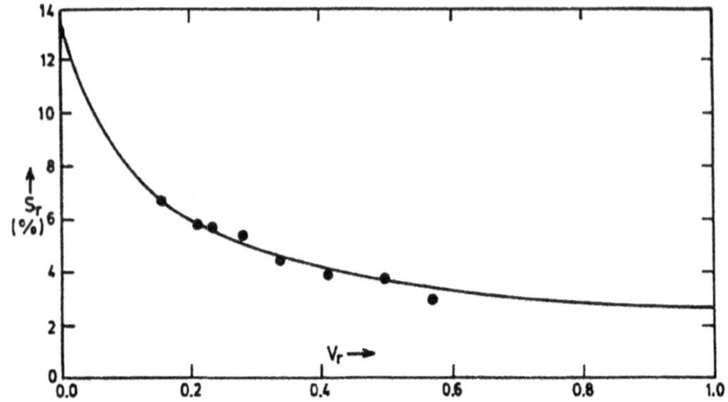

Fig. 4.28. Experimental points and calculated curve of radial shrinkage s_r in relation to fraction V_r of ray tissue, as determined by McIntosh (1955) for beech, assuming a ratio E_0/E_r of 0.124, where E_0 and E_r are the radial Young's moduli of ray-free tissue and of ray tissue, respectively

Table 4.5. Percent radial shrinkage from wet to dry condition of wood rays, rayless wood, and wood containing rays (taken from various sources)

Reference	Kind of wood	Radial shrinkage (%)[a]			
		S_{rr}	S_{r0}	S'_{r0}	S_r
Greenhill 1944	*Casuarina iukmanni*	1.2	–	–	3.3
	Grevillea robusta	1.2	–	–	3.7
	Xylomelum pyriforme	0.7	–	–	2.0
	Quercus sp.	3.2	–	–	4.9
Lindsay and Chalk 1954	*Quercus ilex*	3.1	–	6.0	–
	Cardwellia sublimis	0.9	–	3.5	–
	Helicia terminalis	0.8	–	1.9	–
Morschauser and Preston 1954	*Quercus borealis*	2.5	12.0	–	5.1
McIntosh 1955	*Quercus borealis*	2.6	6.8	4.8	–
	Fagus grandifolia	2.3	12.7	6.7	–
Wijesinghe S. de 1959	*Arctocarpus integra*	1.3	4.8	–	3.8
	Canarium zeylanicum	0.2	4.0	–	3.4
	Acer pseudoplatanus	3.8	4.9	–	4.3
Schniewind and Kersavage 1962	*Quercus kelloggii*	2.1	–	5.8	3.0

[a] S_{rr}, Isolated rays; S_{r0}, ray-free wood; S'_{r0}, wood containing fine rays but no broad rays; S_r gross wood including all rays.

of the rays themselves as listed in Table 4.5. Also shown are data obtained from several other sources which give the radial shrinkage of ray tissues and also of rayless wood and of wood containing rays. The data clearly indicate that the ray restraint mechanism is at least one of the factors responsible for transverse shrinkage anisotropy for these hardwoods.

Crews (1965) found that ray shape was a factor in influencing the T/R shrinkage ratio in five United States hardwoods. Short, wide rays reduced the radial shrinkage more than high, narrow rays. Crews attributed the relative ineffectiveness of narrow rays to their low stiffness in the radial direction. He also concluded that the effective radial shrinkage restraint of ray tissue decreases as the overall specific gravity of the wood increases. This is in agreement with Eq. (4.47), since the terms K_{0r} and X_{r0} would then increase as the specific gravity of the wood increases for a given ray volume, giving an increase in the radial coefficient X_r.

The reduced radial shrinkage of ray tissue may be related to the alignment of fibrils in the ray tissues, althougt this evidence is conflicting. Another possibility is that the hygroscopicity of ray tissue is lower than that of longitudinal tissue, which would reduce the total shrinkage. For example, Morschauser and Preston (1954) report that the fiber saturation point M_r was only 17.63% for ray tissue, compared with 25.57% for longitudinal tissue for *Quercus borealis*, based on radial-shrinkage intersection measurements.

In the above discussion no mention was made of tangential shrinkage or of any possible coupling of tangential and radial or longitudinal shrinkages through Poisson ratio effects. Any reduction in radial shrinkage of the longitudinal tissue caused by ray restraint would tend to increase tangential shrinkage because of Poisson ratio effects in longitudinal tissue. This may be partially reduced by the same effect in the ray itself in that increased radial shrinkage in the ray would be expected to produce decreased tangential shrinkage in the ray. However, this effect should be smaller than the increase in tangential ray shrinkage caused by the great reduction in ray shrinkage parallel to the axes of the longitudinal cells. This is true for two reasons. First, the amount of restraint of the rays parallel to the grain of the wood may be greater than that in the radial direction, since there is essentially no shrinkage of the ray permitted parallel to the grain because of the low longitudinal shrinkage of the wood. Second, the Poisson ratio would be expected to be higher in the longitudinal-tangential plane than in the radial-tangential plane because of the orientation of the cell axes in the radial direction.

Boyd (1974) has concluded that the ray-restraint mechanism is not an important factor in differential transverse hygroexpansion in wood. This may be true for softwoods, as Nakato (1958) has suggested, but for hardwoods the available evidence seems to indicate that it is an important factor.

The *earlywood-latewood interaction* theory for explaining transverse expansion anisotropy has been called the Mörath theory by Pentoney (1953) because Kollmann (1936) attributed it to Mörath. However, Erickson (1955) points out that the general idea behind the theory has been known at least as far back as Wagner (1917) and Forsaith (1926).

This theory (Fig. 4.29) states that the tangential expansion is greater than the radial because of the alternation of earlywood and latewood in the radial direction in many woods grown in the temperate zone, particularly softwoods. Hygroexpansion of latewood is greater than that of earlywood, and the latewood is much stronger than the earlywood. Therefore, the strong tangential bands of latewood force the weak bands of earlywood to expand tangentially to about the same extent as the latewood. In the radial direction, however, the total is nearly equal to the weighted mean expansion of the two components. Actually, it is less than this because of the decrease in radial dimension of the earlywood caused by excessive expansion in the tangential direction through the Poisson effect. There is

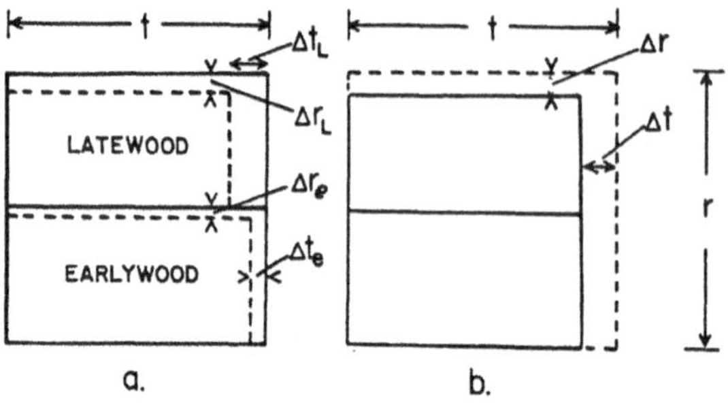

Fig. 4.29. Schematic diagram of model used for earlywood-latewood interaction theory of anisotropic shrinkage showing: **a** Radial and tangential shrinkages of isolated earlywood (Δr_e, Δt_e) and latewood (Δr_l, Δt_l). **b** Radial (Δr) and tangential (Δt) shrinkages of whole wood

also some decrease in the radial expansion of the latewood because its tangential expansion is somewhat reduced by the resistance of the earlywood to excessive tangential expansion.

For the expected tangential expansion coefficient X_t based on earlywood-latewood interaction an approximate equation similar to Eq. (4.47) and with similar assumptions can be derived. It can be written

$$X_t = (X_{te}V_e + X_{tl}V_lK_{le})/(V_e + V_lK_{le}) \ , \qquad (4.48)$$

where K_{le} is the ratio (E_l/E_e) of the Young's modulus (E_l) of the latewood to that (E_e) of the earlywood, V_e and X_{te} are the volume fraction and tangential expansion coefficient, respectively, of the isolated earlywood, and V_l and X_{tl} are the corresponding terms for the latewood.

An approximate expression can also be written for the expected radial-shrinkage coefficient X_r. This expression is more complex because of the Poisson's ratio effect. If it is assumed that the Poisson's ratio U is the same for latewood and earlywood,

$$X_r = X_{re}V_e + X_{rl}V_l + [V_eV_lU(X_{tl} - X_{te})(1 - K_{le})]/(V_e + V_lK_{le}) \ , \qquad (4.49)$$

where the additional terms X_{tl} and X_{te} are the tangential coefficients of the isolated latewood and earlywood. The anticipated ratio X_t/X_r, often designated as the T/R ratio, can be obtained by dividing Eq. (4.48) by (4.49).

Pentoney (1953) gives a more exact equation for the anticipated T/R shrinkage ratio in terms of the total directional shrinkages s_{re}, s_{rl}, s_{te}, and s_{tl} of the isolated earlywood and latewood, rather than the coefficients X. These equations in terms of T/R ratios are essentially similar to those given above if it is assumed that

$$100 - s_{te} \simeq 100 - s_{tl} \qquad (4.50\,a)$$

$$100 - s_{re} \simeq 100 - s_{rl} \ , \qquad (4.50\,b)$$

where the shrinkages are given in percent.

Ylinen and Jumppanen (1967) developed and tested a more rigorous equation for testing the earlywood-latewood interaction theory. They also extended the theory into longitudinal shrinkage on the assumption that such shrinkage is greater for earlywood than for latewood, as is usually observed.

Pentoney (1953) tested the earlywood-latewood interaction hypothesis experimentally by measuring the tangential and radial shrinkages of isolated early-

Table 4.6. Mean percent shrinkages of isolated latewood, earlywood and whole wood of Douglas-fir in tangential (T) and radial (R) directions

Reference	Latewood			Earlywood			Whole wood		
	T	R	T/R	T	R	T/R	T	R	T/R
Pentoney (1953)	7.2	8.9	0.8	4.8	2.4	2.0	7.2	4.4	1.7
Vintila (1939)	9.9	10.9	1.1	5.7	2.9	2.0	8.2	5.2	1.6
Quirk (1984)[a]	14.4	14.2	1.0	6.6	0.6	6.4	8.9	4.7	1.9

[a] Dry to water-soaked swelling.

Table 4.7. Percent shrinkage[a] of isolated earlywood and latewood in tangential (T) and radial (R) directions, and T/R ratios

Reference	Species	Latewood			Earlywood		
		T	R	T/R	T	R	T/R
Vintila 1939	Douglas-fir	10.9	9.8	1.1	5.7	2.9	2.0
	Pine	11.3	8.2	1.4	8.0	2.9	2.8
	Larch	12.2	10.2	1.2	7.1	3.2	2.2
	Fir	8.8	6.2	1.4	5.0	2.4	2.1
Pentoney 1953	Douglas-fir	7.8	9.8	0.8	5.0	2.4	2.1
Browne 1957[a]	Douglas-fir	8.6	12.3	0.7	6.7	5.6	1.2
	S.Y. pine	7.3	13.0	0.6	6.4	7.7	0.8
Ritter and Mitchell 1952	Longleaf pine	9.3	–	–	6.6	–	–
Erickson 1955	Douglas-fir	9.3	–	–	4.3	–	–
	W. red cedar	7.4	–	–	3.9	–	–

[a] Data of Browne (1957) are for percent swelling.

wood and latewood of Douglas-fir, and compared these with those obtained in the gross wood with the two kinds of tissues attached. The results are shown in Table 4.6. It is clear that the tangential shrinkage of the latewood is essentially the same whether isolated or attached to earlywood. The earlywood, when attached to latewood, has been forced to shrink essentially the same tangentially as the strong latewood. It also appears, as the theory suggests, that the radial shrinkage of the earlywood is decreased when it is attached to the latewood. The isolated earlywood shrinks much less than the isolated latewood in both directions, as anticipated. Furthermore, the isolated earlywood shows a T/R ratio of close to 2, while the isolated latewood T/R ratio is slightly less than unity. The high T/R ratio for the earlywood indicates that some mechanism other than the earlywood-latewood interaction may be operating in the earlywood.

Earlier work by Vintila (1939) on four softwoods tends to support the earlywood-latewood interaction theory for anisotropic transverse shrinkage, as do those by Ritter and Mitchell (1952), Brown (1957), and Erickson (1955), some of whose data are tabulated in Table 4.7. It is interesting to note that Vintila found large T/R shrinkage ratios in isolated earlywood compared with those for isolated latewood for all four woods, in agreement with the results of Pentoney. This

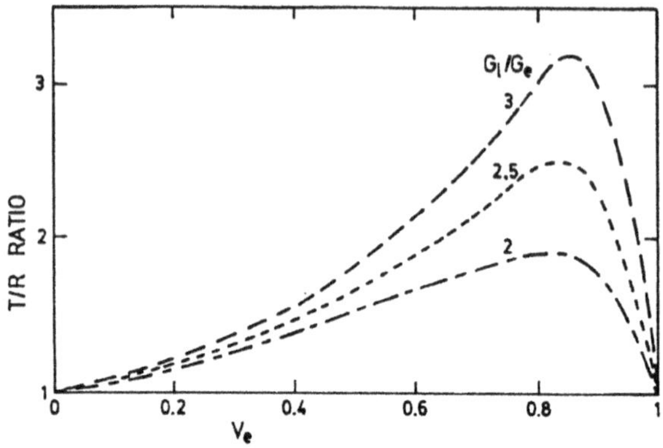

Fig. 4.30. Calculated curves of the T/R shrinkage ratio as a function of earlywood fraction V_e for different ratios of G_l/G_e based on the earlywood-latewood interaction mechanism

follows the general observation that the T/R shrinkage ratio decreases with increasing wood density, first noted by Kollmann (1936).

Quirk (1984) used optometric techniques to measure the densities, cell-wall areas and thicknesses, and cell lumen or cavity areas, of extractive-free Douglas-fir, in both the water-saturated and oven-dry conditions. From these he calculated shrinkages in cell areas, perimeters, tangential and radial dimensions, etc., some of which are summarized in Table 4.6. They tend to parallel the results of Pentoney and Vintilla, although the shrinkages of the isolated latewood seem exceptionally high, while the reverse is true for the radial shrinkage of isolated earlywood. It is not apparent whether these exceptional values occur because the material was extractive-free or because of the measurement technique, or both. Quirk concludes that the earlywood-latewood interaction theory of Mörath (1932) is supported by his study, but that other mechanisms are involved. He noted, for example, that the cell cavities of earlywood cells actually expanded somewhat during drying, while those of the latewood cells contracted. He attributed this to the high ratio of lignified middle lamella to secondary wall in earlywood cells. Presumably the relatively stiff and less hygroscopic middle lamella resisted shrinkage in this case. This tends to support the theory of Bosshard (1963) concerning the role of the lignified middle lamella in differential transverse shrinkage, as discussed later.

Hypothetical transverse expansion coefficients, X_t and X_r and of the corresponding T/R ratios may be calculated for wood consisting of various fractions of earlywood and latewood by use of Eqs. (4.48) and (4.49) if the expansion coefficients and the ratio K_{le} are known. The term K_{le} may be approximated from the relation that $E_l/E_e = (G_l/G_e)^{2.25}$, since Young's modulus across the grain varies approximately as $G^{2.25}$ for clear wood. Figure 4.30 shows curves of the T/R ratio calculated by use of Eqs. (4.48) and (4.49) as a function of the fraction of earlywood V_e, for three different ratios of G_l/G_e, assuming that the T/R ratios in isolated earlywood and latewood are unity, and that $(G_l + G_e)/2 = 0.5$.

The effect of fibril angle on the longitudinal transverse shrinkage anisotropy of wood has already been demonstrated. It has been proposed by Ritter and Mitchell (1952) and by Frey-Wyssling (1940a, b) that fibril angles are greater in the radial than in the tangential walls, the difference ranging up to 15 °C. Ritter and

Mitchell attribute the greater fibril angle in the radial walls to the circular orientation of the crystallities or microfibrils around the pits which are so abundant on these walls, in softwoods particularly.

Pentoney (1953) has calculated that a small fibril angle difference of 15° cannot account for the larger T/R shrinkage ratio observed for wood, based on the lattice model proposed by Cockrell (1947). According to Pentoney, the radial or tangential shrinkage should be proportional to the cosine of the fibril angle. He calculated that if the mean fibril angle θ in the radial walls was 30° and in the tangential walls 15°, the T/R ratio should be equal to (cos 15°)/(cos 30°), or 1.11, much smaller than the commonly observed ratio of 2. However, based on the simplified model of Barber and Meylan (1964) discussed in Section 4.5.1, there should also be a greater reduced expansion in the thickness of the tangential wall than of the radial wall which would tend to increase the T/R ratio to a still greater extent. This is qualitatively clear from Fig. 4.15d which shows that the decrease in transverse expansion in going from a fibril angle θ of 15° (tangential wall) to one of 30° (radial wall) is accompanied by an increase in the thickness of the corresponding wall. The figure also shows the relative dimensional changes in the two walls as anticipated from the model. Barber and Meylan (1964) should be consulted for a more quantitative explanation of this effect.

Boyd (1974) has also considered the effect of differences in fibril angles in the radial and tangential walls. He concluded that this effect can account for only about 15% of the difference in transverse hygroexpansion in earlywood tissue of conifers, and about 2% in the case of latewood. For the combined composite of earlywood and latewood, that is, for the gross wood, the estimated effect is less than 10%.

The third category of theories for explaining transverse shrinkage anisotropy is based on differences in the swelling characteristics and arrangements of the various *cell wall layers.*

Frey-Wyssling (1940a, b) proposed that variations in the compound middle lamella cause the transverse shrinkage anisotropy in wood. He found, for example, that there are more crosswalls in larch (and therefore more layers of middle lamella) per unit length in transverse microtome sections in the tangential than in the radial direction (Fig. 4.31a). Furthermore, he also found (Fig. 4.31b) that the thicknesses of the individual middle lamellae were greater in the radial than in the tangential walls. He proposed that there is more thickness shrinkage in the middle lamella than in the secondary wall and therefore more shrinkage in the tangential than in the radial direction. The high concentration of hygroexpansion pectin substances in the middle lamella region may cause greater swelling in these layers. This theory has not been confirmed, and Matsumoto (1950), according to

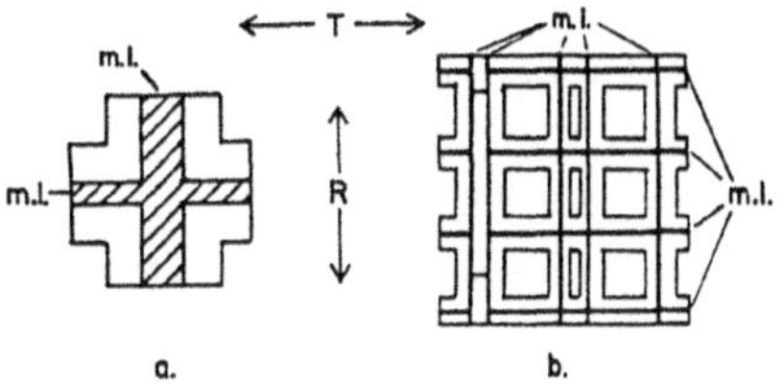

Fig. 4.31. Model used to explain anisotropic transverse shrinkage based on variations in: **a** Thicker middle lamella in radial than tangential walls, and **b** Greater numbers of radial walls and therefore of middle lamellae than tangential walls. (After Pentoney 1953)

Crews (1965), has concluded instead that little shrinkage occurs in the middle lamella and that most shrinkage is confined to the secondary walls.

Bosshard (1956) suggests that there is a higher lignin content in radial than tangential walls of longitudinal wood cells. This therefore restricts radial shrinkage somewhat since the lignin is stiffer and less hygroscopic than the holocellulosic fractions of the cell wall. Subsequent studies by others, including Kato and Nakato (1968), have shown that the radial walls of softwoods are indeed more highly lignified than the tangential walls. Boyd (1974) attributes this to the simultaneous occurrence of lignification and radial growth (of radial walls) during cell growth. He postulates that lignin precursors occupy spaces between microfibrils in the radial walls at this stage and then polymerize to form a permanent bulking agent. This stiffens the radial walls and resists hygroexpansion.

Boyd (1974) gives additional supporting evidence for the lignification theory from Kato and Nakato (1968). They found that the T/R hygroexpansion ratio was nearly equal to the ratio of lignification of the radial and tangential walls of ray-free *Cryptomeria japonica* earlywood, the mean value for the T/R ratio approximating 1.9. A similar analysis for latewood also showed a linear relationship except that the proportionality was considerably less, the mean value for the T/R ratio being close to 0.6. Boyd attributes this low T/R ratio for ray-free latewood to the greater total wall thickness in the radial direction (tangential walls) than in the tangential direction (radial walls). Latewood tracheids have much thicker walls than those of earlywood, primarily because the S2 layer, which is responsible for most transverse hygroexpansion, is much thicker (Fig. 4.32).

Quirk (1984) also gives evidence of the relative effectiveness of the lignin-rich middle lamella in reducing lateral hygroexpansion of longitudinal tracheids of Douglas-fir, using optometric measurements. During shrinkage from the green to oven-dry condition, the cell cavities of earlywood cells actually expanded by 2% while the transverse area of the cell overall shrank by 7%. In contrast, the

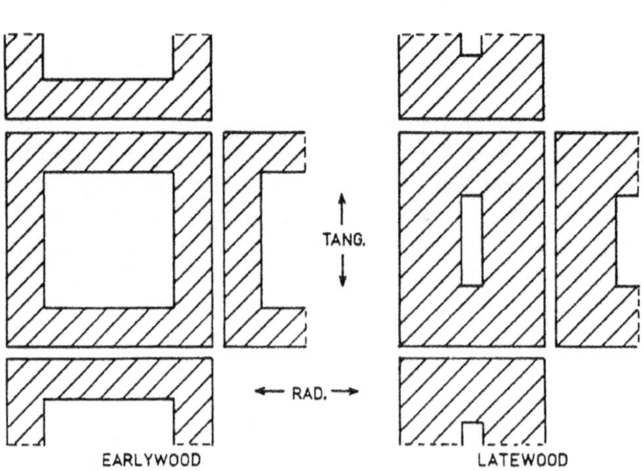

Fig. 4.32. Diagram showing: **a** The thin radial and tangential walls in earlywood tissue. **b** The thick walls in latewood with greater total wall thickness in the radial direction (tangential walls), as noted by Boyd (1974)

latewood cavities shrank by 2% and the overall cell by 27%. Presumably the highly lignified middle lamella reduced the shrinkage of the thin-walled earlywood cells but was less effective in reducing that of the thick-walled latewood cells.

The green-volume specific gravities G_f of the earlywood and latewood of the Douglas-fir used by Quirk were 0.28 and 0.80, respectively. It is of interest to note that the calculated ratios s_{vf}/G_f of the volumetric shrinkage s_{vf} (assuming area shrinkage is equal to volumetric shrinkage) to G_f are 7/0.28 and 27/0.80, or 25% and 34%, respectively, for earlywood and latewood. These are estimated values for M_f based on Eq. (4.23) which assumes constant cell cavity volume during shrinkage, and cell wall shrinkage equal to the volume of water lost.

The volumetric expansion of the cell cavity of the earlywood is probably less than the two percent expansion of the cell cavity transverse area noted above. This is because longitudinal shrinkage of the cell wall, and therefore of the cell cavity, is probably significant for earlywood because of the relatively steep fibril angle in the S2 layer (Sect. 4.5.1).

Boutelje (1973) has proposed that the most important factor govering transverse differential hygroexpansion in *Pinus sylvestris* and *Picea abies* is the "geometrical distribution of the anisotropic cell wall in the tissue". He suggested that hygroexpansion is least in the direction in which the stiffness of the tissue is greatest, and vice versa. He cited the positive relationship observed by Keylwerth (1951) between the ratio E_r/E_t of the directional Young's moduli, and the ratio X_t/X_r of the hygroexpansion coefficients. Thus in one sense this agrees with the relative distribution of lignin in the two directions, although it can be interpreted otherwise.

It seems apparent from the above discussions that several mechanisms may operate to cause transverse anisotropy in wood shrinkage and that these are not necessarily identical for all woods. Furthermore, considerable controversy exists concerning the relative importance of the various mechanisms which have been proposed.

There is evidence that the T/R shrinkage ratio is not constant over the entire hygroscopic moisture range. For example, Kelsey (1956) and Ecklund (1961) have reported that the shrinkage intersection point (the apparent FSP determined by shrinkage measurements during drying) in the radial direction is lower than in the tangential direction. Keylwerth (1962, 1964) has shown that the linear relationship between swelling and moisture content continues to a higher moisture content for the tangential than for the radial direction for a number of species (Fig. 4.33). Goulet and Fortin (1975) reported a similar phenomenon in sugar maple. Hittmeier (1967) found that swelling in the radial direction of Engelmann spruce reached a maximum value at a lower moisture content than in the tangential direction during moisture adsorption. Noack et al. (1973) also reported that, in the region of fiber saturation, tangential dimensional changes of most of the wood species studied were relatively greater than radial changes compared with their behavior at lower moisture contents.

Chomcharn and Skaar (1983a) measured tangential and radial shrinkages of round (2 cm diameter) discs (0.4 cm thick along the grain) of basswood, yellow birch, and cherry, drying from the green condition to equilibrium with air at 25 °C

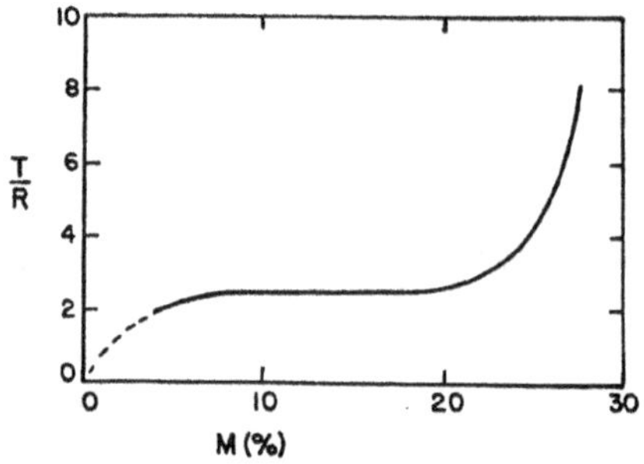

Fig. 4.33. Curve of T/R shrinkage ratio in relation to wood moisture content for European beech. (Keylwerth 1962)

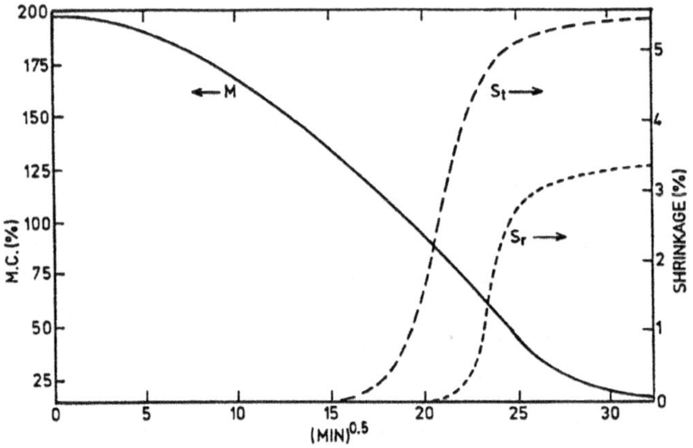

Fig. 4.34. Curves of tangential (S_t) and radial (S_r) shrinkages and of moisture content, against square root of drying time for basswood, observed by Chomcharn and Skaar (1983a)

and 77% humidity. They noted (Fig. 4.34) that radial shrinkage was delayed in time compared with tangential shrinkage for all three woods. They suggested that the apparent fiber saturation point may be lower for ray tissue, as reported by Morschauser and Preston (1954) for oak. They also theorized that the ray tissue dried more slowly than longitudinal tissue in this case, thus delaying radial shrinkage. This seems plausible, based on the findings of Hart (1984) that water is retained in parenchyma cells including ray cells of green hardwood samples at humidities as low as 93%.

In subsequent experiments, Chomcharn and Skaar (1983b) subjected the same samples to sinusoidally varying humidities between 77% and 47% at 25 °C after first equilibrating them at 77% as noted above. They recorded moisture content as well as tangential and radial dimensions during humidity cycling, for several cycles at each of several different cycling periods. They calculated the shrinkage intersection point, M_i, that is the estimated value of the moisture content below which shrinkage begins, for both tangential and radial hygroexpansion. The values of M_i were 36% and 29% for birch, 37% and 30% for basswood, and 65% and 41% for cherry, for tangential and radial shrinkage, respectively. The

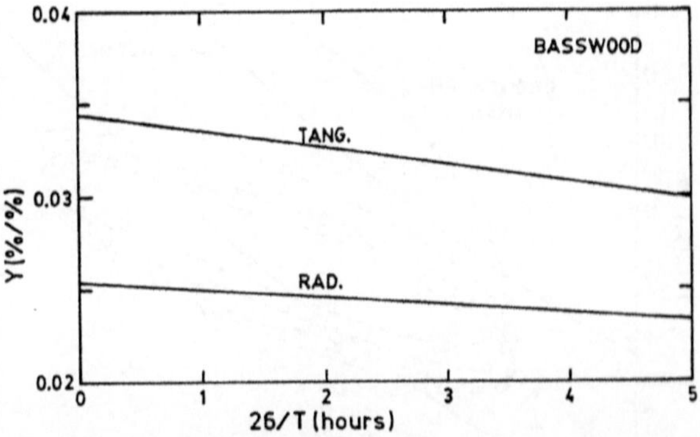

Fig. 4.35. Curves of tangential and radial dynamic humidity expansion coefficients Y against the reciprocal of cycling period for basswood, as reported by Chomcharn and Skaar (1983a)

shrinkage intersection points M_i were consistently higher for tangential than for radial hygroexpansion, in agreement with results of others as noted above. The reason for the exceptionally high values of M_i for cherry are unknown. Perhaps it is related to the high retention of moisture in parenchyma cells as reported by Hart (1984).

Chomcharn and Skaar (1983b) calculated the moisture, X_t and X_r, and humidity expansion coefficients, Y_t and Y_r, for both the static and dynamic conditions. The static values were calculated from the moisture contents, humidities and dimensions at the beginning and completion of the cyclic tests, the dynamic values from these parameters measured during the cyclic tests.

The dynamic *humidity expansion coefficients* Y_{td} and Y_{rd} [Eqs. (4.41) and (4.42)] in the tangential and radial directions, respectively, increased with length of the cycling period, as shown in Fig. 4.35. The values of Y_{td} and Y_{rd} extrapolated to zero (Fig. 4.35) represents the values expected if the cycling periods were sufficiently long that the sample moisture contents were uniform throughout. In other words, there were no moisture gradients and the moisture contents throughout a sample varied sinusoidally but uniformly with time. Similar curves were used to calculate the dynamic values X_{td} and X_{rd}.

The static and dynamic values of X and Y for basswood are tabulated in Table 4.8, together with Z, the ratio Y/X, equivalent to the effective slope, dM/dH, of the sorption isotherm as defined by Eqs. (4.41) and (4.42). The data in Table 4.8 indicate that the static and dynamic values of X_t and X_r are not consistently different. However, the dynamic values of Y_t and Y_r are generally less than half

Table 4.8. Calculated mean values of moisture (X) and humidity (Y) expansion coefficients, and the effective slope (Z) of the sorption isotherm for basswood. (Chomcharn and Skaar 1983b)

	Tangential		Radial	
	Static	Dynamic	Static	Dynamic
X (%/%)	0.246	0.339	0.231	0.182
Y (%/%)	0.0731	0.0346	0.0688	0.025
Z (%/%)	0.298	0.102	0.298	0.137

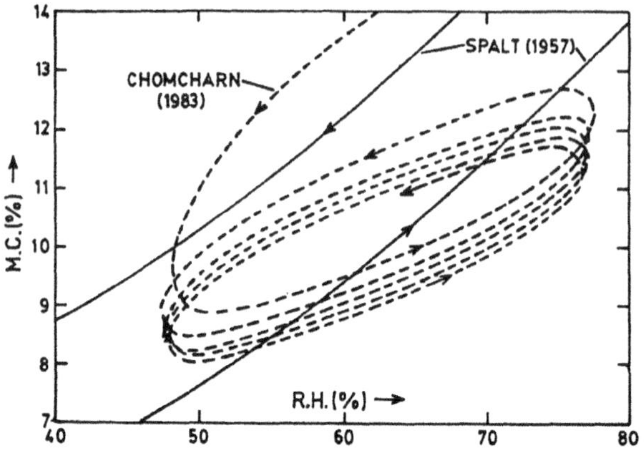

Fig. 4.36. Dynamic curve of moisture content M against relative humidity H for basswood cycled for several cycles at a period of 5.33 h as reported by Chomcharn and Skaar (1983b). The *two solid lines* are portions of the desorption and resorption isotherms as given by Spalt (1957)

those of the static values. It is believed that this results primarily from the strong sorption hysteresis effect which greatly reduces the effective slope dM/dH of the sorption isotherm during cyclic sorption. Figure 4.36, which shows a plot of M against H during cyclic sorption for basswood cycled at 5.33 h and also a portion of the static desorption and adsorption isotherms for basswood taken from Spalt (1957), illustrates this point. The slope of the long axis of the ellipse represents a mean value of dM/dH for the cyclic sorption. This is considerably lower than the slope of either the adsorption or desorption isotherms over the humidity range shown. It is recognized that this dynamic slope would be greater for a longer period T_0 but is still considerably lower than that of the static equilibrium isotherm.

The hygroexpansion of wood subject to cyclic humidity changes as described above have been termed "movement" by Stevens (1963). Movement is considerably smaller for a given range of humidity change than is shrinkage during the initial drying of green wood as Stevens has noted, and has been described above.

4.6 Hygroexpansion and Mechanical Stress

The shrinkage and subsequent hygrodimensional changes in wood are affected to a greater or lesser degree by mechanical stresses. These stresses may arise from a number of causes, including moisture gradients, mechanical restraints, macroscopic tissue swelling anisotropy, and microscopic and submicroscopic anisotropy within the cell wall itself. In any of these cases the effect on dimensional change may be indirect in that EMC itself is changed because of the Barkas (1949) effect (Chap. 2.7), or it may be a direct mechanical deformation either elastic or inelastic or both. It is impossible in practice to eliminate stresses completely during moisture changes in wood, particularly those associated with the third and fourth items mentioned above. We will discuss each of these types of stresses and their effects on dimensional changes, in the order given above.

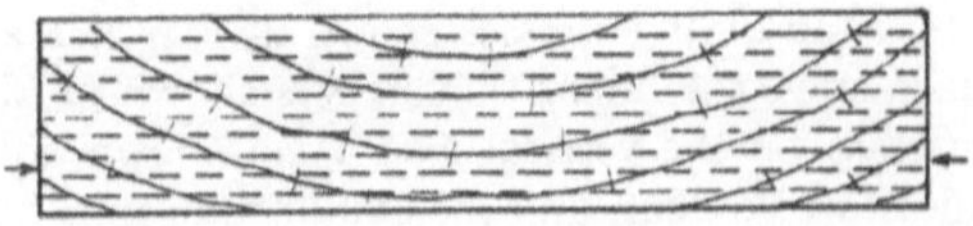

Fig. 4.37. Diagram showing how a cross-section of a drying board is subdivided into layers for determining moisture content and drying stresses

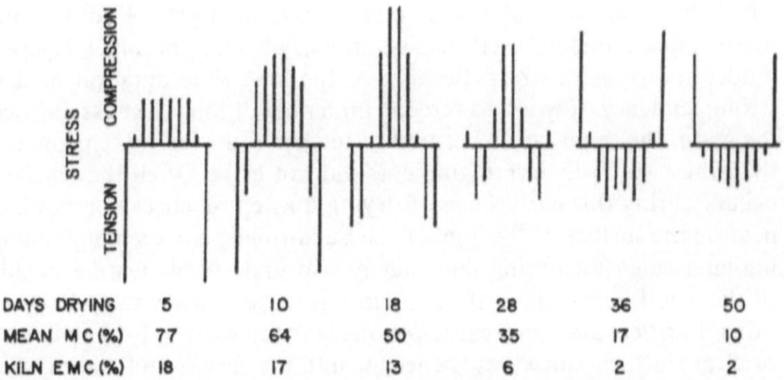

Fig. 4.38. Distributions of residual stresses at various stages of drying throughout the thickness of a drying board. The surfaces are in tension after 5 days of drying, and in compression after 50 days. (After McMillen 1958)

Moisture gradients always exist in wood under normal conditions, not only during initial drying but also during subsequent use, since atmospheric humidity and temperature are subjected to both cyclic and random changes. The stresses associated with drying of lumber have been studied in some detail and actually form the basis for many lumber-drying schedules.

The method commonly used to measure transverse residual stresses in lumber during drying is illustrated in Fig. 4.37, which shows the cross-section of a board drying primarily from the upper and lower surfaces. The cross-section is subdivided into a number of layers, in this case ten, each parallel to the surface. At various stages of drying a section of the board about 2 cm in length along the grain, is cut from the board. The width of the section is measured in situ at each of the ten locations. The section is then immediately sawn into the ten layers, and the width of each individual layer is remeasured, after which its moisture content is obtained gravimetrically.

The difference in the width of each layer measured before and after sectioning, divided by the in situ width represents the in situ strain in the layer. The stress is calculated as the product of the strain and Young's modulus in the width direction.

Figure 4.38, taken from McMillen (1958), shows a time sequence of transverse residual stresses such as are found in lumber at various stages during the normal kiln drying process. The residual stresses shown were obtained by the general method described above. It is instructive to follow the sequence of moisture contents and drying stresses shown in Fig. 4.38 at each of six stages of drying. For each stage the board thickness is shown on the horizontal axis and the residual stresses on the vertical axis. Also shown are the drying time, the mean wood moisture content, and the equilibrium moisture content (EMC) of the air in the kiln.

During the initial stages of normal drying the surface dries below fiber saturation and some hygroscopic shrinkage takes place. However, this is resisted by the wet core which is above saturation. As a consequence, the situation shown in the left diagram in Fig. 4.38 occurs after 5 days of drying. Here the outer layers of the wood are under severe tensile stress, whereas the inner layers are under a milder compressive stress. Because of the long time duration of drying and the strong tendency of wood to "creep" under conditions of stress, particularly across the grain, the outer layers become "stretched" or "set" in tension and shrink less than they would if drying gradients did not exist. Often the tensile stress which occurs during this early stage of drying may cause checks or small cracks to appear on the surface of the lumber. These normally close and are not as visible during later stages of drying, but they are an undesirable feature in subsequent use of the wood, particularly if the surface is to be exposed to cyclic humidity conditions. Furthermore, they cause problems if the surface is to be coated with paint or clear finishes, since these penetrate into the checks and make them visible, particularly with clear finishes.

As drying progresses further, as shown for the 10- and 18-day drying times in Fig. 4.38, the compressive stress in the interior increases, as intermediate layers dry and commence to shrink. The tensile stress in the extreme outer layer decreases, however, because the adjoining layers begin to develop some tensile stress as they try to shrink relative to the inner undried layers. After 28 days the outer layer is under compression and remains so throughout the subsequent drying time. This occurs because the outer layer has been stretched or set severely in tension during the initial drying period compared with adjoining layers. Therefore, as the intermediate layers dry below fiber saturation, they tend to shrink to smaller dimensions parallel to the wide face of the board than the adjoining surface layers, even though the latter may be somewhat drier. Finally, when the wood is nearly uniformly dry after 50 days, the stresses have essentially reversed themselves from those existing after 5 days, as Fig. 4.38 indicates. This final distribution of stresses is that which is characteristic of so-called *casehardened* lumber.

The term "casehardening" as applied to dry lumber does not indicate that the surface is mechanically any harder than wood which is not casehardened. The term is really a misnomer but has probably been adapted from the field of metallurgy because of the analogy between stresses arising from moisture gradients during drying of lumber and stresses caused by temperature gradients near the surfaces of metal objects when they are casehardened as a part of the manufacturing process in order to harden the outer surface.

Casehardening is not a serious problem in lumber unless it is resawn into thinner pieces, when the residual stresses cause excessive cupping of the surfaces. For ordinary construction lumber, for example, it has little importance. Casehardening is removed, if necessary, by subjecting the lumber to air of high humidity and temperature for a short time near the end of the drying process. This causes additional compressive stress in the outer surface sufficient to cause the reverse kind of creep which brought about the original casehardening, and if properly carried out, results in stress-free lumber. If casehardening is over-corrected by application of too high a humidity-temperature combination for too long a time period, the wood may attain the condition known as *reverse casehardening*. As the name implies, this is a state in which the reverse stresses exist in the dry lumber. In other words, the surface layers now have a compression set and are therefore in a state of tension compared with the interior layer of wood when the wood is uniformly dry.

It is clear that the outer surfaces of normally dried wood containing casehardening have a tension set and are in a stretched condition. There thus tends to be less shrinkage across the width of the board in these outer layers than for wood dried without stress. However, the inner layers have been somewhat compressed during the early stages of drying and therefore tend to show greater shrinkage than unstressed wood. There is much more volume in the interior with compression set than exists in the surface layers with tension set. Therefore, although the tension set in the surface layers is usually more pronounced than the compression set, the overall shrinkage across the width of the board is determined largely by the amount of compression set. According to McMillen (1958), the amount of compression set in the board interior, and therefore the extent of shrinkage, is minimized by use of low temperatures and humidities during the drying operation.

Espenas (1971) made a study of the comparative transverse shrinkage of Douglas-fir, western hemlock, and red alder when dried from the green condition at temperatures of 90 °, 150 °, 180 °, 200 °, and 215 °F at relative humidities designed to give normal EMC's of 6%, 9%, and 12%. His results confirmed the fact that greater shrinkage occurs during the drying at higher temperatures. The difference was greatest for the samples dried at 12% EMC conditions. The increase in tangential and radial shrinkages for material dried at 215 °F compared with that dried at 90 °F, for example, was 33% and 51% for Douglas-fir, 65% and 82% for western hemlock, and 130% and 137% for red alder. Only a small portion of these differences could be attributed to the decrease in hygroscopicity of the wood associated with drying at the higher temperatures.

Additional evidence to support the effect of stresses arising from moisture gradients in wood is shown in Fig. 4.39, taken from Hittmeier (1967). This shows curves of the squares of fractional radial and tangential swelling (the ratio of the swelling at time t to the total equilibrium swelling) as functions of time for initially dry wood during soaking in distilled water at 20 °C. Two of the curves are for flat-sawn samples 0.3 cm thick radially by 3×3 cm tangentially and longitudinally, and two curves are for quarter-sawn samples 0.3 cm thick tangentially by 3×3 cm radially and longitudinally.

It is apparent from Fig. 4.39 that the width swelling is delayed compared with the thickness swelling for both flat-sawn and quarter-sawn samples. According to Hittmeier, this is attributed to the fact that most of the diffusion of water into the dry sample takes place in the direction of thickness. During the early stages of soaking, outer layers take up water and tend to swell. They are restrained from

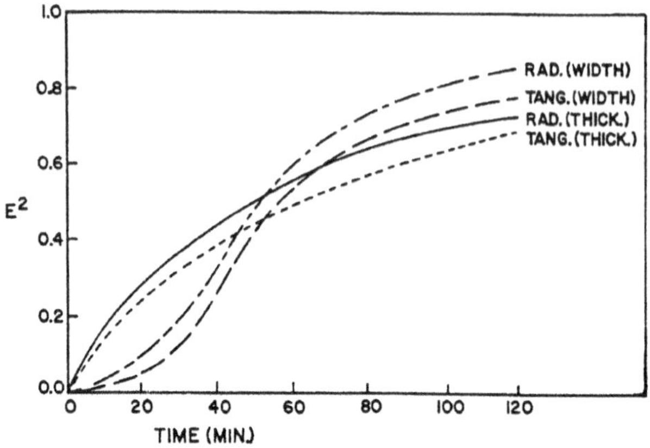

Fig. 4.39. Curves of fractional radial and tangential swelling squared (E^2) in relation to time for both width and thickness directions in each case. Delayed swelling in width is caused by drying stresses in the width direction. (After Hittmeier 1976)

swelling in the width direction, however, because the dry interior wood is stronger than the moist surface layers. Swelling does take place in the thickness direction and may even be somewhat greater than normal because of the Poisson ratio effect in the moist surface layers. Later, as the interior wood finally becomes moist, the width swelling increases more rapidly than the increase in thickness. This stress pattern in the wood is nearly the reverse of that which occurs during the normal drying of lumber. Hittmeier carried out similar experiments with cross-sectional samples. In this case there was little restraining effect because the water penetrated the cell cavities in the center of the samples almost immediately, and the rate of penetration of the cell wall was essentially the same throughout the sample thickness.

The phenomenon described above is very similar to that observed by Hermans (1946), who measured the length and diameter changes with time of soaking in water of model cellulose fibers which were initially dry. He noted that the diameter (analogous to thickness in Hittmeier's experiments) increased with time immediately after immersion of the fiber in water. However, the length (analogous to width of sample in Hittmeier's experiments) remained essentially unchanged for 10 min until some water penetrated to the center of the fiber. Thereafter the length swelling increased more rapidly than the diameter swelling. Hermans was also able to study the penetration of the water boundary into his fiber sample and noted that the increase in length began at nearly the same time that the boundary penetrated to the center of the fibers.

Sadoh and Christensen (1967), in their study of longitudinal shrinkage and swelling of *Araucaria cunninghamii*, noted a hysteresis effect. They attributed this in part to the stresses associated with drying, even in thin microtomed section.

The effect of mechanical restraint on hygroscopic dimensional changes is very marked in wood because of the large component of rheological or inelastic deformation associated with mechanical stress. This, of course, is the cause of casehardening as described previously. The inelastic deformation under mechanical restraint can be advantageous or not, depending on the circumstances. For example, mechanical restraint is used to reduce cupping, twisting, bowing, and crooking of lumber caused by diagonal grain or anisotropic shrinkage during drying (Rasmussen 1961). Koch (1971) has used similar techniques to produce straight 2×4-inch lumber from juvenile wood of southern yellow pine which is prone to crook, bow, and twist due largely to excessive longitudinal shrinkage.

The deleterious effects of mechanical stress on shrinkage are most apparent when wood which is restrained from movement is subjected to cyclic moisture changes. This effect appears, for example, in wood flooring which is used indoors in temperate climates where marked annual changes in indoor humidity occur. If the flooring is fitted tightly during installation at a moisture content lower than the highest which it attains during the subsequent humidity cycles, then it will be subjected to a compressive transverse stress during the high humidity portion of the cycle. This tends to buckle the flooring, but if the flooring is adequately fastened, the wood relaxes and develops a compression set similar to that described in the previous section. The transverse dimension of each board is now smaller at any given moisture content than when it was installed. Therefore, the spaces develop between the boards and the appearance of the floor deteriorates.

The loosening of wood fittings subjected to cyclic moisture changes is also accelerated by the development of compression set due to restrained swelling. One of the more strinking examples of the effect of restrained swelling on subsequent wood dimensions is found in the loosening of wooden axe handles and other tool handles which are soaked in water to tighten them. They are tightened as the wood handle swells against the restraint imposed by the tool head but in time the swelling pressure relaxes causing an excessive compression set. Thus the portion of the handle inserted into the hammer or axe head now shrinks to smaller dimensions than the original dimensions when it is exposed to normal atmospheric conditions again. Figure 4.40 shows the sequence of dimensional changes in the wood during a cycle such as is described above.

Tiemann (1944) has demonstrated a similar effect in which the wood is restrained from shrinkage in one dimension during drying. In this case the wood is "stretched" in the direction of restraint having attained a tension set and is of larger dimensions in this direction than the unrestrained wood.

The effect of macroscopic tissue anisotropy, and the development of mechanical stresses thereby, has already been discussed in connection with the

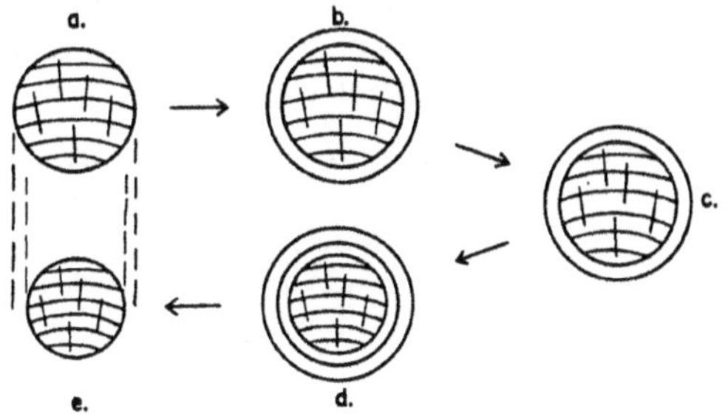

Fig. 4.40. Development of compression "set" in wood dowels restrained from transverse swelling by steel restraining ring. Residual compression set causes the difference in diameters of samples shown in **a** and **e**

theories of anisotropic shrinkage. Both the ray restraint and the earlywood-latewood interaction theories are based on the anisotropy of the respective tissues and the mechanical stresses which arise from this differential shrinkage.

The microscopic and submicroscopic anisotropy of shrinkage within the cell wall itself has also been discussed. It should be noted, however, that if the swelling stresses during water resorption were sufficiently great, the cell wall would completely disperse and go into solution in the water. The swelling would then be infinite at unity relative vapor pressure, as it is for true solutions. This does not occur, fortunately, for two reasons. First, the water apparently does not penetrate into the crystalline portions of the cellulose, and second, the cellulose chains are not stressed to the point where they rupture during the normal course of swelling in water, thus retaining the basic integrity of the cell wall. It is also worth repeating that the S1 and S3 layers of the cell wall, with their microfibrils nearly perpendicular to the cell axis and to the S2 layer, restrict the swelling of this layer and thus reduce the external dimensional changes in the cell and of the gross wood. The stiffening effect of lignin may also reduce the expansion, as noted above.

4.7 Reducing Hygroexpansion in Wood

The sensitivity of wood to changes in atmospheric humidity, particularly with respect to hygroexpansion, causes many practical problems during its initial drying and also in subsequent use. Therefore considerable research has been carried out to find practical and economical methods to reduce the natural tendency of wood to shrink and swell with changes in ambient humidity. The results of these studies are summarized in several publications, a partial list of which includes Stamm (1964, 1977). Rogers (1967), Kollmann et al. (1975), Goldstein (1977), Bariska (1979), Feist and Hon (1984), and Meyer (1984). The present treatment is restricted to a general discussion of the elementary principles involved in stabilizing wood with respect to its natural hygroexpansion characteristics. It is given in terms of reducing shrinkage, but the same principles apply to swelling.

Stamm, in Kollmann et al. (1975), has suggested five different ways by which hygroexpansion in wood can be reduced. These include:

1. Crosslaminating veneers, such as in plywood.
2. Applying moisture-resistant coatings.
3. Heat treatment to reduce hygroscopicity.
4. Chemical treatment to induce crosslinking of wood constituents.
5. Chemical treatment with bulking agents to reduce shrinkage.

These five methods are discussed in the order given above.

Before considering the details of the methods it is desirable to define the parameters used to evaluate the effectiveness of the various treatments. The most common parameter is called the *anti-shrink efficiency* (ASE), the percent reduction in shrinkage, defined as

$$\text{ASE}\ (\%) = 100(S - S_x)/S\ , \qquad (4.51)$$

where S and S_x are the shrinkages before and after treatment, respectively, between any two humidity limits. It is usually most convenient to measure shrinkages between the water-soaked and oven-dry conditions, so the humidity range is 100%. In some cases the ASE is defined in terms of the reduction in hygroscopicity between any two limits of humidity. In this case

$$ASE = 100(\Delta M - \Delta M_x)/\Delta M \ , \qquad (4.52)$$

where ΔM and ΔM_x are the respective moisture changes between the two humidities for the untreated and treated wood. If the mean moisture expansion coefficient $\bar{X}$ is the same for untreated and treated wood, Eqs. (4.51) and (4.52) give the same result over a given humidity range. The shrinkages are usually measured in one dimension, preferably in the tangential direction.

Crosslaminating veneers and gluing them together into plywood, reduces transverse hygroexpansion almost to the level of longitudinal expansion, normally only a small fraction of transverse expansion. This is because the longitudinal Young's modulus is about 20 times greater than the transverse modulus. The reduction in expansion is particularly great for plywood made from rotary-cut veneers, since these would normally expand in the tangential direction.

An approximate equation for calculating the hygroexpansion coefficient X for plywood along the principal axes is

$$X = \sum (X_i E_i d_i)/\sum (E_i d_i) \ , \qquad (4.53)$$

where X_i, E_i, and d_i are the hygroexpansion coefficients, Young's modulus and thickness of the individual plies along either of the two principal surface directions. For example, for three-ply plywood of a balanced construction made from rotary-cut veneers in which the center ply thickness is equal to the sum of the thicknesses of the two surface plies, Eq. (4.53) reduces to

$$X = (X_l E_l + X_t E_t)/(E_l + E_t) \qquad (4.54)$$

for the case in which X is measured either parallel or perpendicular to the grain direction of the surface veneers, X_l and E_l are the hygroexpansion coefficient and Young's modulus, respectively, in the longitudinal direction, and X_t and E_t are those in the tangential direction.

Other hygroexpansion parameters such as percent shrinkage or swelling S, or the humidity expansion coefficient Y, may be substituted for X in Eqs. (4.53) and (4.54). For coast Douglas-fir the tangential shrinkage s_t is 7.6% from green to oven-dry, while the longitudinal is about 0.15% for mature wood. The relative values of E_l and E_t are 2.14 and0.091 (Bodig and Jayne 1982). Substituting these values in Eq. (4.54) gives an estimated shrinkage value of 0.45% from fully swollen to oven-dry. This is three times as great as the assumed longitudinal shrinkage of 0.15%, but only about 1/17 as great as the assumed tangential shrinkage of 7.6%. The reason for the great reduction in hygroexpansion is the large ratio E_l/E_t of 23.5 (2.14/0.091) of the Young's moduli in the two directions.

It should be noted that transverse mechanical stresses are produced in plywood, particularly on surface veneers. These may result in surface checking which is unsightly and may eventually induce failure in surface coatings. Furthermore there is a slight increase in thickness expansion due to the *Poisson ratio* effect.

The second general method for reducing hygroexpansion in wood is to provide *moisture barriers* between the wood and the source of moisture. Two general approaches have been taken. The most common of these is to coat the *external wood surface* with a moisture barrier material which reduces the rate at which moisture can enter or leave the wood surface. The second approach is to coat the *interior surface* of the wood cells with a similar material.

External coatings include paints, varnishes and other finishes normally sprayed or brushed onto the surface. These coatings normally bond to the wood by the same mechanism by which water is sorbed, that is by hydrogen bonding. They usually have some attraction for water and are always somewhat permeable to water vapor. Therefore they do not completely prevent the exchange of water between the wood and ambient atmosphere but retard the rate of exchange. Nevertheless this is adequate for many purposes, particularly if the atmospheric changes are short-lived such as those due to rain or daily humidity fluctuations.

For humidity variations of long duration, such as annual fluctuations, external coatings retard, but do not prevent the accompanying hygroexpansion. Furthermore, they are subject to deterioration due to weathering and/or abrasion and must therefore be renovated or replaced periodically.

Internal coatings are those applied to line the insides of individual cell cavities with one or more layers of moisture repellant material. These include waxes, resins, and toxic materials dissolved in more or less volatile carriers. They are generally applied by dipping the wood into the solution for a sufficiently long time to permit the required depths of penetration of the solution by capillary forces. The depth of penetration is therefore a function of the interfacial surface tension between the wood and the solution and the permeability of the wood. The latter is orders of magnitude greater along than across the grain (Siau 1984), so penetration is also much deeper on surfaces with end grain exposed.

Internal coatings are often used in solid wood millwork products to reduce sorption of liquid water prior to the time of application of external coatings. They reduce the likelihood of fungal activity and of hygroexpansion prior to application of a final exterior coating such as paint.

Exposure of wood to *high temperatures* reduces its hygroscopicity after cooling to normal temperatures. However, there is often a simultaneous reduction in some of its mechanical properties which tends to offset the benefits of this *heat treatment* technique for reducing hygroscopicity.

The rate of reduction in hygroscopicity of wood by heating in the dry condition appears to follow the *Arrhenius* relation in regard to the time-temperature combination required for a given reduction in hygroscopicity. The effect of heating wood in the wet condition results in less well-known changes in hygroscopicity compared with heating in the dry condition. In contrast to the case of thermal degradation, there is only a minor effect on reduction of hygroscopicity whether dry wood is heated in the presence or absence of air or oxygen.

Using data of Stamm et al. (1955), the times in hours required to achieve three different degrees of anti-shrink efficiencies are given in Table 4.9, for heating temperatures of 200 °, 250 °, and 300 °F. According to the table, the high-temperature drying (HTD) schedules currently in vogue, or contemplated, will result in a 5% or lower level of anti-shrink efficiency based on moisture sorption. This is less

Table 4.9. Hours of heating time required to produce anti-shrink efficiencies of 5%, 10%, and 20% for Sitka spruce heated under dry conditions at temperatures of 200 °F, 250 °F, and 300 °F. (Data from Stamm et al. 1955)

Hours to 5% ASE			Hours to 10% ASE			Hours to 20% ASE		
200 °	250 °	300 °	200 °	250 °	300 °	200 °	250 °	300 °
1000[a]	80[a]	4[a]	4000[a]	210	15	15000[a]	1000	50

[a] Estimated.

than has been observed in dry-kiln practice, where wood kiln-dried at higher temperatures has an appreciably lower EMC than wood dried at lower temperatures. For example, using data of Salamon et al. (1975) the mean moisture contents in June for wood samples air-dried, conventionally kiln-dried and dried at 220 °F and 245 °F temperatures were 17.1%, 14.7%, 13.8%, and 13.2%, respectively. Neglecting for the moment the differences between air-dried and kiln-dried material, attributable primarily to hysteresis since the air-dried stock was never subject to very dry atmosphere during drying, we can calculate the anti-shrinkage efficiencies of the two sets of HTD material based on the EMC of conventionally kiln-dried lumber. Using Eq. (4.52), the ASE for lumber dried at 220 °F is 100% (14.7 − 13.8)/14.7, or 6.1%. For that dried at 245 °F the corresponding ASE is 100% (14.7 − 13.2)/14.7, or 10.2%. It is clear from these figures that the effective ASE is higher than might be expected based on the data given in Table 4.9. This may be because of lower EMC conditions in the kiln at the higher temperatures and the consequent increased hyteresis effect which can cause an anti-shrink efficiency based on Eq. (4.52) of about 20%.

Regardless of the cause of reduced EMC for wood dried at high temperatures, the point is that a lower EMC will affect the mechanical properties of the wood in use, or at time of test. There are two methods for evaluating the effect of HTD on the mechanical properties of wood, based on the observed change in hygroscopicity. One is to compare the properties of the wood at equilibrium with a given *humidity* condition, in which case the wood dried conventionally will have a lower EMC. This we will designate as the *constant humidity condition* (CHC). The second method, designated as the *constant moisture conditions* (CMC), Compares the properties of the wood at the same moisture content. Since most testing is done on samples equilibrated to the same humidity condition, it is necessary in this case to adjust the calculated values to the same moisture content for all samples. This is done on the basis of correction procedures established on the basis of known moisture content-strength relationships for wood. For example, the modulus of rupture increases approximately 4% per 1% decrease in wood moisture content (Table 1.7).

Reported studies of the effect of high temperatures on the mechanical properties of wood have used the CMC values in some cases and the CHC values in others. In evaluating the effect of HTD on the mechanical properties of wood, it is important to distinguish between the two methods of evaluation. From the viewpoint of the fundamental mechanism of thermal degradation of wood, it is probably best to use the CMC method, but for practical evaluation of the effect of HTD on the strength of wood in service, the CHC method is probably more realistic. However, in the latter case, it is also important to know the degree of permanence of the reduced hygroscopicity resulting from HTD, although according to Stamm et al. (1955) the stabilization caused by heating is permanent.

In the previous paragraphs we have been discussing essentially the strength of clear defect-free wood. Most lumber, however, contains defects of various kinds and it is not known whether or not HTD accentuates the strength-reducing attributes of these defects. It is conceivable, for example, that the tissues around knots may be degraded more severely by HTD than clear wood insofar as the mechanical properties are concerned. On the other hand, it is also conceivable that the high temperatures used may minimize the effects of discontinuities around knots by virtue of the softening effect and consequent reduction of stresses.

The last two techniques for reducing hygroexpansion both involve chemical treatments, but differ in their basic approach. In one case the chemicals induced *crosslinking* of the structural constituents of wood of sufficient strength and numbers to prevent the dry wood from swelling when exposed to water. In the other case, chemicals are introduced into the wood primarily as *bulking agents*, thus keeping the wood at or near the fully swollen condition.

It has already been noted in Chapter 2.6 that the cell wall of wood shows limited moisture sorption and swelling due to the *crosslinking* and intertwining of the cellulosic network. If additional crosslinks could be incorporated into the cell wall to further reduce the number of moisture sorption sites and to provide additional restraints to swelling, hygroexpansion would be reduced. This is the principle involved in the use of chemical treatments to induce crosslinking.

Formaldehyde has been used successfully as a crosslinking agent for wood, according to Stamm (1964, 1977). The wood must be nearly dry and is best treated with formaldehyde vapor over heated paraformaldehyde in the presence of strong mineral acids. Anti-shrink efficiencies (ASE), based on Eq. (4.51), of as high as 70% were achieved in small wood samples with an uptake of 4% (based on dry wood weight) of formaldehyde. However, toughness and abrasion resistance of the treated wood were reduced by 84% and 95%, respectively. Those characteristics, which give natural wood its high toughness and resilience, are related to its hygroscopicity, and when this is reduced by treatment, the desirable mechanical properties are also reduced, in general.

Two classes of chemicals are used in *bulking treatments* for wood, *water-soluble* and *water-insoluble*. The water-soluble chemicals are usually nonreactive with the wood and essentially displace water in the swollen cell wall and/or reduce the humidity or activity of the water in the cell wall, thereby reducing its humidity expansion coefficient Y. Water-insoluble chemicals, in contrast, are reactive in or with the cell wall, forming permanent bonds due to polymerization in situ. Treatment with water-soluble chemicals will be discussed first.

Several classes of *water-soluble bulking chemicals* have been used to reduce hygroexpansion in wood, including salts, sugars and polyethylene glycol. When wood sections are soaked in concentrated aqueous solutions of these chemicals, they leach into the cell wall, displacing some of the water. Because of the reduction in activity of the remaining water and the bulking effect of the chemicals, the wood remains almost fully swollen at relatively low ambient humidities.

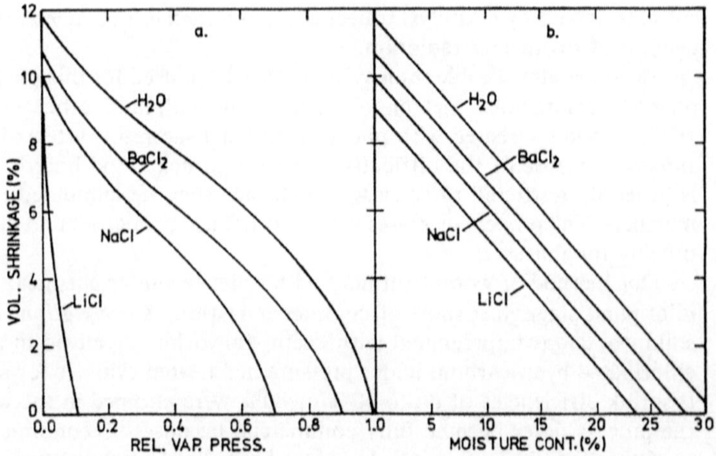

Fig. 4.41. Curves of percent volumetric shrinkage s_v against relative vapor pressure h and against wood moisture content M for this Sitka spruce cross-sections after having been soaked in one of several salt solutions shown, as reported by Stamm (1977)

Figure 4.41 shows curves of percent volumetric shrinkage s_v against relative vapor pressure and against wood moisture content for thin Sitka spruce cross-sections soaked in several different salt solutions. In each case shrinkage begins at an activity equivalent to that of the water in a saturated solution of the individual salt. The problem with this simple salt treatment is that the wood surface remains damp at ambient humidities higher than that at which shrinkage begins in each case, and the salts are corrosive. Surface treatment of lumber with salt has been used to reduce surface checking during the drying of lumber.

Sugar solutions of various concentrations have been used in place of salts. These are noncorrosive but are attractive to biological organisms which deteriorate wood, and are difficult to coat with wood finishes, partly because the wood surface is damp at high humidities. Toxic substances have been added to the solutions to inhibit biological organisms in some cases.

Polyethylene glycols (PEG), of molecular weights near 1000, or lower, have been used successfully to prevent shrinkage in green wood. Stamm (1977) recommends that green wood be soaked in a 30% by weight aqueous solution of PEG of 1000 molecular weight, for various time periods depending on the amount of end-grain exposed and the permeability of the wood. This technique is particularly effective in preventing checking and splitting of tree cross-sections.

Wood treated with PEG remains in the swollen condition unless exposed to liquid water which leaches it out of the wood. It can best be coated with polyurethane type of finishes. It is a suitable technique for treating wood carvings and also archeological artifacts made of wood if they have a tendency to deteriorate when exposed to normal atmospheric conditions.

The final mode of chemical treatment considered here consists of using *water-insoluble chemicals* as bulking agents. The general approach used here is to impregnate the wood with low molecular weights resins dissolved in suitable solvents, including a catalyst if necessary, and then to polymerize the resin in situ, using heat or nuclear radiation.

Various water-soluble resin systems have been used for this purpose, including phenols, resorcinols, melamine, furfuryl alcohol, and other systems (Stamm 1977). Wood so treated with phenol-formaldehyde resin systems has been called *Impreg*. Because of the difficulty of treating thick wood boards, this technique is generally restricted to veneers, which may then be laminated to form thicker members. The process increases some mechanical properties and decreases others, notably toughness.

One method of wood treatment with water-insoluble chemicals, carried to the pilot plant stage, just short of commercialization, is *acetylation*. In this process, solid wood was impregnated with acetic anhydride, dissolved in an aromatic or chlorinated hydrocarbon, under pressure in a heated cylinder for about 12 h. Anti-shrink efficiencies of up to 70% or 80% were attained in the wood. However, the process never became fully commercial because of economic factors.

Other chemical treatments have also been used to stabilize wood. Meyer (1984) has reviewed some of these methods, with particular emphasis on the use of vinyl type monomers. These, of which methyl-methacrylate is representative, are used to make wood-polymer composites, in which the monomer is introduced into the cell cavities as well as the cell wall of the wood, prior to polymerization. Both gamma radiation and heat, using a catalyst, have been used to polymerize the resin. These products have specialized uses such as in musical instruments and sports equipment. In these cases the wood provides the skeleton of a new material, filled with a resin matrix.

Chapter 5

Moisture Movement in the Wood Cell Wall

5.1 Introduction

Wood is continually changing in moisture content both during the initial drying from the green condition and during subsequent use. The latter changes are due to changes in atmospheric humidity and temperature which may be sporadic or somewhat cyclic, or both. Therefore moisture and related gradients are usually present in wood even in its final product form.

It has been shown in Chapter 3 that wood moisture may appear in three general forms, capillary water in the cell cavities and possibly in the cell wall, vapor in the cell cavities, and hygroscopically bound water in the cell walls.

When wood is above fiber saturation, liquid water is present in the cell cavities, as in green wood. In this case capillary forces and related thermodynamic potential gradients operate to cause liquid flow. The flow rate is then determined by the combination of these forces and wood permeability, particularly that between cells. This subject is treated extensively by Siau (1984) and will not be covered here except to note its significance.

Vapor movement through the cell cavities in response to vapor pressure and total pressure gradients may also be important in moisture movement, particularly at high temperatures and in woods of high permeability. High temperature drying of wood may cause steam pressures in the wood to overcome some of the back pressures due to capillary forces, and actually force liquid water out of the wood, particularly along the grain direction. However, at low temperatures, such as in the room temperature range, vapor movement makes only a small contribution to the total transverse movement. This is because bound-water movement through the cross walls between cells is two or three orders of magnitude slower than movement across the cell cavity and therefore controls the overall transport rate (Choong 1965).

Stamm (1946) first proposed an electrical analog model to simulate moisture movement through wood in the hygroscopic range. A simplified circuit of the type he proposed is shown in Fig. 5.1, together with a schematic cross-section of a wood cell showing the corresponding paths of moisture movement. The electrical circuit consists of four resistances as follows: R_1 is the resistance of the parallel cell wall, R_2 is that of the cell cavity, R_3 is that of the cross wall and R_4 is that of the pit openings.

For woods of moderate density R_1 is much greater than the combination of R_2 in series with the parallel combination of R_3 and R_4. Also R_2 is much lower than the combined parallel resistance of R_3 and R_4 because vapor movement is so rapid across the cell cavity.

The resistance R_4, shown as a simple hole, is extremely variable among and within woods. Its value depends on the sizes and numbers of pit openings between

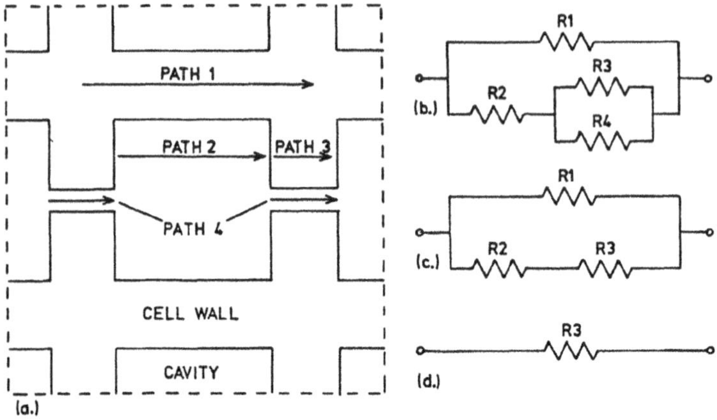

Fig. 5.1. Schematic diagrams showing: **a** The four primary paths for transverse moisture transport through a single wood cell. **b** The analogous electrical resistance circuit. **c** The reduced electrical circuit when pit pores (path 4) are neglected. **d** The approximate circuit when path 3 is the rate determining path, as is assumed by Siau (1984)

cells, the same factors which determine the transverse permeability of wood to liquids or gases (Siau 1984). Because R_4 is so extremely variable and is usually large compared with its parallel path R_3, Siau neglects it in his model for moisture diffusion through wood. He cautions that it may be important in woods of high transverse permeability, particularly at low moisture contents and high temperatures.

Bound-water movement through the cell wall is the most important mechanism of transport in the hygroscopic moisture range. It is also the most controversial, and the present discussion is limited primarily to this mechanism. The interested reader is referred to Stamm (1946, 1964) and to Siau (1984) for more detailed discussion of the other two mechanisms.

5.2 Isothermal Moisture Diffusion

Most experimental studies of moisture movement in wood have been carried out isothermally, that is, under constant temperature conditions. Strictly speaking, there is always some transfer of heat during moisture movement, due to the transport of energy with the water molecules. This causes slight temperature differences, a diffusion thermo-effect considered in Section 5.3, but neglected in the present discussion.

5.2.1 General Mechanism of Bound-Water Diffusion

For historical reasons, most analyses of moisture transport in wood and other materials have used Fick's law or a minor adaptation thereof for quantifying the process. The steady-state form of Fick's law, for one dimension, is

Table 5.1. Tabulation of some moisture transport coefficients used for wood, assumed potentials, and relationships to diffusion coefficient D (all in cgs units)

Assumed potential	Symbol (units)	Transp. coeff. (units)	Relationship to coefficient D
Moisture conc.	c (g/cm^3)	$D = -J/(\partial c/\partial x)$ (cm^2/s)	$D = D$
Fractional M.C.	m (g/g)	$K_m = -J/(\partial m/\partial x)$ (g/cm/s)	$K_m = D(\partial c/\partial m)$
Vapor pressure	P (erg/cm^3)	$K_p = -J/(\partial p/\partial x)$ ($g\text{-}cm^2/erg/s$)	$K_p = D(\partial c/\partial p)$
Water activity	h (ratio)	$K_h = -J/(\partial h/\partial x)$ (g/cm/s)	$K_h = D(\partial c/\partial h)$
Chemical potential	μ (erg/mol)	$K_\mu = -J/(\partial \mu/\partial x)$ ($g\text{-}mol/cm^2/erg/s$)	$K_\mu = D(\partial c/\partial \mu)$
Osmotic pressure	π (erg/cm^3)	$K_\pi = -J/(\partial \pi/\partial x)$ ($g\text{-}cm^2/erg/s$)	$K_\pi = D(\partial c/\partial \pi)$
Spreading pressure	ϕ (erg/cm^2)	$K_\phi = -J/(\partial \phi/\partial x)$ (g-cm/erg/s)	$K_\phi = D(\partial c/\partial \phi)$

$$J = -D(\partial c/\partial x) \;, \qquad (5.1)$$

where J is the steady-state flux of moisture (g/cm^2 s, in cgs units), $\partial c/\partial x$ is the gradient of concentration c (g/cc) of water in the wood and D (cm^2/s) is the coefficient of proportionality normally assumed for Fick's law.

The point of controversy in applying Fick's law is the assumption that the potential for causing movement is the concentration c. Babbitt (1950) suggested that the gradient of a more fundamental potential than concentration is involved. He further proposed that a gradient of spreading pressure is the actual driving force for moisture transport in the cell wall (Sect. 5.2.6). Other potentials have also been proposed such as vapor pressure, chemical potential, etc., most of which are discussed in this chapter. Each of these potentials is associated with a particular transport coefficient K, by analogy with Eq. (5.1).

A number of moisture potentials, together with their gradients and transport coefficients, are listed in Table 5.1, with their relationships to Fick's first law. It should be noted that each of the transport coefficients K can be related to the Fick's law coefficient D if the rate of change of the assumed potential with respect to c is known. They can also be related to each other. One interesting and useful property of the coefficient D is that it appears in unmodified form in Fick's second law $\partial c/\partial t = \partial[D(\partial c/\partial x)/\partial x]$, whereas all other transport coefficients K appear in combination with other parameters.

Before discussing specific models for bound water transport a general model based on Fick's law will be considered which lends itself to general thermodynamic treatment. Other models and their relationship to this classical model will be treated subsequently.

It has already been noted in the discussion of moisture sorption (Chap. 3) that bound water is believed to be sorbed at or near sorption sites in the cell wall which have a strong attraction for water molecules. The attractive forces are attributed mostly to the attraction of polar hydroxyl groups in the wood for the polar water molecules by hydrogen-bonding (Chap. 3).

It is convenient in this discussion to consider that water molecules are attracted to these sorption sites, and that a particular sorption site may contain from one to several water molecules in its vicinity, depending on the moisture content of the wood. For purposes of illustration consider that the number of molecules of water per sorption site is, on the average, one for each 7% moisture content. This means that at 0% all sorption sites are empty. At 7%, 14%, 21%, and 28% there would be 1, 2, 3, and 4 molecules, respectively, per sorption site. These assumptions are for convenience only, since it is not known how many sorption sites there are, nor exactly how many molecules are deployed at each sorption site at any given moisture content (Chap. 3).

Based on the model, Fig. 5.2a illustrates a hypothetical diagram of the moisture distribution in the cell wall of wood where a moisture gradient exists. Five different sorption sites, each at a different moisture level, are shown. Also shown in the figure are the potential energy "wells" corresponding to these sites, and the energy levels of water in vapor, liquid and hygroscopic form at the different moisture contents. Water molecules in these potential energy wells are continually vibrating about a position of equilibrium, the average energy of vibration being proportional to the product RT of the gas constant R, and Kelvin temperature T.

The energy wells for bound water at wood moisture contents of 0%, 7%, 14%, 21%, and 28% are shown in Fig. 5.2a. In this case, if 28% is taken to be fiber saturation, there is essentially no difference in the enthalpies of bound water at M_f and of liquid water.

It is evident from Fig. 5.2a that a water molecule at a 14% site must attain energy equivalent to the difference in the enthalpy levels H_v and H_s in order to

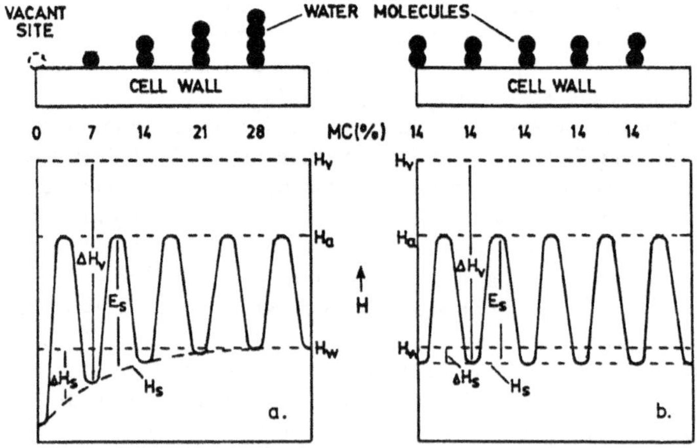

Fig. 5.2. Hypothetical model for interpreting the effect of moisture distribution in the cell wall on the energy levels H_a, H_w, H_s, and activatation energy E_s, for the case of: **a** a non-uniform moisture distribution, **b** an uniform moisture distribution

evaporate from the wood. This energy difference corresponds to ΔH_v calories per mole at 14% M. C. For the same molecule to escape into the liquid form, it must attain the smaller additional energy ΔH_s, equivalent to the difference, $H_w - H_s$.

A water molecule may jump from the 14% to the 7% site by attaining enough energy to overcome the energy "barrier"separating the two wells. The height of this barrier, defined as the activation energy, E_s, on a molar basis, is the energy required for a mole of water molecules to jump from one sorption site to the next. Thus E_s is equal to the energy difference $H_a - H_s$, where H_a is the enthalpy level of the activated molecules, corresponding to the horizontal line joining the energy peaks in Fig. 5.2a.

According to the diagram the same energy barrier E_s must be overcome for a water molecule to jump from the 14% site to the 21% site as from the 14% to the 7% site. However, molecules from the 21% sites are also jumping to other sites, but the energy barriers or activation energies are lower. Therefore more molecules per unit time are migrating from the 21% to 14% sites than in the reverse direction. The same reasoning follows with respect to the 14% and 7% sites so more molecules jump from 14% to 7% sites than the reverse.

The picture represented by Fig. 5.2a gives some insight into the reason why bound-water movement takes place in wood, and the direction in which it moves. Drier wood contains more low energy sites than more moist wood, and therefore molecules tend to migrate toward the drier wood until an equilibrium is attained such that the free energy levels of all the bound water molecule sites throughout the wood are the same. This is not a static equilibrium, because water molecules are continually "jumping" from one sorption site to another because of the random distribution of kinetic energy among them. However, on a macroscopic scale, the average moisture content throughout the wood tends to remain constant.

Figure 5.2b shows the moisture distribution and potential energy levels of the corresponding sorption sites under equilibrium conditions at an uniform moisture content of 14%. Note that each sorption site, according to the model, contains two water molecules, and that the potential energy wells are all of the same energy level, with the same height of barrier between any pair of sites. In this case a given water molecule is as likely to jump in one direction as another. It is improbable that each sorption site contains the same number of molecules, even though the moisture content is uniform in the macroscopic sense. This is because water molecules are continually jumping from one sorption site to another, and the sites probably vary among themselves in their energies or forces of attraction.

Until now we have considered only the direction in which moisture will move under a hypothetical moisture gradient such as is shown in Fig. 5.2. It will move "downhill," that is, in the direction of decreasing free energy level, in analogy with the observation that water in liquid form, such as in rivers, moves toward lower elevations where the potential energy is lower. The rate of movement, on the other hand, is determined by the heights of the energy barrier, E_s. This is because the probability of a given water molecule attaining sufficient kinetic energy to overcome a given barrier is inversely related to the height of the barrier. Therefore the

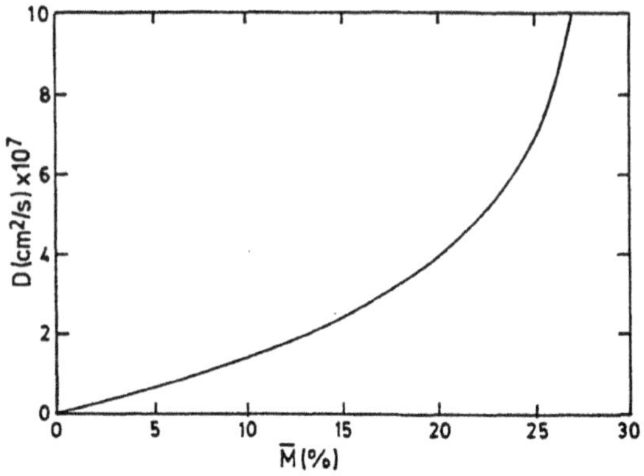

Fig. 5.3. Bound water diffusion coefficient D_b in relation to the mean moisture content for the cell wall of Sitka spruce as measured by Stamm (1959) at 26.7 °C in the longitudinal direction

rate of movement or of jumping from one sorption site to another is expected to be higher if the energy barrier is lower. Experimental observations by Stamm (1959) of bound-water movement through wood under steady-state conditions have shown that the rate of movement under a given concentration gradient is more rapid at high than at low wood moisture contents (Fig. 5.3). Therefore, the model of Fig. 5.2 shows that the energy barriers, proportional to activation energy E_s, are higher at the lower wood moisture contents.

Figure 5.2a shows only a small change in the heights of the energy barriers with changing moisture content, while Fig. 5.3 shows a large change in the diffusion coefficient D with increasing moisture content. The reason for the great difference is the exponential relationship between the height E_s of the energy barrier and the rate of molecular jumps between sites, the latter being directly proportional to the transport coefficient. If a concentration gradient is assumed, as in Fick's law, the diffusion coefficient D (Table 5.1) is the appropriate transport coefficient, proportional to the rate of molecular jumps.

The Arrhenius equation predicts the relationship between D and E_s to be

$$D = D_0 \exp(-E_s/RT) \ , \tag{5.2}$$

where D_0 is taken as a constant, dependent on the concentration and distribution of sorption sites. According to Eq. (5.2), a proportionately small change in E_s may cause a large change in D for the isothermal case.

Equation (5.2) can be used to evaluate E_s if data for D as a function of wood temperature at constant moisture content are available. Rearranged, Eq. (5.2) can be written in two forms to evaluate E_s,

$$E_s = -R\, d(\ln D)/d(1/T) \tag{5.3}$$

$$E_s = RT^2\, d(\ln D)/dT \ . \tag{5.4}$$

Stamm (1964) and Choong (1965) have shown that curves of the logarithm of D against the reciprocal of Kelvin temperature are essentially linear, based on drying studies of a number of workers (Fig. 5.4). Calculation of the activation energy E_s using Eq. (5.3) gives values in the order of 8500 calories per mole. Although the data used are based on gross wood samples, the rate of moisture transport is

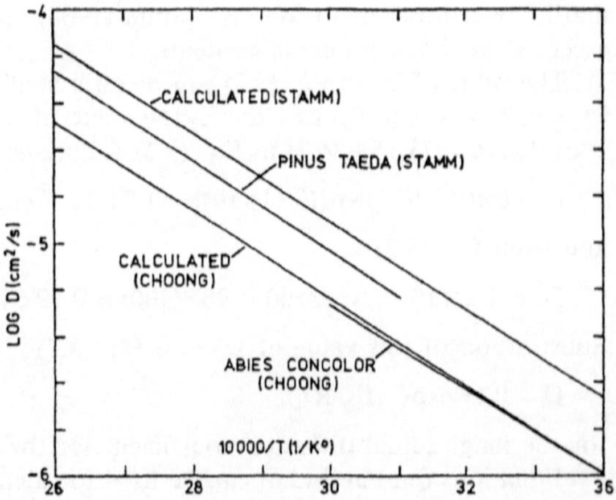

Fig. 5.4. Calculated and measured curves of log (D) as functions of reciprocal of Kelvin temperature, as given by Stamm (1964) for *Pinus taeda* and by Choong (1965) for *Abies concolor*

determined essentially by the rate of bound-water movement through the cross walls, according to Siau (1984). This may not be true in all cases, particularly for woods of high transverse permeability, in which case appreciable flow may take place through pit openings in the cross walls. If bound water transport is limiting, the effect of temperature and/or moisture content on transverse diffusion is essentially the same as their effects on bound-water diffusion.

Kawai et al. (1980) calculated activation energies for longitudinal diffusion in *Chamaeyparis obtusa* wood of Japan, to be in the order of magnitude of 8.1 to 8.7 kcal/mol of water, in good agreement with the values reported above. Apparently even in this case bound-water diffusion in the longitudinal cross walls limited the rate of movement. They discussed a model for activated diffusion similar to that given here, and proposed that the chemical potential (Sect. 5.2.5) should be used as the basic driving potential for bound-water movement. Their method of measuring moisture distributions during drying was a novel nondestructive technique involving the use of "soft" X-ray densitometry.

Figure 5.3, taken from Stamm (1959), shows a strong increase in the longitudinal diffusion coefficient D for bound water in Sitka spruce with increasing moisture content at 26.7 °C. It was noted that D increased exponentially with moisture content over much of the moisture range from 5% to 25%, from which the following empirical equation can be calculated,

$$D = 0.40 \times 10^{-7} \exp(0.11\,M) \ . \tag{5.5}$$

In order to calculate E_s as a function of moisture content M, it is desirable to have measurements of D at each of several constant moisture contents as a function of temperature. Such data are not available, but it is possible to estimate this relationship if D is known as a function of temperature at a single moisture content. For example, Choong (1965) and Stamm (1959) report activation energies of about 8500 cal/mol for wood at a mean moisture content of 10%, based on application of the Arrhenius equation [Eq. (5.2)]. Although Choong's measurements were made on gross wood, the temperature effect should be essentially the same as for the cell wall, since diffusion in the transverse direction is limited primarily by cell-wall diffusion, as Stamm (1964) and Siau (1984) have calculated, except at very low moisture contents.

The value of D_0 in Eq. (5.2) can be estimated by evaluating D in Eq. (5.5) at M = 10%, and substituting for E_s the value of 8500 cal/mol and a temperature T of 300 K (273.15+26.7) in Eq. (5.2). Thus, from Eq. (5.5)

$$D = 0.40 \times 10^{-7} \exp[(0.11)(10)] = 1.2 \times 10^{-7}\ \mathrm{cm^2/s}\ ,$$

and from Eq. (5.2)

$$D_0 = 1.2 \times 10^{-7} \exp(8500/1.987/300) = 0.19\ \mathrm{cm^2/s}\ .$$

Substitution of this value of D_0 into Eq. (5.2) gives

$$D = 0.19 \exp(-E_s/RT) \tag{5.6}$$

for the longitudinal diffusion coefficient for the cell wall of Sitka spruce.

Equations (5.5) and (5.6) can be used to calculate the dependency of the activation energy E_s on moisture content M by equating them for D at the temperature T of 300 K. This results in the empirical equation based on Stamm's (1959) data,

$$E_s = 9600 - 170\,M + 6.95\,M^2 - 0.160\,M^3\ \mathrm{(cal/mol)}\ . \tag{5.7}$$

Siau (1984) used an approximation to Eq. (5.7) which gives reasonable values over the moisture range from 5% to 25%, that is, $E_s = 9200 - 70\,M$. Using this in Eq. (5.6) yields

$$D = 0.19 \exp(-(9200 - 70\,M)/RT)(\mathrm{cm^2/s})\ . \tag{5.8}$$

Equation (5.8) may be used to estimate D for longitudinal diffusion through the cell wall of Sitka spruce. Stamm (1960) has shown that it is approximately 2.5 times greater than D in the transverse cell wall direction. Therefore the pre-exponential term 0.19 is divided by 2.5 to obtain 0.076, close to the value 0.07 given by Siau (1984) for transverse diffusion. Figure 5.5 shows curves for the transverse diffusion coefficient as functions of moisture content and temperature based on Eq. (5.6) and (5.7), and using 0.076 as the pre-exponent term in Eq. (5.6).

It is anticipated that, as the cell wall approaches moisture saturation, the apparent activation energy should approach that for the self-diffusion coefficient

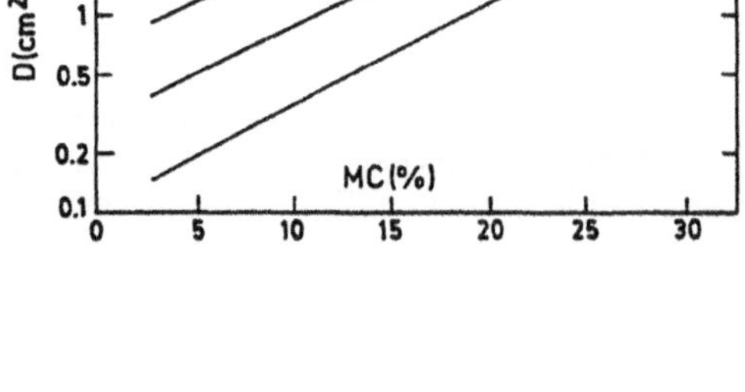

Fig. 5.5. Calculated curves of the transverse bound water diffusion coefficient D (log scale) as a linear function of wood moisture content M for several temperatures, using Eq. (5.8)

D_w for water. In Stamm's (1946) initial attempt to calculate D for bound-water movement through the cell wall of wood he used the viscosity η_w of liquid water and the Einstein diffusion equation

$$D_w = kT/(6\pi r \eta_w) \ , \tag{5.9}$$

where k is the gas constant R divided by Avogadro's number (6.023×10^{23} molecules per mole) and r is the radius of the water molecule, to calculate D_w. Although Stamm (1959) later rejected this model as inadequate, the Einstein equation is useful for the present discussion since it shows that D is inversely related to η, that is, the product $D \cdot \eta$ is nearly constant with temperature except for the small direct effect of T in Eq. (5.9).

Jost (1960) uses the relationship of constant $D \cdot \eta$ for liquids to point out that the activation energy E, obtained by application of the Arrhenius equation to diffusion in liquids (or solids) based on the "hole" theory, is essentially the same as the activation energy obtained by application of the same equation to the variation of the viscosity η of the same liquids. This is because the energy required to form "holes" in a liquid is the most important factor affecting viscosity, as it is for diffusion.

Jost (1960) shows that the relationship of D to T is given essentially by Eq. (5.2), written here as

$$D = D_0 \exp(-E/RT) \ , \tag{5.10}$$

where D_0 is practically constant. If D and η are interchanged in Eq. (5.9) and the equivalent for D given by Eq. (5.10) is substituted for D, the result obtained can be rearranged to

$$\eta = (D_0 kT/6\pi r) \exp(+E/RT) \ . \tag{5.11}$$

A comparison of Eqs. (5.10) and (5.11) shows that, except for T in the kT term, which is essentially negligible compared with the exponential term, the effect of temperature is the same on D and $1/\eta$.

The activation energy E for the viscosity of liquid water is in the range of 4 or 5 kcal/mol, according to Jost (1960). Therefore that for D_w should be similar. Extrapolating the value of E_s from Eq. (5.7) to fiber saturation, say $M = 30\%$, gives an estimated value for E_s of 6.4 kcal/mol. This decreases rapidly to 6.0, 5.6, and 5.0 kcal at moisture contents of 32%, 34%, and 36%, in the range of activation energies for the self-diffusion coefficient of liquid water. According to Nelson (1984), the activation energy E_w for water is 5.3 kcal/mol.

5.2.2 Eyring's Absolute Reaction Rate Theory

The diffusion process was first treated in terms of the absolute reaction theory of chemical kinetics by Eyring (1936) who applied it to diffusion in liquids. An abridged version of Eyring's theory, adapted from Jost (1960), is given here.

In this theory the diffusion coefficient D is given by

$$D = l^2 r \ , \tag{5.12}$$

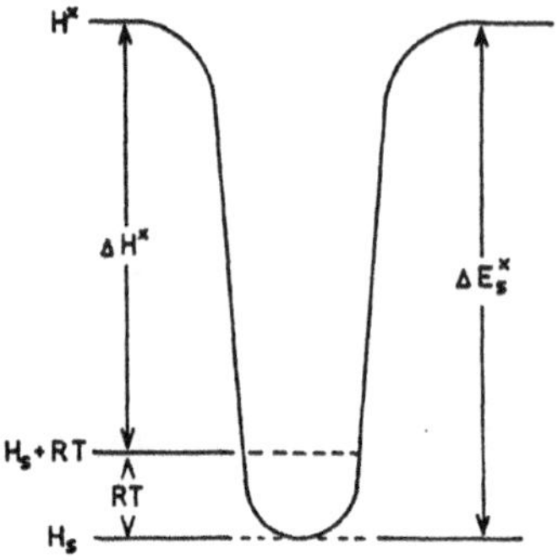

Fig. 5.6. Schematic diagram illustrating the relationships among the terms H_s, ΔE_s^*, ΔH^*, H^*, and RT

where l is the distance between successive equilibrium positions of the diffusing molecule, and r is a rate constant, in this case the number of jumps per second. The rate constant r is given by

$$r = (kT/h)\exp[(T\Delta S^* - \Delta H^*)/RT] , \qquad (5.13)$$

where k is the Boltzmann constant, h is Plank's constant and ΔS^* and ΔH^* are the entropy and enthalpy of activation, respectively. The diffusion coefficient D, from Eqs. (5.12) and (5.13) is given by

$$D = l^2(kT/h)\exp((T\Delta S^* - \Delta H^*)/RT) . \qquad (5.14)$$

The empirical activation energy ΔE_s^* of Eq. (5.6) is equivalent to $\Delta H^* + RT$. Figure 5.6 shows schematically the relationship among the energy terms H_s, H^*, RT, ΔH^* and ΔE_s^* based on this analysis. At zero Kelvin, ΔE_s^* and ΔH^* are the same since $RT = 0$, provided ΔE_s^* is constant with temperature.

The change in Gibbs free energy ΔG^* of the activated molecules is defined as

$$\Delta G^* = \Delta H^* - T\Delta S^* , \qquad (5.15)$$

or, from Eq. (5.14), letting $A = l^2(kT/h)$,

$$D = A\exp(-\Delta G^*/RT) . \qquad (5.16)$$

The free energy change ΔG^* can be calculated if the entropy change $T\Delta S^*$ is known. This is small compared with ΔH^* for other diffusion systems. It can be estimated in the present case if both D and ΔE_s are known. Substituting for ΔH^* in Eq. (5.14) its equivalent, $\Delta E_s - RT$,

$$D = A\exp[(T\Delta S^* - \Delta E_s + RT)/RT] , \qquad (5.17)$$

which can be rearranged to solve for $T\Delta S^*$

$$T\Delta S^* = \Delta E_s - RT[1 - \ln(D/A)] . \qquad (5.18)$$

At 15% moisture content, terms in Eq. (5.18) are evaluated as follows: ΔE_s is taken as 8500 cal/mol, RT as 600 cal/mol at room temperature, D as 2×10^{-7} cm²/s from Fig. 5.3, kT/h is approximately 6×10^{12}/s and l is assumed to be 5×10^{-8} cm; the mean distance between sorption sites based on the Dent sorption isotherm model (Chap. 3.2.1). Substitution of these values into Eq. (5.18) yields a value for $T\Delta S^*$ in the order of 1200 cal/mol, about twice the value of RT.

Based on the calculations given above, most of the increase ΔH^* in enthalpy of the activated molecules is due to the increase ΔG^* in free energy, according to Eq. (5.15). The magnitude of ΔH^* in our example, using

$$\Delta H^* = \Delta E^* - RT \tag{5.19}$$

is approximately 8500 − 1200 or 7300 cal/mol. Therefore ΔG^* is approximately 6100 cal/mol from Eq. (5.15) since $T\Delta S^*$ is 1200 cal/mol.

5.2.3 Moisture Content as a Driving Force

The moisture transport coefficient K_m based on a moisture content gradient is related to the Fick's law coefficient D essentially by a constant. The slopes dc/dm or dc/dM (Table 5.1) decrease somewhat with increasing moisture content due to the swelling of wood. The relationships are given by

$$K_m = K_M(100) = DG_m\varrho_w = Dd_m \ , \tag{5.20}$$

where G_m is the specific gravity of the wood or of the cell wall based on swollen volume at M or m, ϱ_w is the density of water, and d_m is the density index [Eq. (4.17)].

Use of the coefficients K_m or K_M and the associated moisture gradients dm/dx or dM/dx to describe moisture movement in wood is therefore essentially the same as using Fick's law in its original form.

5.2.4 Vapor Pressure as a Driving Force

The assumption of vapor pressure p as the driving potential for moisture movement yields the transport coefficient K_p (Table 5.1). This form of coefficient is used widely to describe transport of vapors through solid non-porous materials such as polymer films used as vapor barriers. Often called the "permeability constant" it is relatively easy to evaluate since only vapor pressures need be measured, and not moisture contents.

It should be noted that the use of the term "permeability" in this case is different from the sense in which it is normally used for wood. In conventional use the term refers to the effectiveness of the porous network in a material such as wood to conduct fluids through this porous system under a total pressure gradient. The fluid is presumed to move through the porous system but not through the cell wall material per se, that is, it is not sorbed and transmitted through the cell wall as bound water.

In the present discussion water vapor is presumed to condense on the cell wall, diffuse through as bound water, and evaporate from the opposite surface. [Darcy's law is not used here, since it refers to flow through porous media (Siau 1984)]. The permeability K_p is calculated from

$$K_p = -J/(dp/dx) \ , \tag{5.21}$$

where J is the measured moisture flux (g/cm^2 s) and p is the partial water vapor pressure. The relationship of K_p and D (Table 5.1) is

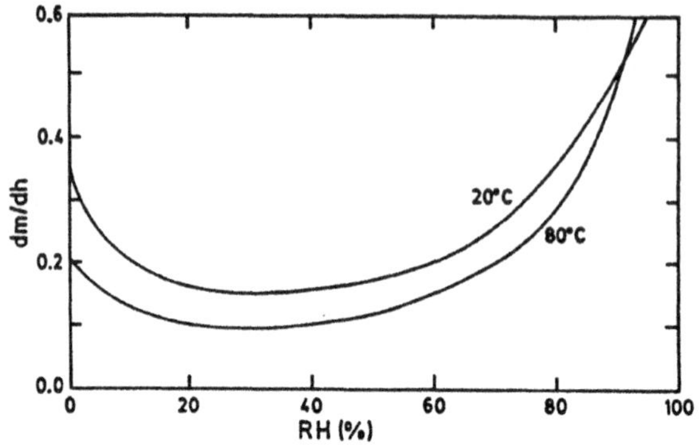

Fig. 5.7. Curves showing the relationship of the slope dm/dh to relative humidity at temperatures of 20° and 80 °C. (After Simpson 1982)

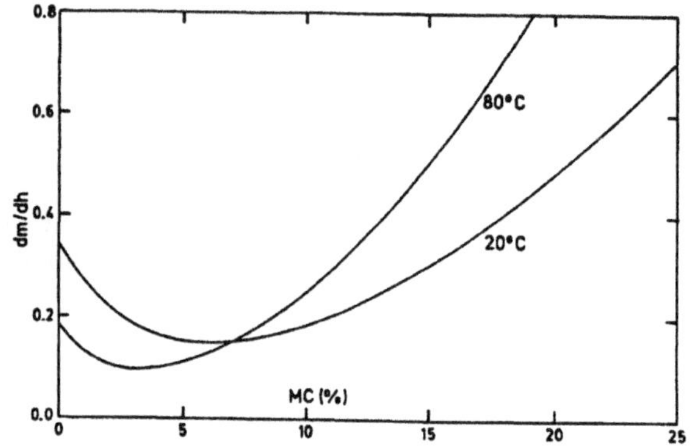

Fig. 5.8. Curves showing how the slope dm/dh varies with wood moisture content at temperatures of 20° and 80 °C. (After Simpson 1982)

$$K_p = D(dc/dp) \ . \tag{5.22}$$

If Henry's solubility law is obeyed, that is, if c is a linear function of p, the term $\partial c/\partial p$ is defined as the solubility S. It is customary among those who study the vapor transport properties of polymer films to use the symbol P for permeability K_p as used here (Rogers 1967). In this case, assuming that the solubility S is a constant, Eq. (5.22) is written

$$P = DS \ . \tag{5.23}$$

The term dc/dp is linearly related to the slope (dm/dh) of the sorption isotherm for wood, as follows

$$dc/dp = (G_m \varrho_w / p_0)(dm/dh) \ . \tag{5.24}$$

Therefore it is not constant since dm/dh varies in a complex manner with relative humidity, wood moisture content, and temperature, as Figs. 5.7 and 5.8, adapted from Simpson (1982), show. Furthermore the specific gravity G_m decreases somewhat with increasing m.

Two general methods are used for measuring K_p. In the first method opposite faces of the sample are exposed to atmospheric air of which water vapor is only one constituent. The partial water vapor pressure p is greater on one face than

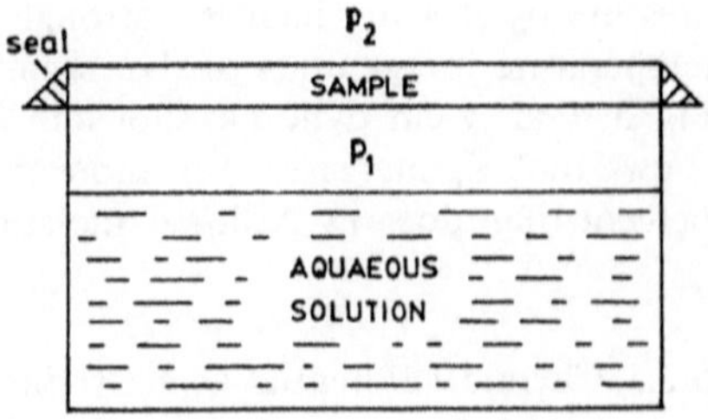

Fig. 5.9. Diagram of a typical vapometer cup

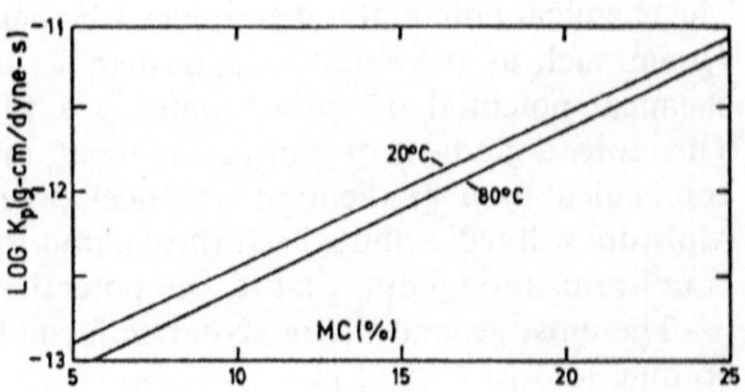

Fig. 5.10. Curves showing the strong increase in log (K_p) with increasing wood moisture content at 20 and 80 °C

on the other, resulting in a vapor pressure gradient through the sample. This is usually accomplished by use of a "vapometer" cup (Fig. 5.9) whose opening is covered by the sample, sealed on the edges to prevent edge losses. The aqueous solution inside the cup maintains a constant vapor pressure p_1, and the outer surface is exposed to air with a different, but constant, water vapor pressure p_2. The flux J is measured simply by weighing the cup at appropriate time intervals, and the mean value K_p of the water vapor permeability K_p is determined from Eq. (5.21) in the form

$$\bar{K}_p = -JL/(p_2 - p_1) \; , \tag{5.25}$$

where L is the sample thickness. It is important to stir the solution and air inside the cup so as to maintain an uniform vapor pressure p_1, particularly for high permeability samples. Sometimes pure water or a dessicant is used inside the cup rather than a solution.

In the second method no atmospheric gases are present except water vapor. This requires an evacuated system on each side of the sample, except for water vapor introduced into one or both sides of the sample. Each side is maintained at constant but different vapor pressures p_1 and p_2. The steady-state flux J is determined on either or both sides of the sample by weighing the loss or gain in weight of the water source or sink, as appropriate.

In either method noted above it is important to maintain close and constant temperature control of both the source and sample, because the vapor pressure of water is so highly sensitive to temperature.

Equation (5.22) relates K_p to the conventional diffusion coefficient D. Therefore if D and dc/dp are known as functions of moisture content the vapor permeability K_p can also be calculated at any moisture content. Figures 5.7 and 5.8 permit calculation of K_p as a function of wood moisture content and temperature, using Eq. (5.24) to relate dc/dp to dm/dh, shown in Fig. 5.8. The vapor pressure p_0 of water increases strongly with temperature. It is calculated at each temperature in cgs units of dynes/cm^2. Therefore the units of K_p as plotted in Fig. 5.10 are g-cm/dyne-s as shown in Table 5.1. Comparison of Figs. 5.5 and 5.10 shows that K_p increases even more rapidly with moisture content above 5 to 8 percent than does D. At lower moisture contents the reverse is true.

5.2.5 Chemical Potential as a Driving Force

The chemical potential μ determines whether equilibrium has been attained in a system such as the wood-water system. It has been shown (Chap. 3) that the chemical potential of sorbed water is a function of wood moisture content. Therefore, a gradient of moisture content, or more specifically of activity, is accompanied by a gradient of chemical potential under isothermal conditions. Moisture will redistribute itself throughout the wood until the chemical potential is uniform throughout, that is, the potential gradient $d\mu/dx$ is zero.

The most general thermodynamic formulation of the diffusion equation according to Jost (1960), is

$$J = -(cu/N_0)(d\mu/dx) \ , \tag{5.26}$$

where J is the moisture flux (g/cm^2 – s), c is the moisture concentration (g/cm^3), u is the mobility of the diffusing water molecule [(cm/s)/(dyne/molecule)], N_0 is Avogadro's number (6.02×10^{23} molecules/mol) and $d\mu/dx$ is the thermodynamic gradient (ergs/mol-cm). The transport coefficient K_μ (Table 5.1) is given by

$$K_\mu = cu/N_0 \ . \tag{5.27}$$

The gradient $d\mu/dx$ under isothermal conditions is related to the gradient of activity a by

$$d\mu/dx = (RT/a)(da/dx) \ , \tag{5.28}$$

and to the concentration gradient by

$$d\mu/dx = (RT/a)(\partial a/\partial c)(dc/dx) \ . \tag{5.29}$$

Thus Eq. (5.26) can be written in terms of dc/dx as

$$J = -(cu/N_0)(RT/a)(\partial a/\partial c)(dc/dx) \ , \tag{5.30}$$

and the diffusion coefficient D is given by

$$D = (cukT/a)(\partial a/\partial c)_T \ , \tag{5.31}$$

where k ($= R/N_0$) is the Boltzmann constant. Moisture content m can be substituted for c in Eq. (5.31) with essentially no error, and h for a, giving

$$D = ukT(m/h)(\partial h/\partial m)_T \ . \tag{5.32}$$

If the sorption isotherm is known, the diffusion coefficient D can be related to mobility u at any temperature. Thus Eq. (5.32) can be rearranged to calculate the ratio D/u

$$D/ukT = (m/h)(\partial h/\partial m)_T \ . \tag{5.33}$$

Fig. 5.11. Curves of D/ukt and of D/u_2kt in relation to wood moisture content, at 26.7 °C

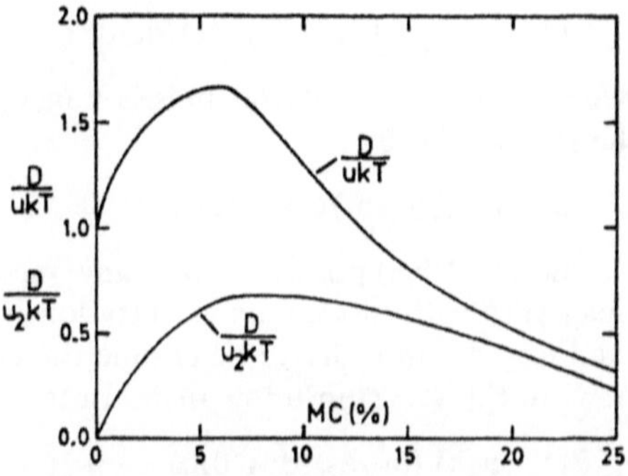

Figure 5.11 shows curves of the ratio D/ukT as a function of wood moisture content at 26.7 °C. In order to evaluate $(m/h)(\partial h/\partial m)_T$, the Hailwood-Horrobin sorption model coefficients (Chap. 3.2.3), calculated by Simpson (1971 a) as functions of temperature from the Wood Handbook sorption data, were used to calculate the constants A, B, and C at 26.7 °C. The evaluation was then completed by use of the relationship, obtainable from Eq. (3.1), that

$$(m/h)(\partial h/\partial m)_T = (A+Bh-Ch^2)/(A+Ch^2) \ . \qquad (5.34)$$

It is evident from Fig. 5.11 that moisture content has a relatively small effect on the ratio D/u, compared with its effect on D as shown in Fig. 5.5. It may be concluded therefore that the primary reason for the large variation of the diffusion coefficient D with moisture content is due to increase in the mobility u. It may be noted that if Henry's Law is obeyed, that is, if the sorption isotherm is linear such that $\partial h/\partial m = h/m$, the ratio D/u is then independent of moisture content, and

$$D = ukT \ . \qquad (5.35)$$

The mobility u in the discussion above refers to the overall mobility of all of the sorbed water molecules. Barrie (1968) proposes that some of the sorbed water molecules in polymers are essentially immobile, that is, they are bound so tightly to the polymer that their mobilities are essentially zero. He suggests that Eq. (5.26) be applied only to the mobile molecules. For example, in the sorption theories discussed in Chapter 3, the molecules of primary water in the BET or Dent models, or of the hydrated water in the Hailwood-Horrobin model, may be considered to be immobile since they are bound with such strong forces. Those in the secondary layers (or the dissolved water) are expected to be relatively mobile. The Dent model will be used here to estimate the effect of moisture content on the diffusion coefficient D, considering the mobility u_2 of the secondary layers only.

Equation (5.26) can be written in terms of the concentration c_2, mobility u_2 and chemical potential μ_2 of the secondary water,

$$J = J_2 = -(c_2u_2/N_0)(d\mu_2/dx) \ , \qquad (5.36)$$

since J_1 is taken to be zero. It can be modified to

$$J = -c_2 u_2 (kT/a_2)(\partial a_2/\partial c)(dc/dx) \ , \tag{5.37}$$

where a_2 is the activity of the secondary water. The diffusion coefficient D may be represented by

$$D = u_2 kT (c_2/a_2)(\partial a_2/\partial c) \ . \tag{5.38}$$

Barrie (1968) points out that any one or more of the three terms in Eq. (5.38) may vary with moisture concentration, that is, the mobility u_2, the ratio c_2/a_2, and $\partial a_2/\partial c$. In order to determine the effect of these terms, it is convenient to rewrite Eq. (5.38) in terms of moisture content m. Thus

$$D = u_2 kT (m_2/a_2)(\partial a_2/\partial m) \ . \tag{5.39}$$

The activity a_2 is equal to bh according to the Dent model where b is the equilibrium constant and $m_2 = mbh$ (Chap. 3), and Eq. (5.39) becomes

$$D/u_2 = kT\,b\,m(\partial h/\partial m)_T \ . \tag{5.40}$$

Comparison of Eqs. (5.33) and (5.40) shows that the mobilities u and u_2 are related by

$$u/u_2 = b\,h = m_2/m \ , \tag{5.41}$$

or in the ratio m_2/m. A curve of D/u_2kT is included in Fig. 5.11. Using the mobility u_2 of the secondary water molecules does improve the model somewhat over that given by Eq. (5.33) because the diffusion coefficient increases to a greater extent with increasing moisture content since it begins with zero at zero moisture content. However, it appears that the mobility u or u_2 must also increase with moisture content to explain the large increase in D with moisture content shown in Fig. 5.4. The increase in mobility is considered further in a later model (Sect. 5.2.7).

Stanish (1986) developed an equation based on a gradient of chemical potential to describe moisture transport through wood. His equation included both bound-water and water-vapor diffusion. For isothermal diffusion, the chemical potential gradient $d\mu/dx$ was expressed in terms of vapor pressure gradients, essentially derivable from Eq. (5.28), although Stanish used a more general thermodynamic relationship, which included the effect of temperature gradients. He also assumed that the chemical potentials of the bound water and water vapor were at equilibrium at any location in the wood.

Stanish's (1986) final equation included both vapor and bound water transport through wood under nonisothermal conditions. Furthermore, it was expressed in terms of the vapor pressure and temperature in the wood rather than in terms of moisture content and temperature gradients. In order to use the equations on experiments in which moisture contents rather than vapor pressures are measured, it is necessary to know the sorption isotherms. On the other hand, for analyzing data based on vapor pressure measurements Stanish's equations are most suitable. He has noted that his equations gave closer agreement with experimental data than other equations proposed to explain nonisothermal moisture transport in wood, such as those proposed by Siau et al. (1986) and by Skaar and Siau (1981).

5.2.6 Osmotic Pressure as a Driving Force

Osmotic pressure π is equivalent to the ratio of the chemical potential to the molar volume. Thus, for water [see Eq. (3.74)]

$$\pi = \mu/\bar{V}_w = (RT/18)\ln(a) \;, \tag{5.42}$$

where $\bar{V}_w$ is the partial molar volume, 18 cm^3/mol for liquid water, and π is in ergs/cm^3 or dynes/cm^2.

Osmotic pressure has been equated to swelling pressure (Chap. 2.6) and is relatively easy to visualize as a driving force for moisture movement in wood. For example, in the model shown in Fig. 5.2, there is a gradient of activity a or relative vapor pressure h associated with the moisture gradient. For a typical sorption isotherm the equilibrium relative vapor pressures h_2 and h_1 are approximately 0.7 and 0.3, for moisture contents of 14% and 7%, respectively. In this case the difference in osmotic pressure $\pi_2 - \pi_1$, of the sorbed water is given by

$$\pi_2 - \pi_1 = (RT/18)\ln(h_2/h_1) \;, \tag{5.43}$$

where h is taken to be equal to activity a. At room temperature $\pi_2 - \pi_1$ is the order of 1150×10^6 dynes/cm^2, about 1150 atm. This is equivalent to the difference in pressure at the bottom of a column of water about 7 miles (11 km) high, and that at the earth's surface. However, the resistance to flow in the cell wall is also extremely great, so the rate of movement is slow, despite the high gradient.

Osmotic pressure and chemical potential are the same potential expressed in different dimensions, as Eq. (5.42) indicates. The dimensions of osmotic pressure are force per unit area or energy per unit mass. Therefore, the same factors which affect chemical potential as a driving force also apply to osmotic pressure.

5.2.7 Spreading Pressure as a Driving Force

Babbitt (1950) proposed that the migration of sorbed gases through solids is along internal surfaces. He suggested that the driving force is proportional to the space gradient of the spreading pressure ϕ, a surface parameter having the same dimensions as the surface tension σ associated with capillary forces. Therefore, the dimensions of ϕ are force per unit length or energy per unit area. This distinguishes spreading pressure from osmotic pressure π, which has dimensions of force per unit area or energy per unit volume, and also from the related chemical potential μ.

Skaar and Babiak (1982) adapted Babbitt's transport model to the Dent (1977) modification of the BET sorption theory (Chap. 3.2.1). Nelson (1986a, b, c) extended Babbitt's model further to include nonisothermal moisture transport through wood. The present discussion essentially summarizes these treatments.

In the model proposed by Babbitt (1950), moisture is considered to be condensed as a film of one or more molecular layers on internal surfaces in the cell wall, as in the BET and Dent models. The condensed or surface-sorbed water migrates as individual water molecules in the water film. Figure 5.12 shows a schematic diagram of such a surface with uniformly distributed sorption sites,

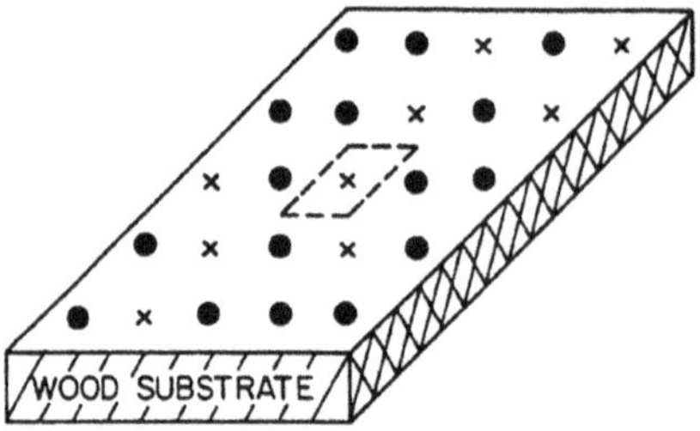

Fig. 5.12. Schematic diagram showing sorption sites on an internal wood surface, some of which are occupied (*dark circles*) and some unoccupied (*marked X*). The area per sorption site a is shown as the area within the parallelogram in the center of the surface. (Skaar and Babiak 1982)

each of area a. Some sites are occupied by one or more water molecules, and some are empty.

The spreading pressure ϕ, as defined by Babbitt (1950)

$$\phi = -(\partial G/\partial a) \tag{5.44}$$

is equal to the rate of change of free energy G per unit area a, defined as the area per water molecule when there is one molecule on each site, or complete monolayer coverage. Equation (5.44) is analogous to that for osmotic pressure π, defined as

$$\pi = -(\partial G/\partial v) \ , \tag{5.45}$$

where G is the free energy, and v the specific volume, of a water molecule.

The spreading pressure increases as more sites are occupied, that is, as the ratio m_1/m_0 increases. For example,

$$\phi = (kT/a) \ln\{1/[1-(m_1/m_0)]\}$$

$$\phi = (kT/a) \ln [m_0(m_0-m_1)] \tag{5.46}$$

where k is Boltzmann's constant, and m_0 and m_1 are the moisture contents corresponding to complete and partial monolayer coverage, respectively. The partial monolayer coverage m_1 increases with total moisture content m as is shown by Eq. (3.34). The spreading pressure ϕ also increases with activity h, according to the following equation, obtained by substituting for m_1 in Eq. (5.46) its equivalent given by Eq. (3.18), and rearranging to give

$$\phi = (kT/a) \ln [(1+b_1h-b_2h)/(1-b_2h)] \ , \tag{5.47}$$

where b_1 and b_2 are the Dent coefficients (Chap. 3.2.1). Table 5.2 lists a number of equations relating spreading pressure ϕ to several other potentials used to describe moisture transport in wood.

The relative spreading pressure ϕ/ϕ_0 is analogous to relative vapor pressure or activity h, where ϕ_0 is the value of ϕ at unit activity (h = 1.0), following Nelson (1986a, b, c). Equation (5.47) can be written in terms of relative spreading pressure as follows

$$\phi/\phi_0 = \frac{\ln [(1+b_1h-b_2h)/(1-b_2h)]}{\ln [(1+b_1-b_2)/(1-b_2)]} \ . \tag{5.48}$$

Figure 5.13 shows curves of the ratios m/m_0, m_1/m_0, and ϕ/ϕ_0, each plotted against activity h, using m_0, b_1, and b_2 as 0.0594, 10.26 and 0.777, respectively.

Table 5.2. Relationship of spreading pressure ϕ to several other potentials for describing moisture transport

Potential	Relationship to spreading pressure
c	$\phi = (kT/a) \ln [c_0/(c_0 - c_1)]$
m	$\phi = (kT/a) \ln [m_0/(m_0 - m_1)]$
p	$\phi = (kT/a) \ln [(p_0 + b_1 p - b_2 p)/(p_0 - b_2 p)]$
h	$\phi = (kT/a) \ln [(1 + b_1 h - b_2 h)/(1 - b_2 h)]$
μ	$\phi = \mu/(N_0 a)$ [a]
π	$\phi = \pi (v/a)$

[a] N_0 = Avogadro's number (molecules/mol).

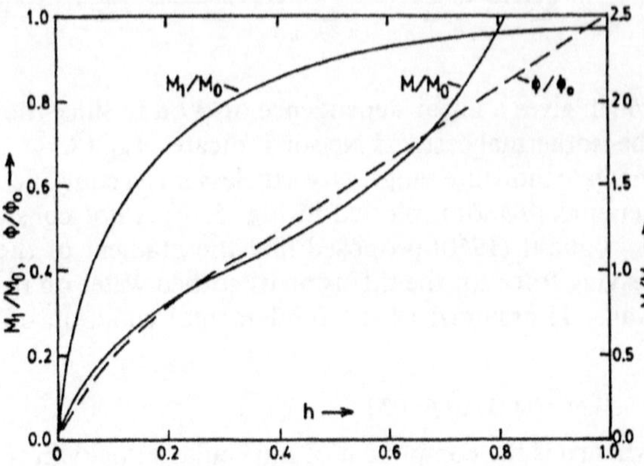

Fig. 5.13. Calculated curves of M/M_0, M_1/M_0 and ϕ/ϕ_0 as functions of vapor activity h

The curve for m/m_0 is on a different scale than the other two. For the Dent model, the ratio m/m_0 is equivalent to n, the *average* number of layers of water, and m_1/m_0 is the fraction of sorption sites occupied by at least one sorbed water molecule. It is of interest that the shape of the curve for spreading pressure ϕ in Fig. 5.13 resembles that for total moisture content m except at the higher activities.

Nelson (1986a, b, c) approximated Eq. (5.48) by an empirical equation of the following form

$$\phi/\phi_0 = K h^z \ , \tag{5.49}$$

where ϕ_0 is the value of ϕ at h = 1.0, K is a constant ranging from 0.94 to 0.96, and the exponent z varies between 0.65 and 0.85. He then combined Eq. (5.49) with the differential equation for the spreading pressure, or free energy of adsorption of a film of moisture, given by

$$d\phi = [RTm/(18A)] [d(\ln p)] \ , \tag{5.50}$$

where A is the surface area per gram of dry wood. The resulting equation may be written

$$\phi = [RT/(18Az)] m \ , \tag{5.51}$$

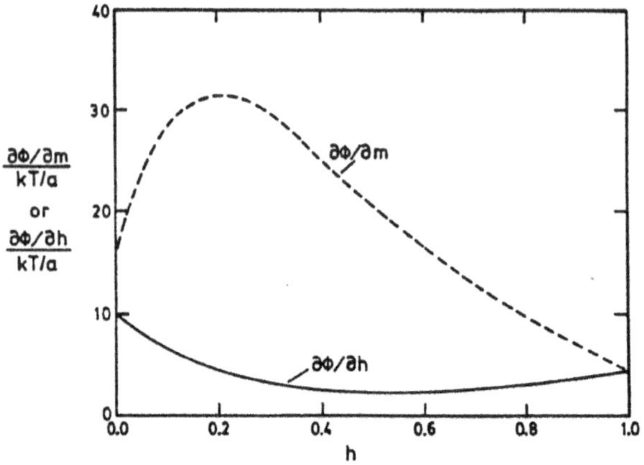

Fig. 5.14. Curves of $\partial\phi/\partial m$, and $\partial\phi/\partial h$ as functions of vapor activity h, calculated from the equations in Table 5.2 relating ϕ to m and h

which gives a linear dependence of ϕ on m since the other terms are constant for the isothermal case. As Nelson indicates, Eq. (5.51) is not valid except in the intermediate moisture range. Nevertheless it is a convenient approximation, but not exact since $\partial\phi/\partial m$, plotted in Fig. 5.14, is not constant.

Babbitt (1950) proposed that the gradient of the spreading pressure ϕ is the driving force for the diffusion of sorbed water on the internal surfaces of the cell wall. He assumed as the fundamental equation of diffusion for unidirectional flow

$$u = -(1/f)(\partial\phi/\partial x) \ , \tag{5.52}$$

where u is the component of molecular velocity in the x-direction, and f is a coefficient of frictional resistance. The driving force $\partial\phi/\partial x$ will be considered first, followed by discussion of the frictional resistance f.

The gradient $\partial\phi/\partial x$ can be related to the conventional moisture gradients, $\partial m/\partial x$ or $\partial c/\partial x$. For the latter the derivative of $\partial\phi/\partial x$ of Eq. (5.46), with c_1 and c_0 substituted in place of m_1 and m_0, can be taken and rearranged into the following form

$$\partial\phi/\partial x = \{(kT/a)/[(c-c_0)^2+4cc_0b_2/b_1]^{0.5}\}(\partial c/\partial x) \ . \tag{5.53}$$

From this it can be seen that $\partial\phi/\partial c$ is given by

$$\partial\phi/\partial c = (kT/a)/[(c-c_0)^2+4cc_0b_2/b_1]^{0.5} \ . \tag{5.54}$$

There are analogous relationships $\partial\phi/\partial U$ relating ϕ to other potentials U, several of which are listed in Table 5.2. Figure 5.14 shows curves of $\partial\phi/\partial m$ and $\partial\phi/\partial h$ plotted against activity h, based on these relationships.

According to Eq. (5.52), the driving force $\partial\phi/\partial x$ is equal and of opposite sense to the resistive forces uf opposing diffusion, or

$$uf = -(\partial\phi/\partial x) \ . \tag{5.55}$$

The flux J (mass of water diffusing per unit time per unit cross-sectional area of the cell wall) is given by the product uc. Multiplying both sides of Eq. (5.55) by c and equating this to J gives

$$J = uc = -(c/f)(\partial\phi/\partial x) = -K_\phi(\partial\phi/\partial x) \ . \tag{5.56}$$

The term c/f is equivalent to the moisture *conductivity* K_ϕ based on the spreading pressure gradient, and its reciprocal f/c is the moisture resistivity. Therefore, the resistance to movement is inversely proportional to the concentration c, or moisture content m, as Babbitt (1951, 1958) proposed.

The commonly used transport coefficients D, K_m, K_h, etc., listed in Table 5.1 may be calculated if K_ϕ, defined as

$$K_\phi = c/f \tag{5.57}$$

is known. For example, from Table 5.1, $D = K_\phi(\partial\phi/\partial c)$ may be combined with Eq. (5.57) to give

$$D = (c/f)(\partial\phi/\partial c) \ . \tag{5.58}$$

The moisture concentration c in the expression c/f may be converted to the appropriate variable by use of the relationship between c and the desired variable. For example, in terms of moisture content m, c is replaced by m/V_m where V_m is the moist volume of the cell wall at moisture content m, per unit dry mass.

The frictional resistance term f in Eqs. (5.57) or (5.58) is difficult to evaluate. It is related to the activation energy E_s for diffusion (Eq. 5.2) by

$$f = f_0 \exp(E_s/RT) \ , \tag{5.59}$$

where f_0 is a constant, increasing slightly with temperature T. Skaar and Babiak (1982) related f to the effective viscosity η_s of the sorbed water, using the relationship

$$f = \eta_s/\delta \ , \tag{5.60}$$

where δ is the mean thickness of the sorbed water film. They postulated that η_s is in the order of 2000 times greater than the viscosity η_w of liquid water. This is based on the assumption that the ratio of the viscosities of liquid to sorbed water is inversely related to the ratio of the self-diffusion coefficient D_w of liquid water to D_s for sorbed water. Using the Arrhenius relation, therefore

$$\eta_s/\eta_w = D_w/D_s = \exp[(E_s - E_w)/RT] \ , \tag{5.61}$$

where E_w is the activation energy for the viscosity, taken to be equal to that for the self-diffusion of liquid water. They assumed a value of 4000 and 8500 cal/mol for E_w and E_s, respectively, and assumed that E_s was invariant with moisture content. Figure 5.15 compares a plot of the diffusion coefficient D_s, calculated as a function of activity h for their model, compared with the curve measured by Stamm (1959) at 25 °C. The general shape is similar, but Stamm's values are approximately three times greater than those calculated from the model. Equation (5.61) may also be written

$$D_s = D_w \exp[-(E_s - E_w)/RT] = D_0 \exp(-E_s/RT) \ , \tag{5.62}$$

since

$$D_w = D_0 \exp(E_w/RT) \ , \tag{5.63}$$

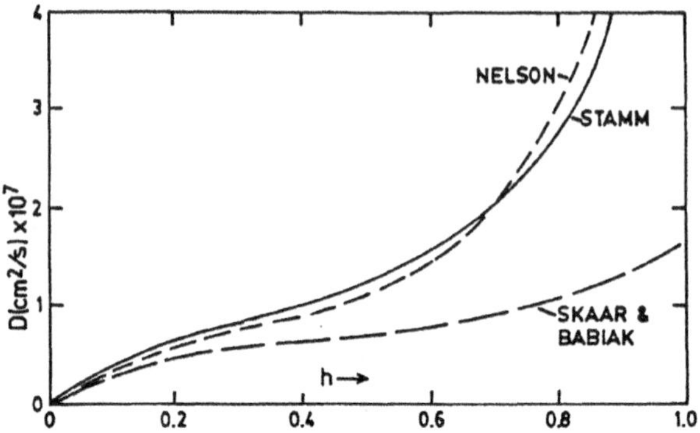

Fig. 5.15. Curves of the diffusion coefficient D against vapor activity h as measured by Stamm (1959), and as predicted by the models of Skaar and Babiak (1982), and also by Nelson (1986a, b, c), all at 25 °C

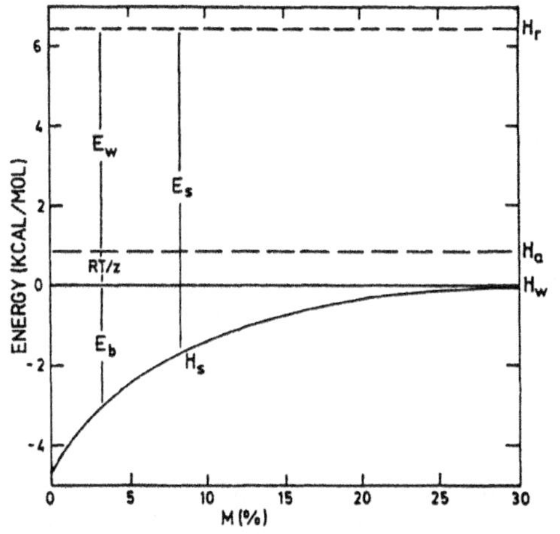

Fig. 5.16. Relative energy levels, compared with that of liquid water H_w, of sorbed water H_s, dissociated water H_a, and fully activated water H_r, in relation to wood moisture content M. Also shown are the relative energy differences among the various energy levels. (After Nelson 1986a, b, c)

where the constant D_0 is approximately 0.2 cm^2/s at room temperature, increasing slightly with temperature.

Nelson (1986a, b, c) separated the activation energy E_s into three components,

$$E_s = E_w + E_b + E_z \ . \tag{5.64}$$

The component E_w, the self-diffusion coefficient for water is essentially the same as in Eq. (5.63). The component E_b is equal to the molar heat of sorption, ΔH_s, defined by Eq. (2.40). The term E_z is the small energy change associated with the expression RT/z in Eq. (5.51). It is approximately 800 cal/mol at room temperature, since z ranges between 0.65 and 0.85. Nelson (1986a, b, c) defines it as "the energy change from the liquid state to the activated state." As indicated above, it varies somewhat with moisture content.

Nelson also introduced a *tortuosity factor* S into his equation to account for the fact that only a portion of the cell wall is accessible to water. Thus his version of Eq. (5.62) is

$$D_s = (D_0/S) \exp(-E_s/RT) \ , \tag{5.65}$$

where S increases from approximately 2.0 for the dry cell wall to 1.7 at moisture saturation. His predicted values of D_s, included in Fig. 5.15, agree better with Stamm's (1959) data than do those of Skaar and Babiak (1982).

Figure 5.16 shows the relative levels of energy for water in four states, sorbed, liquid, dissociated, and activated, based on the analysis by Nelson (1986a, b, c). Also shown are the enthalpy and/or activation energies separating these levels. The mean enthalpy of liquid water is taken as a reference level of zero in the diagram. The vapor enthalpy level (H_v), not shown on the diagram, is about 10 kcal/mol higher than that of liquid water (H_w).

5.3 Irreversible Thermodynamics and Nonisothermal Diffusion

5.3.1 Introduction

The preceding discussion has been confined to isothermal moisture diffusion in which temperature gradients have been neglected. In most wood-drying processes temperature gradients are present and there is therefore some coupling of heat and moisture transport. One such coupling phenomenon is known as *thermal diffusion* or the *Soret* effect. This is the process by which moisture diffuses through wood under the influence of a temperature gradient. In other words, when wood is subjected to a temperature gradient it will not remain at an uniform moisture content but will approach a nonuniform moisture distribution such as is shown in Fig. 5.17, adapted from Johnson (1977). This coupling of moisture and temperature distributions has been confirmed by a number of workers including Voigt et al. (1940), Choong (1963), Siau and Babiak (1983) and Siau et al. (1986).

Siau (1984) has discussed nonisothermal moisture transport in wood in his book *Transport Processes in Wood* and has derived some of the equations used to describe this process. One of the approaches to this problem is that of irreversible thermodynamics which will be given here in order to supplement the treatment given by Siau. The discussion will be confined to steady-state conditions.

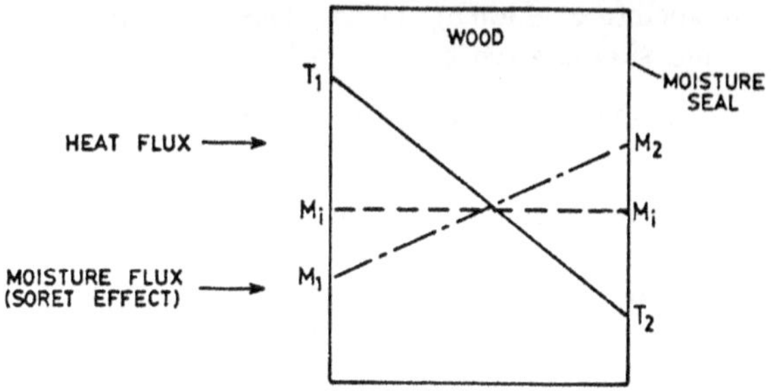

Fig. 5.17. Moisture and temperature distribution curves in a wood sample subjected to a temperature gradient. Heat flows from the hot toward the cool surface and moisture tends to move in the same direction due to the Soret effect. (After Johnson 1977)

5.3.2 Derivation of Coupled Flow Equations

Johnson (1975) defined the coupling coefficients relating heat and moisture transport in wood in terms of the conventional transport coefficients normally used

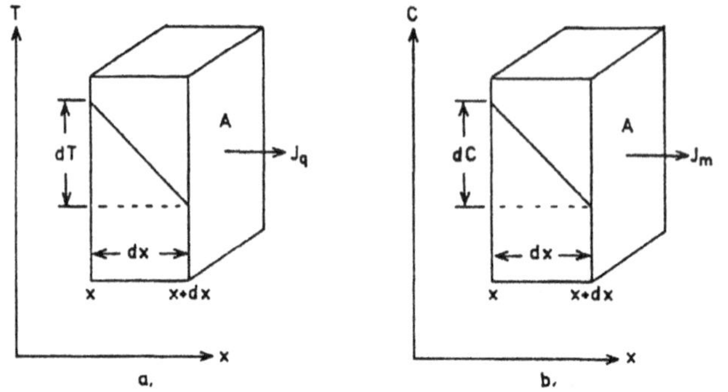

Fig. 5.18. Volume elements for one dimensional flow of: **a** Heat due to a temperature gradient. **b** Moisture due to a moisture concentration gradient

to describe these. He also noted that these coefficients are not necessarily the same as those which arise from the principles of irreversible thermodynamics, because the driving forces assumed are usually different. However, they can be related to those used in the classical descriptions of heat and mass transport in wood.

The equations derived here, and their theoretical foundations, are treated more extensively by several authors, including DeGroot (1950), Van Rysselberghe (1963) and Prigogine (1961). Transports of heat and moisture are not reversible processes even under *steady-state* conditions such as are considered here, because true thermodynamic equilibrium does not exist. Prigogine defines such steady-state conduction as *stationary nonequilibrium state* processes, or simply *stationary-state processes.* In such processes entropy is produced continually at a steady rate.

Entropy production under stationary-state conditions is most readily visualized in the steady-state formulation of Fourier's law of heat conduction. Consider the volume element dV (Fig. 5.18a) of cross-sectional area A and thickness dx through which a steady-state heat flux J_q is flowing under the influence of the temperature gradient dT/dx. The usual formulation of Fourier's law in one dimension is given by

$$J_q = (1/A)(dq/dt) = -K_q(dT/dx) \ , \tag{5.66}$$

where K_q is the thermal conductivity of the medium. For the present it will be assumed that there is no moisture gradient, although such gradients normally occur in wood.

Under stationary-state conditions the rate of heat energy flow dq/dt is the same entering the element at x as leaving it at x+dx. The total energy or enthalpy flow does not change. However, the energy has been degraded since the temperature T_{x+dx} at the exit face is lower than that T_x at the entering face by dT. There is therefore an increase dS in the entropy within the volume element due to the degradation of the energy q.

The rate of entropy change per unit volume dV(= A dx) is given by

$$dS/dt = (1/A)(dq/dt)[d(1/T)/dx] \ . \tag{5.67}$$

The product TS of temperature and entropy has dimensions of energy and therefore the rate of energy degradation or dissipation ϕ_q in the element dV due to heat flow is

$$\begin{aligned}\phi_q &= (T/A)(dq/dt)[d(1/T)/dx] \\ &= TJ_q[d(1/T)/(dx)] \ . \qquad (5.68)\end{aligned}$$

A similar relationship can be derived for the rate of energy dissipation due to steady-state moisture flow. Consider Fig. 5.18b, which shows a steady-state moisture flow of dn/dt moles per unit time passing through a volume element dV of area A and length dx. Using Fick's first law

$$J_m = (1/A)(dn/dt) = -D(dc/dx) \ , \qquad (5.69)$$

where D is the diffusion coefficient and c is the molar *concentration*.

It can be shown that the rate of entropy production dS/dt per unit volume of the wood associated with a steady state flux (1/A)(dn/dt) of moisture is given by

$$dS/dt = (1/A)(dn/dt)[d(-\mu/T)/dx] \ , \qquad (5.70)$$

where μ is the molar free energy or chemical potential of the water.

Therefore, the rate of energy dissipation ϕ_m due to mass flow is

$$\phi_m = (T/A)(dn/dt)[d(-\mu/T)/dx] = TJ_m[d(-\mu/T)/dx] \ . \qquad (5.71)$$

Equations (5.68) and (5.71) are special cases of a more general expression for the rate of energy dissipation ϕ per unit volume. This may be written

$$\phi = \sum_k (J_k)(X_k) \ , \qquad (5.72)$$

where J_k is the energy flux due to the gradient X_k of the particular thermodynamic potential causing J_k. For example, if both heat and moisture are flowing through the volume element Adx, that is, if both temperature and chemical potential gradients are present, the total value of ϕ is the sum of ϕ_q and ϕ_m. Using Eqs. (5.68) and (5.71), and assuming no coupling between heat and mass flow,

$$\phi = \phi_q + \phi_m = TJ_q[d(1/T)/dx] + TJ_m[d(-\mu/T)/dx] \ . \qquad (5.73)$$

The relationships of the gradients $d(1/T)/dx$ and of $d(-\mu/T)/dx$ to the conventional gradients used in Fourier's and Fick's laws are discussed below in connection with coupled heat and mass flow.

Equations (5.68) and (5.71) have been derived on the assumption that heat and mass flow are independent of each other, that is, there is no coupling between the two processes. A more general relationship for these two processes can be written as

$$J_q = L_{qq}X_q + L_{qm}X_m \qquad (5.74a)$$

$$J_m = L_{mq}X_q + L_{mm}X_m \ , \qquad (5.74b)$$

where J_q represents the heat flux (1/A)(dq/dt), and J_m is the moisture flux (1/A)(dn/dt). The coefficients L_{qq} and L_{mm} are transport coefficients related to the thermal conductivity K_q and moisture diffusion coefficient D, respectively. The terms X_q and X_m are the gradients of the thermodynamic potentials or *affinities* which cause the fluxes J_q and J_m. For example in this case,

$$X_q = Td(1/T)/dx = -(1/T)(dT/dx) \tag{5.75 a}$$

$$X_m = Td(-\mu/T)/dx \; . \tag{5.75 b}$$

The coefficients L_{qm} and L_{mq} in Eqs. (5.75 a, b) are the *interference* or *coupling* coefficients. These relate the flux J_q of heat to the thermodynamic gradient X_m, and the flux J_m of moisture to the thermodynamic gradient X_q, respectively. According to Onsager (1931), the coupling coefficients L_{qm} and L_{mq} are equal to each other. This identity is a specific case of a more general theorem known as the *Onsager reciprocity relations* which can be expressed as

$$L_{ik} = L_{ki} \; . \tag{5.76}$$

Thus, when the flux J_i of an irreversible process caused by a thermodynamic gradient X_i is influenced by a different gradient X_k through the coupling coefficient L_{ik}, the flux J_k caused by the gradient X_k is influenced by the gradient X_i through the identical coupling coefficient L_{ik}.

The total rate of energy degradation resulting from the combined thermodynamic potential gradients X_q and X_m can be obtained by combining Eq. (5.72) with Eqs. (5.74, 5.75). Thus, for combined transport of heat and moisture, Eq. (5.72) becomes

$$\phi = J_q X_q + J_m X_m \; . \tag{5.77}$$

Using Eqs. (5.74) to eliminate J_q and J_m, together with Onsager's reciprocity relation, Eq. (5.76), yields

$$\phi = L_{qq} X_q^2 + 2 L_{qm} X_q X_m + L_{mm} X_m^2 \; . \tag{5.78}$$

Thus the total rate of energy degradation can be calculated for any combination of gradients of $X_q [= Td(1/T)/dx]$ and of $X_m [= Td(-\mu/T)/dx]$ if the three phenomenological coefficients L_{qq}, L_{mm}, and L_{qm} ($= L_{mq}$) are known.

It should be noted that the energy degradation rate ϕ given by Eq. (5.78) is not simply equal to the sum of ϕ_q and ϕ_m as given by Eqs. (5.68) and (5.71), unless the coupling coefficients L_{qm} and L_{mq} are zero. It can also be shown (Prigogine 1961) that the magnitude of ϕ is a minimum when the values of the potentials X_q and X_m are such that either J_q or J_m are equal to zero.

The coefficients L_{ik} are really heat and/or mass transport coefficients which relate the heat and moisture fluxes J_q and J_m to gradients X_q and X_m. These differ from the empirical gradients which are generally used to describe heat and moisture movement through wood, although the fluxes J_q and J_m are essentially the same as those normally measured. The transport coefficients L_{qq}, L_{qm}, and L_{mm} are different from the conventional heat and moisture transport coefficients, K_q and D, but may be calculated from them. In order to do so, it is desirable to redefine the driving force for moisture transport X_m. This is because X_m, as defined in Eq. (5.75 b), contains the temperature T as well as the chemical potential μ as part of the space derivative. It is possible to remove T from the gradient using the procedure given by DeGroot (1950).

This procedure results in defining a new set of coefficients J'_q, J'_m, X'_q, and X'_m in place of J_q, J_m, X_q, and X_m as defined in Eqs. (5.68) to (5.77). The system of coupled equations given by Eqs. (5.74a, b) are replaced by the following equations in terms of the new variables,

$$J'_q = L'_{qq} X'_q + L'_{qm} X'_m \tag{5.79a}$$

$$J'_m = L'_{mq} X'_q + L'_{mm} X'_m \ , \tag{5.79b}$$

where

$$J'_q = J_q - H J_m \tag{5.80a}$$

$$J'_m = J_m \tag{5.80b}$$

$$X'_q = X_q = -(1/T)(dT/dx) \tag{5.80c}$$

$$X'_m = -(\partial\mu/\partial x)_T \ . \tag{5.80d}$$

It should be noted that the term X'_m does not contain a temperature term, in contrast to X_m of Eq. (5.75b).

Combining Eqs. (5.79) and (5.80) gives

$$J'_q = -(L'_{qq}/T)(dT/dx) - L'_{qm}(\partial\mu/\partial x)_T \tag{5.81a}$$

$$J'_m = -(L'_{mq}/T)(dT/dx) - L'_{mm}(\partial\mu/\partial x)_T \ . \tag{5.81b}$$

The relationship of the coefficient L'_{mm} to the moisture diffusion coefficient D can be obtained for the *isothermal* case by setting $dT/dx = 0$ in Eq. (5.81b), and combining it with Eq. (5.69), since $J_m = J'_m$. Thus,

$$D = L'_{mm}(\partial\mu/\partial c)_T \tag{5.82a}$$

$$D = L'_{mm}(\partial\mu/\partial h)_T(\partial h/\partial c)_T \tag{5.82b}$$

where h is the activity of the sorbed water, essentially equal to the equilibrium relative vapor pressure, and $(\partial h/\partial c)_T$ is the inverse slope of the sorption isotherm.

The term $(\partial\mu/\partial h)_T$ is equal to (RT/h), and Eq. (5.82b) can be written

$$D = L'_{mm}(RT/h)(\partial h/\partial c)_T \tag{5.83a}$$

or

$$D = L'_{mm}(RT)[\partial \ln(h)/\partial c]_T \ . \tag{5.83b}$$

Similar equations can be written to relate the moisture conductivity coefficient K_m to L'_{mm}. The analogous isothermal equation for Eq. (5.83a) for example is

$$K_m = L'_{mm}(RT/h)(\partial h/\partial m)_T \ . \tag{5.84}$$

The relationship of the thermal conductivity coefficient K_q to the coefficient L_{qq} is obtained by combining the definition of K_q [Eq. (5.66)] with that for L_{qq} as given by Eq. (5.74a) for the case where $X_m = 0$, and setting $X_q = -(1/T)(dT/dx)$ as given by Eq. (5.75a). Thus

$$L_{qq} = J_q/X_q = -J_q T/(dT/dx) \ . \tag{5.85}$$

Eliminating J_q and (dT/dx) between Eqs. (5.66) and (5.85) results in

$$L_{qq} = TK_q \ . \qquad (5.86)$$

A similar relationship exists between L'_{qq} and the thermal conductivity K'_q, defined in terms of the energy flux J'_q,

$$J'_q = -K'_q(dT/dx) = J_q - HJ_m \ . \qquad (5.87)$$

Thus

$$L'_{qq} = TK'_q \ . \qquad (5.88)$$

There are therefore two slightly different thermal conductivities, K_q and K'_q. In thermal conductivity measurements in which wood is subjected to a constant temperature gradient, the energy flux through the wood will consist of two contributions initially. These are the direct heat flux J_q, and the indirect energy flux HJ_m, associated with the moisture flux J_m. This extra energy flux HJ_m is in the same direction as J_q, so the initial apparent thermal conductivity K_q should be greater than K'_q. After moisture equilibrium is attained (Fig. 5.17), the heat flow J'_q is lower than J_q and the conductivity is somewhat reduced. This is a special case of the *Dufuor* effect, or *diffusion thermo-effect*, or *thermomechanical* effect (Jost 1960, DeGroot 1950). It is probably so small that it cannot be measured conveniently, because heat diffuses so much more rapidly through wood than does moisture.

It has already been indicated that the coupling or interference coefficients L'_{qm} or L'_{mq} are identical, based on the Onsager reciprocity relationships. They can be derived from Eqs. (5.79a, b) for special conditions. For example, for the isothermal case the temperature gradient dT/dx and the thermodynamic potential X'_q are zero, and the ratio of J'_q to J'_m becomes

$$(J'_q/J'_m)_T = L'_{qm}/L'_{mm} = L'_{mq}/L'_{mm} \ . \qquad (5.89)$$

The ratio L'_{qm}/L'_{mm} is called the *heat of transfer*, symbolized by Q*, or

$$Q^* = L'_{qm}/L'_{mm} \ . \qquad (5.90)$$

A different quantity of transfer can be obtained from Eq. (5.74) based on the ratio L_{qm}/L_{mm}, equivalent to the ratio of J_q to J_m for the isothermal case. Thus

$$(J_q/J_m)_T = L_{qm}/L_{mm} \ . \qquad (5.91)$$

This ratio is designated as the *energy of transfer* U*, by Katchalsky and Curran (1966). Thus

$$U^* = L_{qm}/L_{mm} \ . \qquad (5.92)$$

The relationship between Q* and U* can be obtained from Eq. (5.80a) by dividing both by J_m and rearranging. For the isothermal case,

$$(J'_q/J'_m)_T = (J_q/J_m)_T - H \ , \qquad (5.93)$$

since $J'_m = J_m$. Thus from Eqs. (5.89) to (5.92) it can be shown that

$$Q^* = U^* - H \ . \qquad (5.94)$$

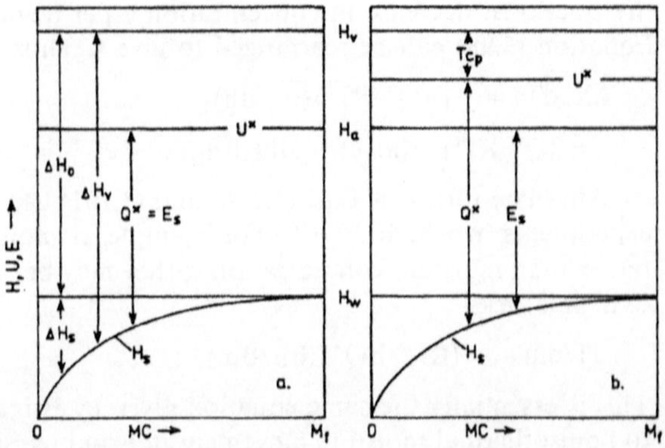

Fig. 5.19. Diagram showing the variation with wood moisture content M of enthalpies H_s, H_w, and H_v, and of ΔH_s, Q^*, and U^*

The term H in the case of the cell wall of wood is equal to the molar enthalpy of the sorbed water, given by

$$H_s = H_w - \Delta H_s \ , \tag{5.95}$$

where H_w is the molar enthalpy of liquid water, and ΔH_s is the molar differential heat of sorption of liquid water by wood at a given moisture content [(Eq. (2.38)].

Figure 5.19a shows how the terms H_s, H_w, H_v, ΔH_v, and ΔH_s vary with wood moisture content at room temperature (25 °C). Also shown are curves of U^* and Q^* based on the assumption that Q^* is equivalent to the activation energy E_s for the diffusion of sorbed water in wood, defined by Eq. (5.3). Therefore, the enthalpy H_a of the activated molecules is assumed to be equal to the energy of transfer U^* in this case.

Nelson (1986a, b, c), based on his analysis of nonisothermal moisture diffusion in wood, concluded that the heat of transfer Q^* is greater than the activation energy E_s and that U^* therefore is greater than H_a. He proposed that the interrelationships among U^*, Q^*, E_s and H_a are as shown schematically in Fig. 5.19b. The other terms retain the same relationships as shown in Fig. 5.19a. The difference $H_v - U^*$, according to Nelson, is equal to the product Tc_p, of the Kelvin temperature and molar specific heat c_p of the water vapor.

Another interesting relationship, the *Soret* effect, can be derived from Eq. (5.81 b) by setting the moisture flux J'_m equal to zero. In this case

$$dT/dx = -(T/Q^*)(\partial\mu/\partial x)_T \ . \tag{5.96}$$

Substituting for $(\partial\mu/\partial x)_T$ its equivalent in terms of moisture concentration gradient dc/dx, Eq. (5.96) becomes

$$dT/dx = -(T/Q^*)(\partial\mu/\partial h)_T(\partial h/\partial c)_T(dc/dx) \tag{5.97}$$

or

$$dT/dx = -(RT^2/hQ^*)(\partial h/\partial c)_T(dc/dx) \ . \tag{5.98}$$

The Soret coefficient s, is defined (Jost 1960) as

$$s = -(1/c)(dc/dT) \ , \tag{5.99}$$

the fractional decrease in concentration c per degree increase in temperature T. Equation (5.98) can be rearranged to give s. Thus

$$(dc/dT) = -(hQ^*/RT^2)(\partial c/\partial h)_T \tag{5.100}$$

$$s = \{(Q^*/RT^2)(\partial[\ln(c)])/\partial[\ln(h)]\}_T \ . \tag{5.101}$$

Alternate forms of Eqs. (5.100) and (5.101) can be used to evaluate Q* from experiments in which $J_m = 0$. For example, if moisture content m is measured, rather than moisture concentration c, they may be written with m substituted for c. In this case

$$dT/dm = -(RT^2/hQ^*)(\partial h/\partial m)_T \ . \tag{5.102}$$

This is essentially the same equation given by Briggs (1967). It was first applied to nonisothermal moisture movement in wood by Siau (1980). He has since used several other equations for describing moisture movement through wood under conditions where temperature gradients exist (Siau 1984). As noted earlier, Nelson (1986a, b, c) proposed that Q* is somewhat greater than the activation energy E_s and decreases in a similar manner with increasing wood moisture content (Fig. 5.19b).

Chapter 6

Electrical Properties of Wood

6.1 Introduction

It has been noted (Chap. 1.2.1.5) that two general types of electrical moisture meters have been used to estimate the moisture content of wood. These are *resistance* meters which measure essentially the direct current (DC) electrical resistance, and *dielectric* moisture meters which measure some function of the complex dielectric constant or electrical permittivity of wood, usually by means of alternating current (AC).

In this chapter the electrical properties of wood are discussed, including the effects of wood moisture content and other parameters, as well as their applications in electrical moisture meters. The electrical resistance of wood is treated first, followed by consideration of the dielectrical properties.

6.2 Electrical Resistance

6.2.1 Introduction

Measurement of the electrical resistivity and/or resistance of wood has been used industrially for many years as a technique for estimating its moisture content. The initial development of electrical resistance moisture meters was based largely on the work of Stamm (1927), who first described quantitatively the tremendous decrease in the DC electrical resistivity of wood with increasing moisture content. These moisture meters are essentially megohmmeters which measure the resistance of wood between electrodes usually consisting of one or more pairs of needle-like probes driven into wood. The megohmmeter scales normally read moisture content directly, based on an empirical calibration of resistance R against moisture content for a particular electrode configuration in a given kind of wood at a specific temperature. For example, Table 6.1 shows the calibration data for Douglas-fir, at 80 °F, of resistance R against percent moisture content M, using two pairs of needle electrodes spaced 1.25 inches apart along the grain, and driven to a depth of 5/16 inch into wood of uniform moisture content. Most

Table 6.1. Electrical resistance (Megohms) at various wood moisture contents

M (%)	8	10	12	14	16	18	20	22	24
Doug. fir (80 °F)	4780	630	120	33	11.2	4.6	2.14	1.10	0.60
Eq. (6.1) (20 °C)	4380	731	156	41	13.1	4.9	2.08	1.00	0.53

resistance moisture meters made in the United States use these calibration data to relate indicated wood moisture content to measured resistance.

Before discussing moisture meters based on measurement of the electrical resistance of wood, it is desirable first to consider some of the factors in addition to moisture content which affect this important property of wood. Vermaas (1975) has listed more than a dozen different factors which affect the measured values of the DC resistance of wood. He divides these into two categories. The first, designated here as *inherent* variables, that is properties of the wood itself, include moisture contact and distribution, temperature, grain orientation, density, species, and chemical constituents, including principal constituents as well as extractives and ash contents. The second or external group of variables Vermaas designates as *experimental* variables. These include kinds of electrodes and their configurations, contact pressure, sample shape and dimensions, electrolytic effects at electrodes, and the magnitude and time duration of applied voltage.

Some of the variables mentioned above can be listed in either of the two categories, but the classification given will be followed here. However, the effect of moisture distribution and differences among species will be discussed primarily in connection with resistance moisture meters, following consideration of the other variables. There is some discussion of variation among kinds of woods in the sections on the effects of moisture content, temperature, and other inherent variables.

Normally the resistance R of a sample is measured directly. If the sample is in the form of a rectangular solid with electrodes of face area A and spacing L, the resistivity r is equal to RA/L, assuming the contact resistance between electrode and wood is zero. The reciprocal of resistance is the conductance k and that of resistivity is the conductivity K. In the following discussion any one or more of these terms will be used, depending on convenience and context.

6.2.2 Inherent Variables Affecting Resistivity

The inherent variables as listed above will be discussed in the following order: moisture content and species, temperature, structural orientation, chemical constituents and extractives, and wood density.

6.2.2.1 Effect of Moisture Content and Species

Dry wood is an excellent electrical insulator, with a resistivity in the order of 10^{17} ohm-cm at room temperature. However, the resistivity r decreases dramatically as moisture content increases in the hygroscopic range (Fig. 6.1). The decrease is exponential with increasing moisture content to approximately 10^{11} ohm-cm at 7% moisture content. This is equivalent to a reduction in resistivity, or increase in conductivity, of five times for each percent increase in wood moisture content in this range.

From about 7% moisture content to fiber saturation, the rate of decrease is less dramatic but still decreases about four to two times for each percent increase

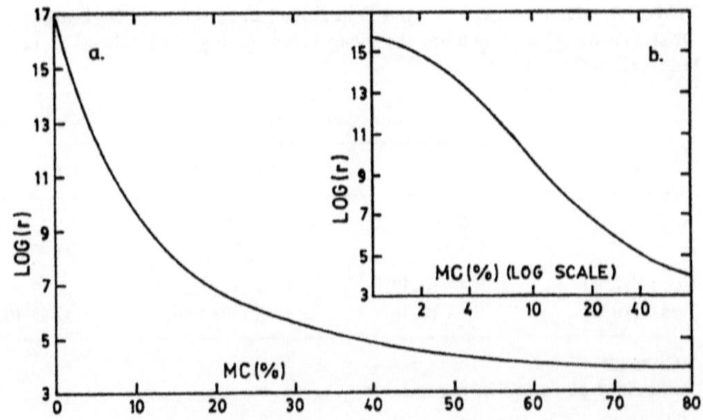

Fig. 6.1. Curves of log (r) (ohm-cm) against wood moisture content on both linear (**a**) and logarithmic (**b**) scale (after Lin 1967)

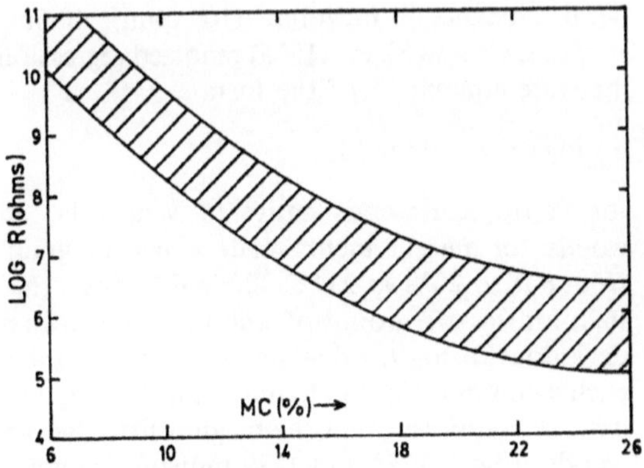

Fig. 6.2. Curves of log R against moisture content. The *shaded area* represents the limits between which curves for 90% of US woods fall. (Courtesy USDA Forest Products Laboratory)

in moisture content, the decrease becoming less pronounced with increasing moisture content. In this moisture range the logarithm of the resistance is essentially a linear function of the logarithm of moisture content (Fig. 6.1 b). This is the moisture range over which resistance moisture meters are most useful. It is well known that the resistivity moisture content relationship varies among different woods even in the useful range from 6% or 7% to fiber saturation. Figure 6.2 shows two curves of log R against moisture content M. The shaded area between these curves gives the range between which 90% of the experimental data points for North American woods lie. It is apparent from the curves that moisture meter readings may vary by as much as 2% or 3% among these woods with the same moisture content.

Keylwerth and Noack (1956) report that the calibration data used by the European resistance moisture meter (Siemens and Halske) is based on data obtained on European oak at 20 °C. The empirical equation which holds between 3% to 23% wood moisture content is given by them as

Table 6.2. Mean values of A and B in Eq. (6.2) as reported by Langwig and Skaar (1975) and by Okoh (1976). Also shown are the mean values of log r[a] calculated at 12% and 18% moisture contents (all at 21 °C)

	Langwig and Skaar (1975)		Okoh (1976)	
	US	Venezuela	US	Ghana
A	17.972	17.248	20.236	20.339
B	9.109	8.979	10.904	10.595
Log r (12%)	8.14 (0.39)[b]	7.56 (0.69)	8.47 (0.26)	8.82 (0.68)
Log r (18%)	6.54 (0.36)	5.98 (0.46)	6.55 (0.32)	6.95 (0.68)

[a] Ohm-cm.
[b] Standard deviation in parenthesis.

$$\log[\log(R)-4] = 1.009 - 0.0322\,M \quad , \tag{6.1}$$

where R is the resistance in ohms. Table 6.1 includes the calculated values of electrical resistance in megohms (10^6 ohm) obtained by use of Eq. (6.1).

Langwig and Skaar (1975) reported regression equations of resistivity r against moisture content M of the form

$$\log r = A - B \log M \tag{6.2}$$

for 18 tropical woods native to Venezuela, and also for 13 North American woods, for measurements made along the grain over the moisture content range from 8% to 20% at 21 °C. Table 6.2 shows the mean values of A and B of Eq. (6.2) for the two groups of woods. Also shown are the mean values and standard deviations of log r, calculated by applying the individual values of A and B for each wood into Eq. (6.2), and evaluating log r at 12% and 18% moisture in each case. It is apparent that the mean resistivities are higher for the North American woods at both 12% and 18% moisture content but that the standard deviations are greater for the Venezuelan woods. The differences may be related to the greater variation in the extractive and mineral contents of the Venezuelan woods.

Okoh (1976) measured the longitudinal resistivities of 22 US hardwoods and 10 hardwoods from Ghana, using a technique similar to that of Langwig and Skaar (1975). His results at 21 °C are also summarized in Table 6.2. The mean coefficients A and B in Eq. (6.2) from Okoh's data are both substantially higher, but the mean resistivities calculated at 12% and 18% moisture contents are similar for the two sets of data. Again the standard deviation for the tropical woods from Ghana was greater than for the temperate zone US hardwoods.

Figure 6.3 shows how the mean resistivities of 22 US hardwoods vary with wood moisture content at 25 °C, based on results of Okoh (1976). Also shown is a curve for the 10 Ghana woods and for the inner barks of the 22 US hardwoods. The curve for inner bark is essentially the same as that for the corresponding woods. However, the bark samples were measured radially and the wood longitudinally. The radial resistivities of the woods would be expected to be greater than the longitudinal values shown, from which it may be concluded that

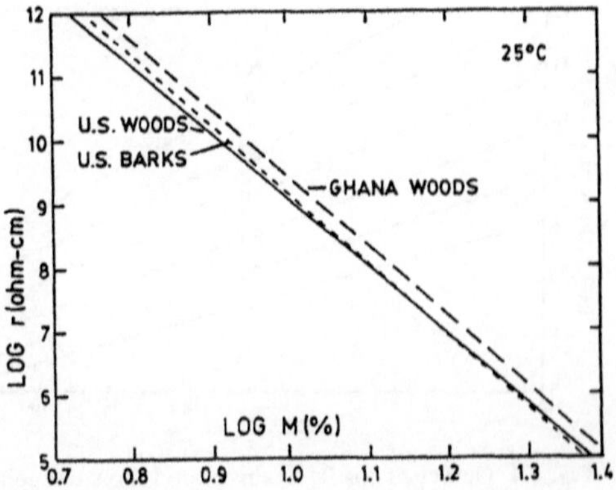

Fig. 6.3. Curves of log (r) against log (M) for mean values of US woods and barks, and for Ghana woods, at 25 °C. (Okoh 1976)

the bark resistivity measured radially is somewhat lower than the corresponding wood values measured radially.

Above fiber saturation, the decrease in resistivity with increase in moisture content is much less pronounced, and its sensitivity to moisture content change is much reduced (Fig. 6.1). Furthermore, the resistivity is much more sensitive to electrolyte content at high moisture contents because it is essentially a function of the numbers and kinds of ions in the cell-cavity water, which varies greatly among samples. In addition, moisture distribution is generally more erratic above fiber saturation so that electrical conduction paths tend to be discontinuous and circuitous.

6.2.2.2 *Effect of Temperature on Resistivity*

Increasing the temperature of wood has the same effect on its resistivity as increasing its moisture content. Therefore when estimating the moisture content M of wood by means of a resistance moisture meter the wood temperature T must be known, as well as the temperature correction factor dM/dT. This empirical factor gives the apparent increase in wood moisture content, as read on the moisture meter, per degree increase in wood temperature T (Kelvin) above the calibration temperature for the meter.

Figure 6.4 shows the logarithm of resistivity plotted against wood moisture content (log scale) from 6% to 24% moisture content, at each of several temperatures. It is possible from this diagram to calculate the mean value of dM/dT at any given moisture content and temperature range. For example, if the wood is at 10% moisture content and 40 °C, log r is approximately 8.8. This resistivity corresponds to a moisture content of approximately 12% at 20 °C. If the calibration temperature for the meter is 20 °C, the mean value of dM/dT is (12 − 10)/(40 − 20) or 0.1% per °C.

James (1975 a, b) has given a convenient diagram, shown in Figure 6.5, which can be used to adjust resistance moisture meter readings made on wood at various temperatures to the true wood moisture content based on the meter calibration temperature, in this case 70 °F. For example, if the wood temperature at time of

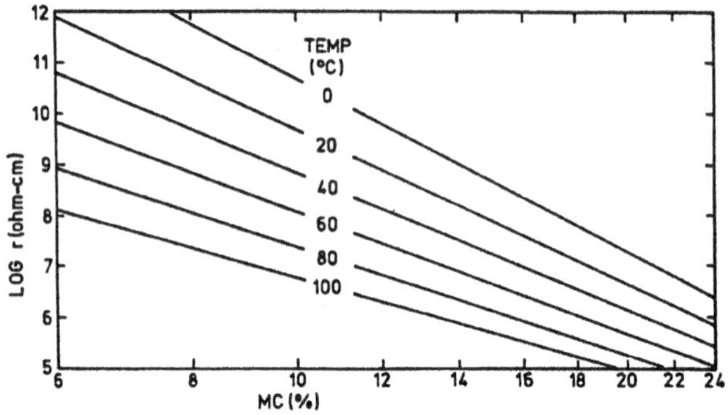

Fig. 6.4. Curves of log (r) against wood moisture content (log scale) for several temperatures

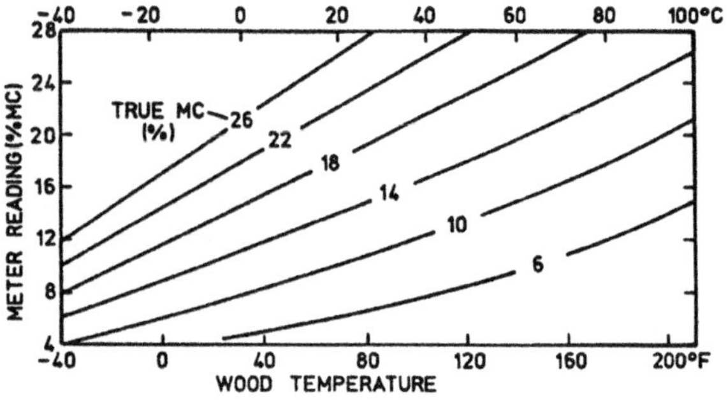

Fig. 6.5. Temperature calibration curves showing resistance meter reading against wood temperature for wood at several different moisture contents. (Courtesy USDA Forest Products Laboratory)

measurement is 120 °F, and the moisture meter reads 13%, the true wood moisture content obtained from Fig. 6.5 is 10%. The mean value of dM/dF (the apparent change in wood moisture content M) per degree Fahrenheit, therefore, is $(13-10)/(120-70) = 0.06\%$ or $(1.8)\times(0.06) \simeq 0.11\%$ per °C or in Kelvin, close to the value of 0.1 obtained by use of Fig. 6.4 over a similar temperature range.

The slopes of the true moisture content curves shown in Fig. 6.5 can be used to calculate the rate of increase (dM/dF) of moisture meter reading per unit increase in wood temperature in °F above the calibration temperature of 70 °F. For example, at 70° the magnitude of dM/dF increases from approximately 0.040 at $M = 6\%$, to 0.144%/°F at $M = 28\%$. This increase in dM/dF with increasing moisture content is approximated by

$$dM/dF \simeq 0.015 + 0.0047\,M \ . \tag{6.3}$$

The change in dM/dT with moisture content at 21 °C (70 °F) is therefore approximated by the empirical equation

$$dM/dT \simeq 0.027 + 0.0085\,M \ . \tag{6.4}$$

Before continuing the discussion of the factor dM/dT it is convenient first to consider the effect of temperature on electrical resistivity in terms of the "activation energy" E, defined below. This is an useful concept, particularly in relation to the mechanism of electrical conduction in wood as discussed in Section 6.2.4, as well as for determining the temperature adjustment factor dM/dT.

J. D. Clark and Williams (1933) first pointed out that the effect of temperature on electrical conduction in dry wood is such that the logarithm of resistivity increases linearly with the reciprocal of the Kelvin temperature. They applied the *Arrhenius* theory of rate processes to calculate the activation energy E for oven-dry wood, using the relationship

$$E = R\,[d \ln (r)]/d(1/T) \ , \qquad (6.5)$$

where E is in calories per mole if R is the gas constant in calories per mole degree and T is Kelvin temperature. They found E values of 22.9, 30.6, 28.4 and 32.0 kcal/mole for oven-dry birch, mahogany, red oak, and Sitka spruce, respectively, over the temperature range from 65° to 110°C.

Subsequent studies have shown that the activation energies calculated by use of Eq. (6.5) decrease as the wood moisture content increases. The slope $d \ln (r)/d(1/T)_m$ changes somewhat with temperature at higher moisture contents according to Fig. 6.6, taken from Lin (1965). However, over small temperature ranges, say from 0° to 50°C, the slopes are essentially linear in the hygroscopic range of moisture contents according to the results of Davidson (1958), Brown et al. (1963), Langwig and Skaar (1975), Okoh (1976), and others. Therefore it is an useful measure of the temperature effect on electrical resistance.

The activation energy E for electrical conduction decreases strongly with increasing wood moisture content. Figure 6.7, taken from Brown et al. (1963), shows how it varies with moisture content for several North American woods tested over the temperature range from 5° to 50°C, within which log K is essentially linear with 1/T. It is apparent that there are considerable variations among woods. Included among the curves is one for *Pinus radiata* (*RP*) taken from Vermaas (1983).

Langwig and Skaar (1975) also reported on the variation of activation energy E with moisture content (range from 6% to 22%), based on measurements made between 10° and 50°C, for 13 North American woods and 18 tropical woods from Venezuela. They tested several kinds of regressions relating E to wood moisture content and reported that results for 16 of the 31 woods were best satisfied by a regression of the form

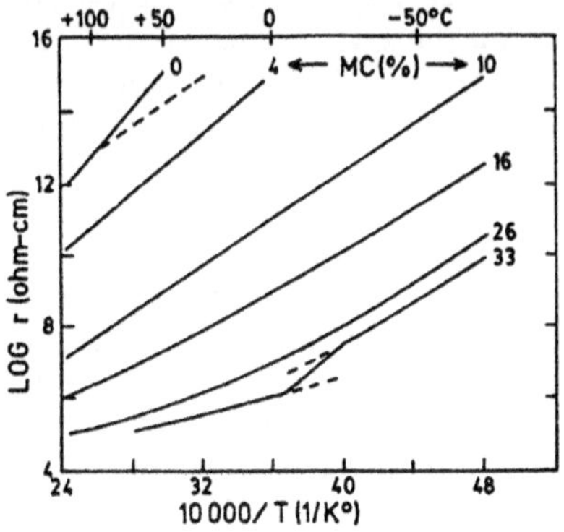

Fig. 6.6. Curves of log (r) against reciprocal of Kelvin temperature at several moisture contents for yellow-poplar. (Lin 1965)

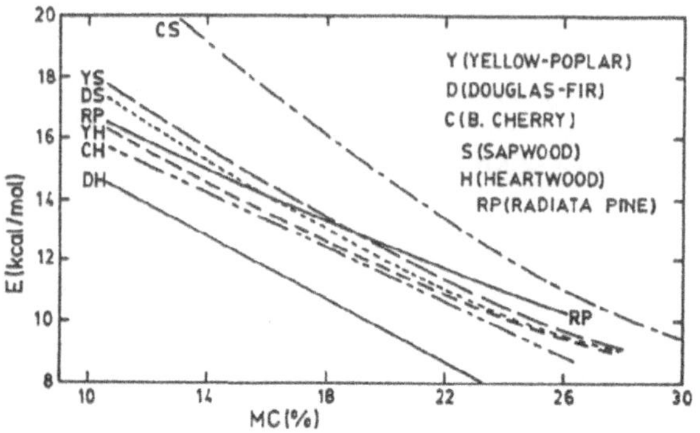

Fig. 6.7. Curves of activation energy E against moisture content for several different woods (Brown et al. 1963, Vermaas 1983)

$$E = A - BM \ . \tag{6.6}$$

For nine woods the best equation was of the form

$$\log E = A - BM \ , \tag{6.7}$$

and for the remaining six woods the best equation was of the form

$$\log E = A - B \log M \ , \tag{6.8}$$

where A and B are empirical constants, different in each of the three types of regression equations.

As was indicated above, the results obtained on more than half of the woods yielded regression equations of the form given by Eq. (6.6). The value of A in this case is the predicted value of E for oven-dry wood. These values ranged from 21.4 to 29.6 with a mean for the 16 woods of 26.2 kcal/mol, generally within the range of those reported above as given by J. D. Clark and Williams (1933), of 22.9 to 32.0. Lin (1965) reported values of E for oven-dry wood ranging from 22.1 to 26.2 kcal/mol for seven woods, six of which were North American.

The values of activation energy at zero moisture content calculated from Eqs. (6.7) and (6.8) by Langwig and Skaar (1975), were considerably higher than those obtained by use of the simple linear model of Eq. (6.6). This is because of the exponential nature of the regression equations, resulting in a strong increase in E when the equations are extrapolated to moisture contents appreciably lower than those over which actual data were obtained.

Vermaas (1983) measured the resistance of *Pinus pinaster* and *Pinus radiata* over a moisture and temperature range from 3% to 25% and 20° to 100°C, respectively, as part of a moisture meter calibration program. He used pin type electrodes in wood samples of uniform moisture content. He noted that the apparent activation energies, calculated from the Arrhenius equation [Eq. (6.5)] at constant moisture contents, were temperature-dependent since the curves of log R vs. 1/T were generally somewhat curvilinear. This effect had also been reported by James (1968) on several woods from North America, as well as by Lin (1965) at wood moisture contents higher than 10% (Fig. 6.6).

The apparent activation energies calculated by Vermaas (1983) as well as their variations with moisture content were temperature-dependent, as Fig. 6.8 in-

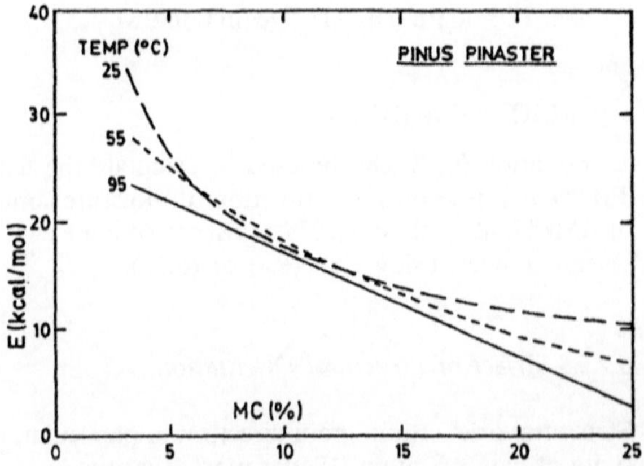

Fig. 6.8. Curves of activation energy E against wood moisture content for *Pinus pinaster* at three temperatures. (Vermaas 1983)

Table 6.3. Values of dM/dT (%/°C) at moisture contents of 12%, 18%, 24% (at or near 20 °C), from various sources[a]

MC (%)	A[a]	B	C	D	E	F	G	H	I
12	0.129	0.131	0.123	0.126	0.114	0.112	0.106	0.118	0.153
18	0.180	0.155	0.121	0.117	0.118	0.127	0.138	0.170	0.178
24	0.231	0.160	–	–	–	–	0.157	–	–

A[a] James 1965: US woods.
B Skaar 1964: Five US woods.
C Langwig and Skaar 1975: 13 US woods.
D Langwig and Skaar 1975: 18 Venezuelan woods.
E Okoh 1976: 22 US woods.
F Okoh 1976: Ten Ghana woods.
G Vermaas 1983: Two South African woods.
H James 1954: US woods.
I Keylwerth and Noack 1956: European woods.

dicates. He reported also that the form of the equation which described the dependency of E on moisture content was different for each of the three temperatures. At 25 °, 55 °, and 95 °C, for example, the relationships were best fitted by empirical curves of the forms given by Eqs. (6.8), (6.7), and (6.6), respectively.

If the conductivity K or resistivity r, and the activation energy E, are known as functions of wood moisture content, the temperature adjustment factor dM/dT for resistance moisture meters discussed earlier can be determined at any moisture content and temperature. To show this, Eq. (6.5) can be rearranged into the form

$$\{\partial[\ln(r)]/\partial T\}_m = -E/(RT^2)\ . \tag{6.9}$$

It can be shown from Eq. (6.2) that

$$\{\partial[\ln(r)]/\partial M\}_T = -B/M\ . \tag{6.10}$$

Therefore

$$dM/dT = \{\partial\,[\ln(r)]/\partial T\}_m / \{\partial\,[\ln(r)]/\partial M\}_T\ , \tag{6.11}$$

and

$$dM/dT = EM/(BRT^2)\ . \tag{6.12}$$

Equation (6.12) can be used to calculate the temperature adjustment factor dM/dT if E is known as a function of moisture content. Table 6.3 lists the values of dM/dT at 12% and 18% moisture content at 21 °C (70 °F) calculated from several sources, using Eqs. (6.4) or (6.12).

6.2.2.3 Effect of Structural Orientation

Structural orientation, grain direction in particular, affects the electrical conductivity of wood. Stamm (1960) reports that the longitudinal conductivity of eight US woods was 1.9 to 3.2 times greater than in the radial direction, and 2.1 to 3.9 times greater than in the tangential direction.

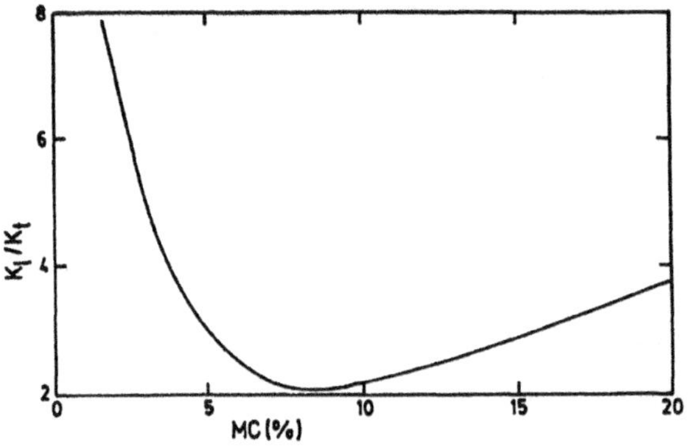

Fig. 6.9. Ratio of longitudinal (K_l) to tangential (K_t) electrical conductivity in relation to wood moisture content. (Kuroda and Tsutsumi 1982)

Kuroda and Tsutsumi (1982) noted that wood moisture content affects the degree of conductivity anisotropy. Figure 6.9, for example, shows how the ratio of longitudinal to tangential conductivity varies with wood moisture content for Sugi wood (*Cryptomeria japonica*). They attributed the minimum in the conductivity ratio near 8% moisture content to the change in the nature of moisture sorption in the cell wall above and below this moisture content.

6.2.2.4 Effect of Chemical Constituents

It has been shown that moisture content and temperature are the two most important factors affecting the electrical conductivity of wood. It has also been indicated that considerable variation exists among woods with respect to the effects of moisture content and temperature. Much of this variability may be attributed to variations in extractive and mineral constuents, as well as to density. However,

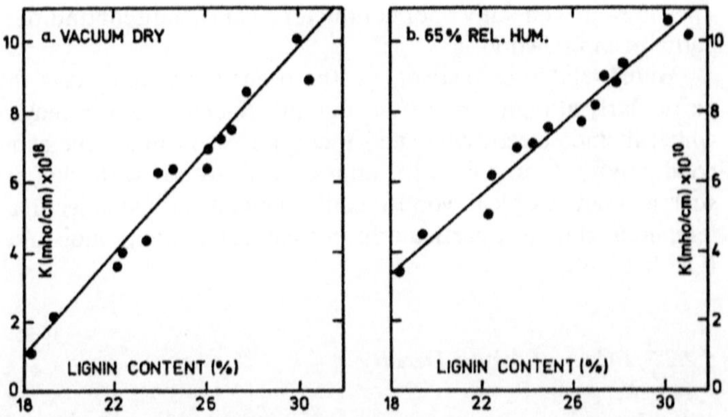

Fig. 6.10. Electrical conductivity in relation to percent lignin content for **a** vacuum-dry wood, **b** wood at equilibrium with 65% relative humidity. (Venkateswaran 1972)

one would also anticipate that variability among woods with respect to their principal constituents such as cellulose, hemicelluloses, and lignin should have some affect, as it does on moisture sorption.

Venkateswaran (1972) noted that the electrical conductivity of isolated lignin ($\simeq 10^{-13}$ mho/cm) is substantially greater than that of vacuum-dried wood ($\simeq 10^{-15}$ to 10^{-16} mho/cm). He reported furthermore that the DC conductivities of 14 North American species increased linearly with their respective lignin contents, both for vacuum-dried woods as well as for the woods conditioned to equilibrium with 65% relative humidity (Fig. 6.10).

Vermaas (1983) measured the DC resistivities and lignin contents of 20 woods of South Africa, dried over P_2O_5. He noted that those woods with higher lignin contents tended to have higher conductivities, as reported by Venkateswaran. However, the relationship was curvilinear rather than linear and there was considerably more scatter of data about the curve.

The charge carriers in wood are believed to be primarily monovalent ions such as sodium, potassium, etc. (Langwig and Meyer 1973). The mineral content varies considerably among woods, both with respect to kind and quantity (Langwig 1971). Therefore, it is anticipated that different woods should also display considerable variation in their electrical conductivities. Langwig has concluded, based on measurements of electrical conductivity and of mineral constituents (using neutron activation analysis), that electrical conductivity is directly proportional to ion content.

Treating wood with wood preservatives containing electrolytes generally increases its conductivity. If logs or timbers are stored in salt water for extended time periods, the electrolyte content may be increased significantly, and resistance moisture meter readings tend to be high. Some organic preservatives such as creosote have the opposite effect in that they decrease the wood resistivity, thus giving low estimates of wood moisture content.

Organic extractives should also affect the conductivity of wood. In general, nonwater-soluble extractives are expected to decrease the conductivity of wood since they are generally poor conductors, causing discontinuities in the conductive paths in moist wood.

Water-soluble extractives, on the other hand, may have the opposite effect, particularly at higher moisture contents. Such extractives may contain complexes which include electrolytes, and therefore act in much the same way to increase conductivity. In any event, it appears that woods with high extractive contents, such as many tropical woods, exhibit greater individuality than woods grown in temperate climates, particularly in their conductivity-moisture content behavior (Okoh 1976).

6.2.2.5 *Effect of Wood Density*

It has been generally assumed that the electrical conductivity of wood should increase with increasing wood density. This is based on the analogy with heat conduction through wood. There is an essentially linear relationship between the longitudinal thermal conductivity and wood density below fiber saturation, the cell wall having a much higher conductivity than the air in the cell lumens. For the simple parallel model (Siau 1984) in which the conductive cell wall is considered to be in parallel with the nonconductive cell cavity, a relationship of the following form is obtained for the electrical conductivity K_l along the grain

$$K_l = K_l'(\varrho/\varrho') , \tag{6.13}$$

where K_l' is the longitudinal conductivity of the cell wall, and ϱ and ϱ' are the densities of the wood and of the cell wall, respectively.

The model for transverse conductivity (Fig. 6.11) is somewhat more complex than the simple longitudinal model. This is because the cross walls, that is, those portions of the cell wall perpendicular to the direction of the electric field, are relatively ineffective in conducting electrical current compared with the parallel walls. If conduction due to fringing of the current into the cross walls is neglected, the transverse electrical conductivity K_t based on the model is approximated by

$$K_t = K_t'(1-(1-(\varrho/\varrho'))^{0.5}) . \tag{6.14}$$

Figure 6.12 shows how the ratios K_l/K_l' and K_t/K_t' vary with the ratio ϱ/ϱ', based on Eqs. (6.13) and (6.14). The actual curve for K_t/K_t' should lie somewhere

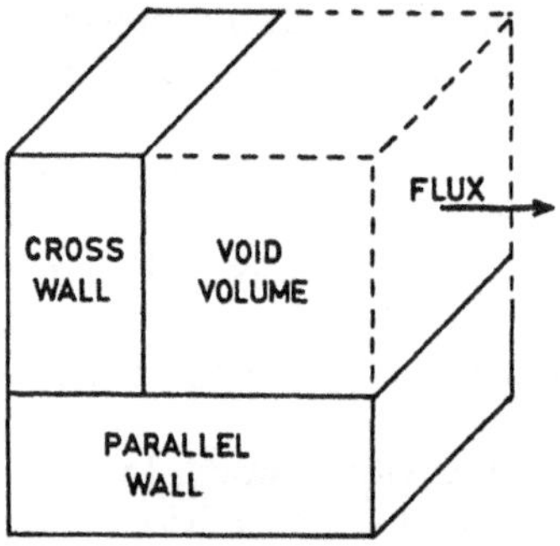

Fig. 6.11. Transverse electrical conductivity model

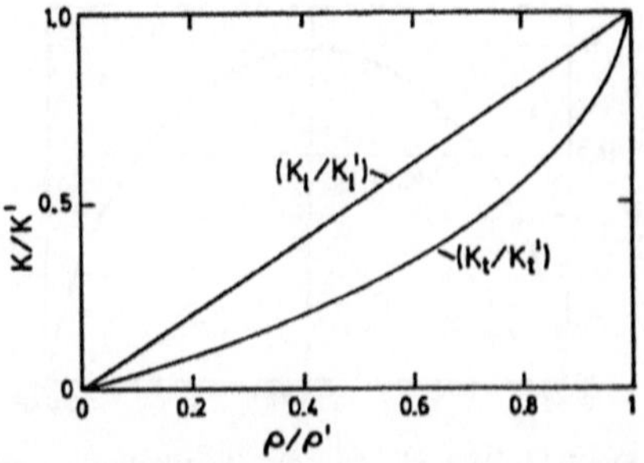

Fig. 6.12. Curves showing the expected effect of wood density on longitudinal and transverse electrical conductivities (see text)

between the two curves shown because of the fringing effect of current flow through the cross walls, as was pointed out by Hart (1964).

Based on the discussion above, it is surprising that there is no consistent relationship between the electrical conductivity of different woods and their densities. Venkateswaran (1972) attributes this to the fact that other factors, such as lignin content and other variables, are more important. This is confirmed by data obtained by Langwig (1971) and Okoh (1976) who reported on measurements of electrical resistivities of a number of woods of varying densities. Their data show no consistent density effect among woods.

Venkateswaran (1973) found that increasing the density of wood by compressing it reduced its resistivity. He proposed an empirical equation, based on his study, which can be written as

$$K = A \exp(-B/\varrho^3) , \qquad (6.15)$$

where A and B are constants and ϱ is the wood density after compression. Equation (6.15) shows a much stronger effect of density on conductivity than are indicated by Eqs. (6.13) and (6.14). He attributes this to the crushing of some wood cells by compression, thus providing additional conduction paths not accounted for in the simple model. Improvement in contact resistance with pressure may account for some of the effects he noted.

6.2.3 External Variables Affecting Resistivity

The second group of variables affecting the measured resistance R or resistivity r of wood are those not treated above as inherent variables. These external variables, according to Vermaas (1975) include: kinds and configurations of electrodes, contact pressure, sample shape and dimensions, electrolytic effects at electrodes, and the magnitude and duration of the applied voltage.

6.2.3.1 Electrode Configuration

Resistance moisture meters measure the resistance R rather than the resistivity r of wood. For an orthogonal hexahedral sample, R and r are related by L and A,

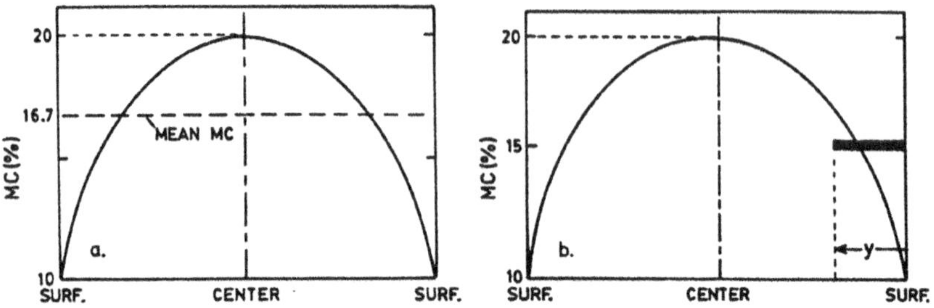

Fig. 6.13. Parabolic moisture distribution curves. **a** Showing mean moisture content. **b** Showing a pin electrode at a distance *y* from the surface

the effective spacing and cross-sectional area of two parallel electrodes, respectively. Thus

$$R = r(L/A) . \tag{6.16}$$

The simplest kind of electrodes used for resistance meters consists of a pair of flat electrodes with large area A and small spacing L. However, these are not practical for field use unless the wood between electrodes is of uniform moisture content throughout. This is rarely the case, since resistance moisture meters are often used to estimate the moisture content of wood while it is drying. Here the wood surface is generally drier than the interior, and resistance measurements on parallel electrodes on opposite surfaces of a drying board are dominated by the resistance of the surface layers. A simple numerical example will be given to illustrate this effect.

Assume that a 2 cm thick board, drying from both surfaces, has a parabolic moisture distribution throughout its thickness such that its surface and center moisture contents are 10% and 20%, respectively (Fig. 6.13a). In this case the moisture distribution is given by

$$\bar{M} = 20 - 10\,x^2 , \tag{6.17}$$

where x is the distance (cm) from the center of the board. The mean moisture content $\bar{M}$ of the board for a parabolic distribution, obtained by integrating M from x = 0 to 1, is 16.7%, thus

$$M = 20 - 10/3 = 16.7 . \tag{6.18}$$

The logarithms of resistivity r and moisture content M are linearly related over the moisture range from 10% to 20% according to the appropriate curve of Fig. 6.4 such that

$$\ln(r) = \ln(a) - b\ln(M) , \tag{6.19}$$

or,

$$r = a\{\exp[-b(\ln(M)]\} , \tag{6.20}$$

where a and b are approximately 10^{20} and 10.4, respectively, with r in ohm-cm. The electrical resistance R between the two parallel electrodes, of 1 cm^2 area and separated by the board thickness of 2 cm, is given by

$$R = 2a \int_0^1 \langle \exp\{-b\,[\ln(M)]\}\rangle dx \ . \tag{6.21}$$

The moisture content M in Eq. (6.21) can be replaced by its equivalent given by Eq. (6.17), so that x becomes the independent variable. Equation (6.21) then becomes difficult to solve analytically but its solution can be approximated by iterative methods. For example, the half-thickness (1 cm in this case) may be divided into N thin sections, each parallel to the surface electrodes and of thickness $\Delta x = (1/N)$ cm. The resistance R_i of each element in the x direction is calculated from

$$R_i = r_i \Delta x = r_i/N \ , \tag{6.22}$$

where r_i is a function of the moisture content M_i of the element i as given by Eq. (6.20).

The total resistance R is then estimated by summing up the individual values of R_i over the total thickness (2 cm). Thus

$$R/2 = \sum_{i=1}^{N} (r_i/N) = (1/N) \sum_{i=1}^{N} (r_i) \ . \tag{6.23}$$

Each value of r_i can be evaluated by use of Eq. (6.20) in the form

$$r_i = a\langle \exp\{-b\,[\ln(M_i)]\}\rangle \ , \tag{6.24}$$

where M_i is the value of M at x_i, the midpoint of section i, given by

$$x_i = (i-0.5)/N \ . \tag{6.25}$$

Thus, from Eq. (6.17)

$$M_i = 20 - 10(x_i)^2 \ . \tag{6.26}$$

Evaluation of Eq. (6.23) using N = 100 gives a value for R/2 of 2.32×10^8 ohm-cm, the apparent resistivity r of the wood with the given moisture distribution. This resistivity corresponds to that for a sample with an uniformly distributed moisture content M of 13.2%, approximately 3.5% lower than the mean moisture content of 16.7% calculated from Eq. (6.18) based on a parabolic desorption moisture distribution.

It was shown above that the use of flat plate electrodes on opposite surfaces of a drying board results in underestimating the wood moisture content because of the high electrical resistance of the dry surface layers. Most resistance moisture meters use pairs of pin electrodes driven into the wood to various depths. In this case the layer of highest moisture content through which the pin electrodes pass tends to dominate the meter reading. This is because the moisture isostere layers are parallel to the direction of current flow rather than perpendicular to it, as is the case for flat-plate electrodes.

The advantage of using electrodes which penetrate the wood is illustrated by the following hypothetical example, based on the same drying board used in the previous example. Assume that two cylindrical electrodes, each of diameter D and center-to-center spacing S, both in cm, penetrate the wood to a depth of y cm (Fig. 6.13b). It will be assumed for convenience that the wood is isotropic with respect

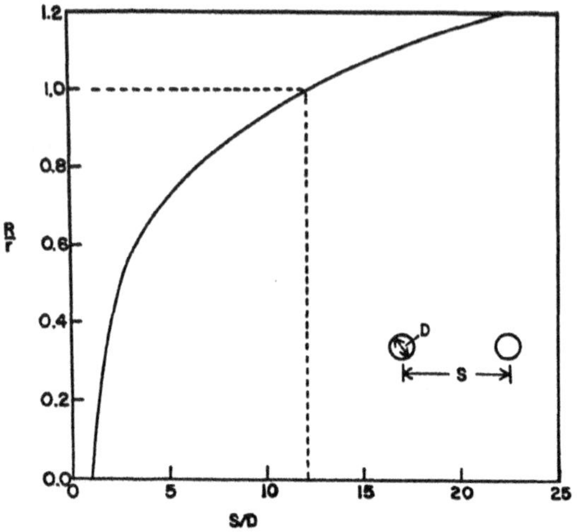

Fig. 6.14. Ratio of resistance R to resistivity r, in relation to the ratio of spacing S to diameter D, of cylindrical electrodes 1 cm long. (Skaar 1964)

to its electrical resistivity. Furthermore, end effects are neglected, that is, there is no component of current flow in the thickness direction.

Figure 6.14 shows R/r, the resistance-resistivity ratio, for two cylindrical electrodes of length y, as a function f(S/D) of the ratio S/D for a homogeneous isotropic medium. In general

$$R/r = f(S/D)/y \ . \qquad (6.27)$$

According to Fig. 6.14, the ratio R/r is essentially equal to 1/y when S/D is approximately 12. This ratio of S/D will be used here for convenience and simplicity, although it is not a necessary assumption so long as the ratio is constant.

Subject to the assumptions given, the resistance R at any penetration depth y can be evaluated for the drying board. In this case it is more convenient to calculate the conductance k, equivalent to the reciprocal of resistance,

$$k = 1/R \ , \qquad (6.28)$$

because the layers of constant moisture content are parallel to the board surface and therefore to each other in the direction of current flow. The effective conductance dk of a layer of thickness dy is

$$dk = (1/r)\, dy \ . \qquad (6.29)$$

Substituting for r its equivalent given by Eq. (6.20)

$$dk = (1/a)\langle \exp\{b\,[\ln(M)]\}\rangle\, dy \ . \qquad (6.30)$$

Equation (6.30) may be integrated analytically over the limits from y = 0 to the depth of penetration y of the electrodes to obtain the total conductance k and its reciprocal R, in which case an equation analogous to Eq. (6.21) is obtained. It may also be integrated numerically, in which case dy, y and dk are replaced by i/N, n/N and Δk, respectively, where n is the value of i at y. Thus

$$\Delta k = (1/r_i)(1/N) \qquad (6.31)$$

and

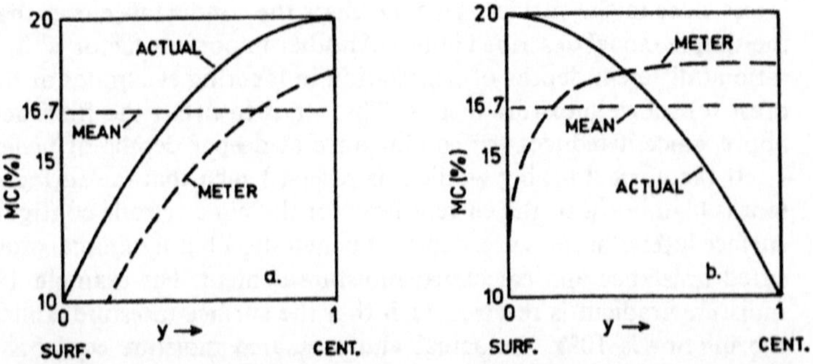

Fig. 6.15. Comparison of actual to meter-estimated moisture contents for **a** parabolic drying distribution, **b** parabolic resorption distribution (see text)

$$k = 1/R = (1/N) \sum_{i=1}^{N} (1/r_i) , \qquad (6.32)$$

where r_i is the resistivity of the layer i, whose mean distance y_i from the wood surface is given by

$$y_i = (i - 0.5)/N , \qquad (6.33)$$

and whose moisture content M_i is given by

$$M_i = 20 - 10(1 - y_i)^2 . \qquad (6.34)$$

The wood moisture content M_e measured electrically is calculated by use of Eq. (6.30), relating the conductance k from Eq. (6.31) and moisture content in the form

$$M_e = \exp [\ln (ak)/b] , \qquad (6.35)$$

since r = R in this case. Figure 6.15a shows the measured moisture content M_e as a function of depth of penetration y of the electrodes based on Eqs. (6.32) and (6.35) calculated for N = 100. Also shown is the curve of the actual moisture distribution as given by Eq. (6.34), as well as the mean value of 16.7%. According to the curve, the measured moisture content is equivalent to the mean moisture content at a depth of penetration y of 0.75 cm, or 3/4 of the distance from the surface to center of the drying board.

Actual wood moisture meters have one or two pairs of needle or pin type electrodes which are spaced about one inch apart and penetrate the wood to a depth of 5/16 inches. Early empirical studies showed that resistance readings obtained from probes penetrating 1-inch lumber to this depth are similar for lumber with normal drying gradients and a given average moisture content, to those obtained on lumber with the same average but uniform moisture content.

The 5/16-inch depth cited is 5/8 of the distance into the center of a 1-inch board, not as deep as the 3/4 distance calculated above. The most important reason for this discrepancy is probably because end-effects and transverse current flow are neglected in the simple model. There is a tendency for current to flow into the deeper portions of the wood where the conductivity is higher than in the layers close to the surface. This increases the conductance over that predicted by the simple model described above. Another important factor is the changing S/D ratio at different depths of penetration in tapering electrodes of the type used in most resistance moisture meters. This tends to offset the first factor mentioned above, since it reduces the conductance at deeper depths of penetration.

If the normal drying gradient is reversed such that the surface moisture content is high, most of the current flow for the pin electrode configuration is in the surface layers. In this case, depth of penetration has a minimal effect on the measured resistance and calculated moisture content. For example, if the parabolic moisture gradient is reversed such that the surface moisture content is 20% and the interior is 10%, the actual and measured moisture contents shown in Fig. 6.15b are obtained. In this case the meter reads the mean moisture content at a penetration depth of 0.2 of the distance to the center.

Some resistance meter manufacturers supply pin electrodes provided with electrical insulation around the pins except for the tip. The insulation minimizes the effects of high moisture contents near the surface and thus assures readings more representative of the moisture contents at the location of the uninsulated tips.

6.2.3.2 Contact Pressure

The effect of increasing contact pressure is to reduce the contact resistance between the electrodes and wood surfaces. Vermaas (1975) has summarized the findings of various workers who have used metal surface electrodes on flat wood surfaces. He notes that most workers have found a strong decrease in contact resistance with increasing pressure up to a certain value, beyond which there is little effect.

Vermaas (1975) noted that the resistance of *Pinus pinaster* samples measured in the radial direction were substantially lower when conductive rubber rather than nickel-plated brass electrodes were used. The measured resistance decreased with increasing contact pressure in both cases (Fig. 6.16), but more so for the metal than rubber electrodes.

The contact resistance of pin electrodes tends to increase with time. This is because the force between the electrode and wood relaxes with time, particularly

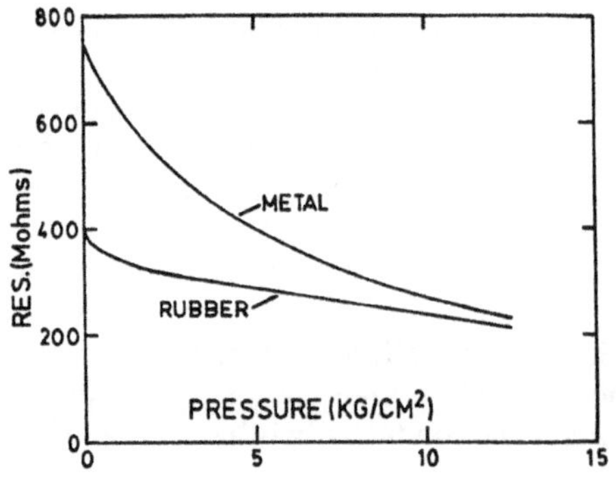

Fig. 6.16. Effect of contact pressure on measured resistance for metal and conductive rubber electrodes. (Vermaas 1975)

if drying occurs after the electrodes have been driven into the wood. The effect is most pronounced when permanent electrodes are used to monitor the progress of lumber drying in a kiln. As the wood dries, the electrode contact resistance increases, giving a higher apparent wood resistance, such that the wood appears to be drier than it actually is. Freshly inserted electrodes have a lower contact resistance, yielding a lower total resistance and higher apparent wood moisture content.

As a corollary to the effect of contact pressure or stress on the electrical resistance of wood, Vermaas (1975) found that a voltage was induced in wood samples under stress. He attributed this to a piezoelectric effect and noted that it was most pronounced in wood conditioned to higher relative humidities. For example, under a stress of about 15 kg/cm^2 the measured piezoelectric voltage was essentially 0, 40, and 90 mV for samples conditioned to 20%, 65%, and 81% relative humidities, respectively. He noted that this could have some effect on the measured resistance of wood, particularly when low voltages are used.

6.2.3.3 Sample Shape and Dimensions

The shape and dimensions of the wood sample determine the kinds of electrodes which should be used in measuring resistance. For example, thin veneer samples may be measured with flat plate electrodes or roller electrodes located on opposite faces of the veneer, because the moisture distribution is essentially uniform in the thickness direction. This would be particularly advantageous for use on face veneers where appearance is important and electrode pin holes would be unsightly. If appearance is not important, sometimes fine multiple pin electrodes may be used, in which case each electrode consists of several pins connected together electrically.

In the case of pin electrodes, particularly those with sharp points, the lines of current flux are highly concentrated in the vicinity of the electrodes. The lines of equal potential are more closely spaced near the electrodes and the current density is therefore greater, as is shown in Fig. 6.17. They are more closely spaced in Fig.

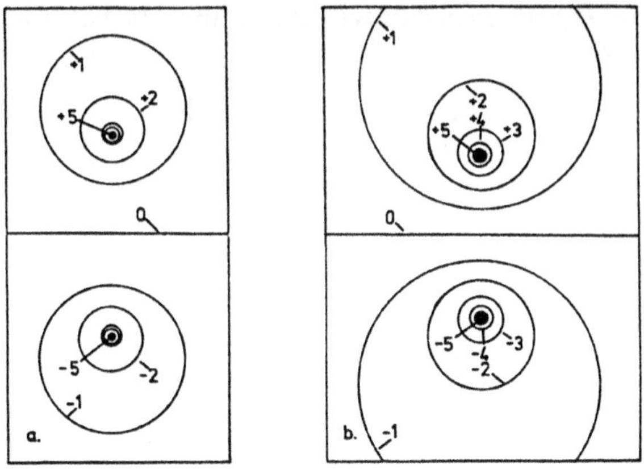

Fig. 6.17. Calculated equipotential curves for pin electrodes in an isotropic material in which the upper electrode in each case is at +5 and the lower at −5 V. The S/D ratio is 28 in **a** and 11 in **b**. (Skaar 1964)

6.17a than 6.17b because the diameter D of the electrode pins is smaller in the former case. These figures show that the current, which flows perpendicular to the circular equipotential lines, is not confined to the area between the electrodes, but spreads into areas behind the electrodes.

The resistance between any pair of adjoining equipotentials is equal to that between any other pair. Therefore the resistance is highly concentrated in the vicinity of the electrodes. The size of a sample block therefore makes little difference so long as the electrodes are completely surrounded by wood. Even sawing out a substantial portion of the wood between the electrodes or changing their spacing has minimal effect on the total resistance between the pins (Skaar 1964).

6.2.3.4 Electrolytic Effects

Direct current electricity inevitably causes electrolytic effects at junctions between dissimilar materials because of polarization. The effect occurs at the interface between electrodes and hygroscopic materials such as wood. It results in an increase in the effective "contact" resistance with time after a DC voltage is initially applied, because of the back electromotive force (EmF) caused by electrolysis.

Langwig and Meyer (1973), using neutron activation analysis, showed that potassium and sodium ions migrate toward the cathode under the influence of a DC voltage. In this case the back EmF produced at the cathode negated some of the applied voltage, thus reducing the voltage gradient through the bulk of the wood.

Duran-Pulido (1977) also used neutron activation analysis to demonstrate that copper and arsenic ions migrated toward the cathode and anode, respectively, when subjected to a DC voltage, but chromium did not migrate. He suggested that this was because chromium formed complex ions of low mobility. When copper, chromium, and arsenic were all present in a wood sample, they tended to form fixation products which resisted migration. In this case other ions, including hydrogen and mobile metallic ions such as potassium, were believed to provide charge carriers for current flow.

Welsh (1978) noted that long-term application of a DC voltage to moist wood of *Populus tremuloides* resulted in migration of electrode material from the anode toward the cathode. The phenomenon is similar to commercial electroplating in which an anode material such as silver is electrodeposited on the cathode.

Figure 6.18 shows the kind of curve obtained by Welsh (1978), of current against time. He postulated that two mechanisms were operating, a short-term mechanism causing the decrease of current, and a long-term mechanism with the opposite effect. The general model he proposed for the total process is described by

$$I = A \exp(-Ct) + Bt^D \,, \tag{6.36}$$

where I is current, A, B, C, and D are constants, and t is time.

The first term in Eq. (6.36) is related to the short time mechanism which Welsh associates with the back EmF due to polarization at the electrodes. It causes the current to decrease and the resistance to increase, both exponentially with time.

Fig. 6.18. Curves of current against time (log scale), showing the short-term and long-term components described in Eq. (6.36). (Welsh 1978)

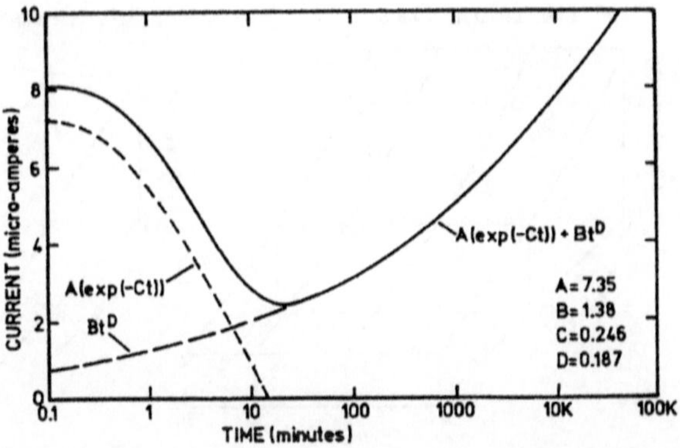

The second term, a long-time effect, he ascribes to the reduction in effective path length with increasing time as cations are deposited at the cathode, which "grows" toward the anode. In most practical applications of moisture meters, only the short time or polarization effect is important, since DC voltages are rarely applied continuously for long time periods.

6.2.3.5 Voltage Magnitude and Duration of Application

It has already been shown that the DC electrical conductivity of wood changes with time (Fig. 6.18) due to polarization at the electrodes. This causes a back EmF, or voltage, which tends to negate part of the voltage applied at the electrodes. One would anticipate therefore that the apparent resistance of a given wood sample should decrease with increasing voltage, for a given back EmF. This nonohmic behavior, termed the Evershed effect, was first observed in cotton by Evershed in 1913, according to Morton and Hearle (1975).

Kuroda and Tsutsumi (1979) investigated the effects of magnitude and duration of application of DC voltage on the longitudinal resistivity at 20 °C of *Cryptomeria japonica* wood, conditioned to three different moisture contents, 2.8%, 10.5%, and 18.9%. Their results, shown graphically in Fig. 6.19, indicate that the Evershed or voltage effect was not noticeable at the low moisture content of 2.8%. However, it became increasingly evident as moisture content increased, particularly in the voltage range below 10 to 30 V, and at longer voltage application times. They interpreted their results to indicate that polarization was most important at low moisture contents and ionic conduction at the higher moisture contents. They applied voltages from 0.1 to 50 V for times up to 100 s on 0.38-cm-thick samples using guarded stainless steel electrodes.

A number of studies have demonstrated the occurrence of the Evershed effect in wood. It is most pronounced in moist wood and appears to increase with duration of voltage application. Vermaas (1975) reported that the radial resistance of 0.5-cm-thick *Pinus pinaster* wood generally decreased with increasing voltage gradient over the range he used (up to 3500 V/cm). However, samples conditioned to lower moisture contents (below approximately 7% or 8%) exhibited some

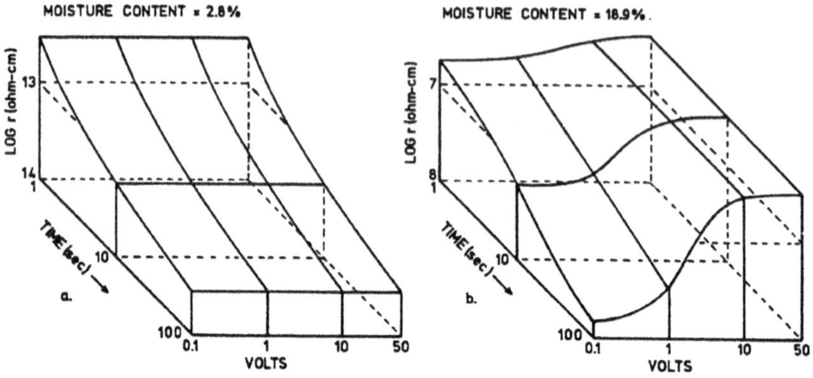

Fig. 6.19. Curves showing the effects of voltage magnitude and duration of application on the logarithm of resistivity at two moisture contents **a** 2.8%, **b** 18.9%. (Kuroda and Tsutsumi 1979)

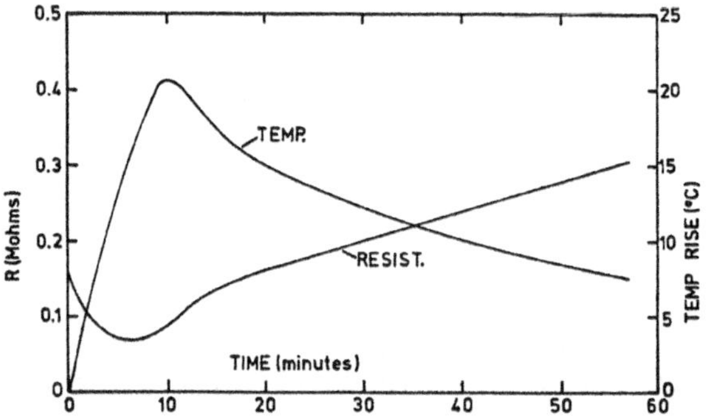

Fig. 6.20. Curves of temperature rise and resistance R against time, for constant applied voltage. (Skaar 1964)

anomalies at lower voltage gradients, where the apparent resistance showed a reversal to this trend before resuming the general trend. He noted that the apparent resistance, using the higher voltage gradients, decreased to about one-third of the values using the low gradients. He also observed the strong decrease in apparent resistance with increasing time of voltage application.

Some ohmic heating always occurs in any material when it conducts current. However, the effect is generally so small as to be negligible in wood using direct current because the current is generally too small to cause measurable effects. However, under certain conditions where high voltages are applied to wood of high moisture content, appreciable heating may occur. For example, Fig. 6.20, taken from Skaar (1964), shows the combined effect of ohmic heating and time-dependent polarization effects on the electrical resistance of wood. One curve shows wood temperature and the other electrical resistance, both as functions of time. A DC voltage of 300 V was applied continuously, across the grain, to a 0.4-cm-thick red oak sample of 2 cm^2 cross-section, using silver paint electrodes. The sample was sealed after having been previously conditioned to 20% moisture content.

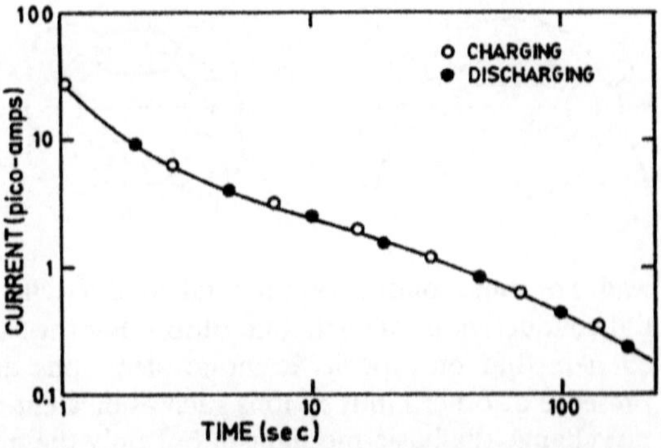

Fig. 6.21. Relationship of charging and discharging current (log scale) to time (log scale) for a wood sample at 4% moisture content. Applied voltage was 3 V for a total charging time of 1000 s. (Kuroda and Tsutsumi 1979)

The initial decrease in resistance was due to ohmic heating of the wood, as is indicated by the sharp initial temperature rise, sensed by a copper-constant thermocouple located near the surface. The polarization effect, causing an increase in resistance, predominated after about 5 min. The maximum temperature was attained at about 10 min, after which it decreased because of the reduction in ohmic heating associated with increasing resistance.

A back electromotive force or residual voltage usually results from continuous application of a DC voltage to wood. This residual voltage decays with time after removal of the applied voltage. At low wood moisture contents, the wood acts like a capacitor in series with a resistance. When a voltage is applied, the charging current decreases with time. When the voltage is removed, the discharge current as a function of time is essentially equal to the charging current, as would occur in a simple resistance-capacitance circuit. Figure 6.21, taken from Kuroda and Tsutsumi (1979), shows this reversible effect for wood conditioned to 4% moisture content and charged with 3 V for 1000 s, prior to discharge. Stainless steel electrodes were used with voltage applied along the grain.

The relationship was more complex at higher moisture contents, where the effect was not perfectly reversible. Kuroda and Tsutsumi (1979) attributed this to the dominance of ionic conduction at higher moisture contents.

6.2.4 Theory of Electrical Conduction in Wood

Several theories have been proposed to explain the strong moisture dependency of electrical resistivity in hygroscopic materials such as wood. Hearle (1953) has reviewed most of these theories for explaining this behavior in natural textile materials such as cotton and wool, which behave similarly to wood in this respect. He also proposed a simple theory in which the increase in conductivity K is attributed primarily to the increase in dielectric constant ε with increasing moisture content. This simple theory, which has been applied to wood by Brown et al. (1963) and Vermaas (1974), is treated first, followed by consideration of modifications suggested by Lin (1965) and Venkateswaran (1971).

Hearle's (1953) model for explaining the tremendous change in electrical resistance of hygroscopic textiles assumes that conduction is by ions in the cell

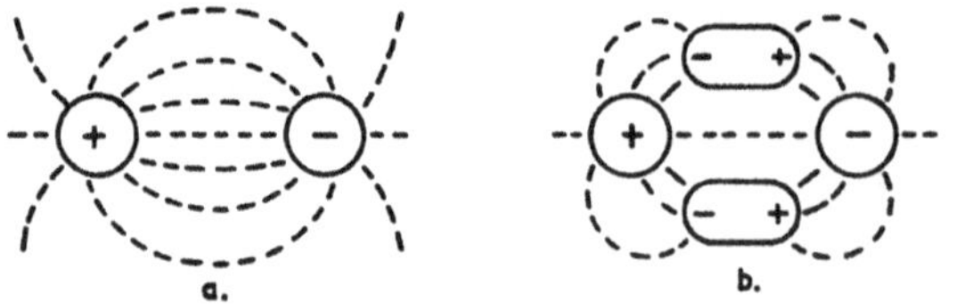

Fig. 6.22. Lines of force (*broken lines*) between two oppositely charged particles. **a** Water molecules absent. **b** Water molecules (*oblong shapes*) present and weakening the lines of force

wall. For ionic conduction the total conductivity K is equal to the sum of the partial conductivities of each kind of ion. For the sake of simplicity it will be assumed here that only positive, monovalent ions are involved in conduction. The presence of other kinds of ions such as bivalent or trivalent or negative ions does not change the basic mechanism but only the magnitude of the effect. The conductivity K is then given by

$$K = N \cdot d \cdot u \cdot e = n \cdot u \cdot e \ , \tag{6.37}$$

where K is in Siemens, N is the total number of potential charge carriers (associated plus dissociated ions) per cm^3 of cell wall, d is the degree of dissociation of these potential charge carriers (the product of $N \cdot d$ therefore is equal to n the number of available charge carriers or dissociated ions per cm^3), u is the mobility (cm^2/s-V) of the available charge carriers, and e is the charge (coulombs per ion) on the carriers.

The electrolyte concentration N remains essentially constant with moisture or temperature changes in the wood unless hydrogen or hydroxyl ions are directly involved as charge carriers. Therefore increases in the degree of dissociation d or of the mobility u, or both, cause the great increase of conductivity with moisture content. Hearle (1953) (see also Morton and Hearle 1962, 1975) considered that of these two factors, the degree of dissociation is most important over most of the hygroscopic moisture range for textile materials. The basis for his dissociation theory follows.

Most inorganic salts dissociate almost completely in aqueous solutions, presumably because the highly polar water molecule weakens the strong interionic bond between the atoms in the molecule, and such solutions are highly conductive (Fig. 6.22). In nonaqueous solutions dissociation does not occur because the solvent molecules are nonpolar. Such solutions are relatively poor conductors of electricity.

In dry wood, the extent of dissociation of potentially conductive ions or charge carriers is small, according to the Hearle theory and therefore few conductive ions are present. The degree of dissociation d can be calculated from the law of mass action, or

$$d^2/(1-d) = (A) \exp(-I/RT) \ , \tag{6.38}$$

where A is a constant, I is the ionization energy per mole of ions, R the gas constant and T the Kelvin temperature.

The ionization energy I is equal to I_0/ε, where I_0 is the ionization energy in a vacuum and ε is the dielectric constant of the medium, in this case the cell wall of wood. According to Hearle when ε is less than about 20, $d^2/(1-d) \simeq d$, and Eq. (6.38) can be written, recalling that $I = I_0/\varepsilon$,

$$d \simeq (a) \exp(-I_0/2\varepsilon RT) \ , \tag{6.39}$$

where $a = A^{0.5}$. Combining Eqs. (6.37) and (6.39) results in

$$K = (N \cdot u \cdot e \cdot a) \exp(-I_0/2\varepsilon RT) \ . \tag{6.40}$$

Taking the logarithm to base ten of both sides of Eq. (6.40), letting $\log(N \cdot u \cdot e \cdot a) = B$, assumed to be a constant, and rearranging, gives,

$$\log K = B - 0.434\ (I_0/2\varepsilon RT) \ , \tag{6.41}$$

where 0.434 is the ratio of decimal to natural logarithms.

Equation (6.41) predicts a linear relationship between $\log K$ and $1/\varepsilon$ at constant temperature T, assuming I_0 is constant. Hearle (1957) noted that, for a variety of textile materials, there was a linear relationship between the logarithm of the DC electrical resistance ($\log R$) and $1/\varepsilon$, where ε is the dielectric constant measured at high audio frequencies or at radio frequencies. Similar relationships have been shown for leather and for silica gel (J. H. Anderson and Parks 1968). Since resistivity r and conductivity K are reciprocal relationships, $\log K = -\log r$. Therefore curves of $\log r$ or of $\log R$ should be linear functions of $1/\varepsilon$, as predicted by Eq. (6.41) but with reversed signs.

Brown et al. (1963) applied the Hearle theory to wood. Equation (6.41) can be written, for the isothermal case, as

$$\log K = B - C/\varepsilon \ , \tag{6.42}$$

where $C = 0.434\ I_0/2RT$, and is therefore constant at constant T. The dielectric constant ε increases with both wood moisture content and density, and decreases with increasing frequency (Sect. 6.3). Since electrical conduction is confined to the cell wall, the dielectric constant for the cell wall should be used in Eq. (6.42). The dielectric constant of wood increases essentially linearly with wood density (Sect. 6.3.3). The mean extrapolated value of ε for the cell wall for frequencies of 2 and 15 MHz, according to Brown et al. (1963), can be expressed by the empirical equation

$$\log \varepsilon = 0.595 + 0.0242\ M \ . \tag{6.43}$$

Eliminating ε between Eqs. (6.42) and (6.43) and rearranging, gives

$$\log K = B - 0.254C\ [\exp(-0.0557\ M)] \ . \tag{6.44}$$

The constants B and C can be evaluated from experimental data in which K or r is related to wood moisture content M. Figure 6.23, taken from Brown et al. (1963) and Vermaas (1974), shows two such curves. One is plotted from the data of Davidson (1958) for Douglas-fir at 21 °C, represented by the empirical equation

$$\log K = 10.24 \log M - 19.81 \ . \tag{6.45}$$

The other is based on measurements of Vermaas on *Pinus pinaster*, from which he obtained the empirical equation, at 20 °C,

$$\log K = 10.05 \log M - 20.10 \ . \tag{6.46}$$

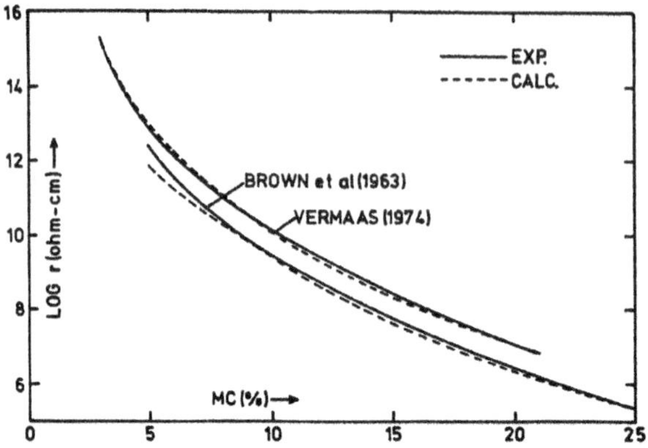

Fig. 6.23. Comparison of experimental (*solid lines*) and calculated (*broken lines*) curves of log resistivity against wood moisture content

Evaluating Eq. (6.45) at moisture contents of 10% and 20% gives values for log K of −9.57 and −6.49 for Davidson's curves. The values of constants B and C in Eq. (6.44) therefore are −2.34 and 49.7, respectively. Substituting these values into Eq. (6.44) gives the predictive equations for log K against M as

$$\log K = -2.34 - 12.6\,[\exp(-0.0557\,M)] \ . \tag{6.47}$$

This is plotted in Fig. 6.23, together with the experimental curve of Davidson. The agreement between experimental and predicted curves for wood at moisture contents between 6 or 7% and 25% is fairly good.

It was noted above that Vermaas (1974) measured the DC conductivity of *Pinus pinaster* at 20 °C, from which he obtained Eq. (6.46). He also measured the dielectric constant ε of the same wood at radio frequencies of 10, 30, and 50 MHz at 20 °C. He calculated the dielectric constant of the cell wall as a function of wood moisture content at each of three frequencies, assuming a linear dependence of dielectric constant on wood density in each case. In a later paper, Vermaas (1976) noted that it would have been more correct to have assumed an exponential dependence of dielectric constant on wood density, as Uyemura (1960) found. Based on his assumed linear relation between ε and wood density, Vermaas (1974) fitted a parabolic curve for ε against M, rather than the exponential form shown in Eq. (6.43), for each of the three frequencies.

Vermaas calculated the empirical constants B and C in Eq. (6.44) in the same manner as Brown et al. (1963) at 10% and 20% moisture contents, using Eq. (6.46) and the calculated values of ε at these moisture contents, for each frequency. He found the best agreement between measured and calculated values of conductivity over the moisture range from 6% to 20% when values of ε measured at 10 MHz were used. His curves, also plotted in Fig. 6.23, showed better agreement than those of Brown et al. (1963), particularly at moisture contents below 10%.

Hearle (1953) has calculated that I_0, the ionization energy in a vacuum as calculated from the constant C of Eq. (6.42), that is

$$I_0 = 2.303\,(2\,RT\,C) \tag{6.48}$$

is in the right order of magnitude based on calculations made from data on textile materials. Brown et al. (1963) have calculated that I_0 is 135 kcal/mol using the value for C of 49.7 obtained at the temperature of Davidson's measurements (294 K). Vermaas (1974) calculated I_0 to be 178 kcal/mol from his value for C of 66.3 at 293 K. These values of I_0 are in the order of magnitude of the dissociation energies of alkali halides, which range from about 160 to 180 kcal/mol. Since $I = I_0/\varepsilon$, from Eq. (6.48),

$$I = 4.6(RTC/\varepsilon) \ . \tag{6.49}$$

R. E. Barker and Thomas (1964) applied the Hearle model to electrical conductivity data obtained with cellulose acetate films "doped" with alkali-halides. They found that the effective value of I_0 was approximately 140 kcal/mol, close to the values noted above. They suggested that the dielectric constant measured macroscopically, as given by Eq. (6.43) for example, was proportional to, and of the same order of magnitude as, the microscopic values within the cell wall in the vicinity of the charge carriers.

Brown et al. (1963) extended the Hearle hypothesis to include the effect of temperature on electrical conductivity. They noted that Eq. (6.41) predicts a linear relationship between log K and 1/T, if B, I_0 and ε are constant with respect to temperature. For example, if B is constant, the rate of change of log K with respect to 1/T is given by

$$d(\log K)/d(1/T) = -0.434(I_0/2\varepsilon R)[1 + (T/\varepsilon)(d\varepsilon/dT)] \ . \tag{6.50}$$

It has been shown (Fig. 6.6) that an approximately linear relationship exists between the logarithm of conductivity and the reciprocal of Kelvin temperature at constant moisture content. Based on this relationship, the Arrhenius equation has been applied to calculate the apparent activation energy E at constant wood moisture content. Equation (6.5) is repeated here in terms of conductivity K, as

$$E = -R[d(\ln K)/d(1/T)] = -2.303\,R[d(\log K)/d(1/T)] \ . \tag{6.51}$$

Comparison of Eqs. (6.50) and (6.51) shows that E and I are related by

$$E = I(D+1)/2 \ , \tag{6.52}$$

where $D = (T/\varepsilon)(\partial\varepsilon/\partial T)_m$. It is apparent that E and I are identical if D = 1, that is if the logarithm of ε increases linearly with increase in the logarithm of T at constant moisture content.

It has been noted (Fig. 6.7) that the apparent activation energy E decreases markedly with increasing wood moisture content. Table 6.4 lists the mean values and standard deviations based on reported results of Brown et al. (1963) for six North American woods, three hardwoods and three softwoods at moisture contents of 12%, 18%, and 20%. They also calculated the ionization energy I for each wood at the same three moisture contents, using Eq. (6.44) to find the constant C, and Eqs. (6.48) and (6.49) to calculate I_0 and I for each wood at each moisture content. The mean values of E, I_0 and I as so calculated are listed in Table 6.4, together with the values of D+1 obtained from E and I by use of Eq. (6.52). The variation of E and I with percent wood moisture content M, using the data in Table 6.4 are

Table 6.4. Summary of results given by Brown et al. (1963) for activation energy E, ionization energies I_0 and I, and the value (D + 1) for six woods at each of three moisture contents at 21 °C

M.C. (%)	E (kcal/mol)	I_0 (kcal/mol)	I (kcal/mol)	(D + 1)
12	16.91 (1.75)[a]	115 (10)	14.96 (1.32)	2.46
18	13.11 (1.20)	115 (10)	10.68 (0.95)	2.46
24	10.29 (1.15)	115 (10)	7.66 (0.68)	2.69

[a] Standard deviation shown in parenthesis.

$$\log E = 4.443 - 0.0180\,M \tag{6.53}$$

$$\log I = 4.465 - 0.0242\,M\ . \tag{6.54}$$

One would expect that the activation energy E should be equal to or somewhat lower than the ionization energy I because it should require the same or less energy to "activate" a molecule from a sorption site than to completely ionize it. However, the values for I given in Table 6.4 are higher than those for E at the moisture contents shown. It seems probable that the values of E are more valid than those for I, since E values were calculated directly from measurements of electrical conductivity K as a function of temperature based on Eq. (6.5), while those of I were based on values of ε calculated for the cell wall at an arbitrary frequency. Furthermore, even if the correct value of ε for the cell wall is used, it is the local dielectric constant in the vicinity of the conducting ions which determines the ionization energy I. It is possible however, as a matter of interest, to calculate how ε should vary with wood moisture content M in order for I to be identical with E. In making this calculation it will be assumed for convenience that D, which gives the temperature dependence of ε, is equal to unity, that is that $\partial \ln(\varepsilon)/\partial \ln(T) =$ unity. With this assumption E = I in Eq. (6.52).

Setting $I = I_0/\varepsilon$ equal to E as given by Eq. (6.53) and rearranging to solve for ε, gives

$$\log \varepsilon = \log I_0 - 4.443 + 0.0180\,M\ . \tag{6.55}$$

The value of I_0 depends on the values of ε originally chosen at moisture contents of 10% and 20% by use of Eq. (6.42). Therefore it will obviously be different than the calculated value of 115 kcal/mol as listed in Table 6.4. However, the rate of change of ε with respect to M as given by Eq. (6.55) is not affected by I_0, and for convenience I_0 will be taken as 115 kcal/mol as listed in Table 6.4. Using this value in Eq. (6.55) it can be shown that

$$\log \varepsilon = 0.618 + 0.0180\,M \tag{6.56}$$

$$\varepsilon = 4.146(10^{0.0180\,M})\ . \tag{6.57}$$

Comparing Eq. (6.56) with Eq. (6.43), the empirical equation used by Brown et al. (1963) to test the Hearle theory, it is clear that the differences in calculated values of ε are not very large. For example, at moisture contents of 12%, 18%, and 24% Eq. (6.43) gives values for ε of 7.68, 10.73, and 14.99, respectively, where

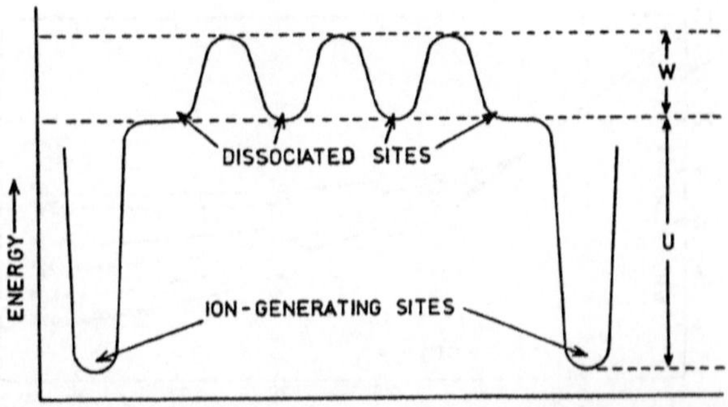

Fig. 6.24. Schematic representation of the relative potential energy levels of ion-generating sites and of dissociated sites as postulated by Lin (1965)

as Eq. (6.56) gives corresponding values of 6.81, 8.73, and 11.18. The same value of ε is obtained from Eqs. (6.43) and (6.56) at a wood moisture content near 4%.

Lin (1965) further extended the Hearle (1957) model by accounting for the effect of wood moisture content and temperature on the mobility u of the charge carriers, as well as the degree of dissociation d. He compared the mechanism of electrical conduction in wood to that in ionic crystals containing impurities. He proposed that the charge carriers in wood, assumed to be ions, may be located in two kinds of sites, each of a different energy level, as shown schematically in Fig. 6.24.

The sites of lowest potential energy he designates as *ion-generating sites.* Here ions are held by strong forces of electrostatic attraction, to ions or to groups in the basic wood structure, of opposite polarity. The second kind of sites are *dissociated sites* of considerably higher energy levels. The dissociated ions on these sites are relatively mobile compared with those bound to the low-energy ion-generating sites.

The difference in energy levels of the dissociated and ion-generating sites is the amount of energy U which must be supplied to an ion for it to move from the ion-generating site to the dissociated site. It is taken to be equal to the energy of dissociation I of Eq. (6.38).

In order for an ion to pass from one dissociated site to another, and thus carry charge, it must overcome the relatively small but significant potential barrier W between dissociated sites (Fig. 6.24). This additional energy, W, Lin calls the *energy of mobility.*

In Lin's model the degree of dissociation d is essentially the same as given by Eq. (6.39), repeated here with U in place of I. Thus

$$d = (a)\exp(-U/2RT) \quad . \tag{6.58}$$

However, the mobility u, taken to be constant in the Hearle model [Eq. (6.40)] is now assumed to be dependent on the energy barrier W between sites, and the temperature T. Thus

$$u = (b)\exp(-W/RT) \quad , \tag{6.59}$$

where b is proportional to the ratio of the distance between dissociation sites to the electric field strength applied, that is, it is inversely proportional to the voltage between sites.

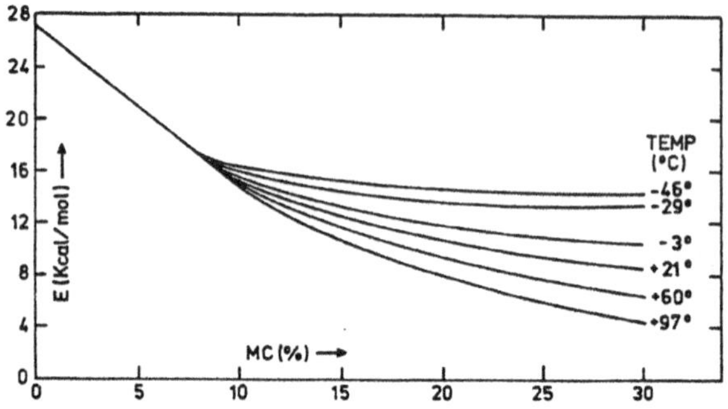

Fig. 6.25. Curves of apparent activation energy E against wood moisture content at several temperatures for Engelmann spruce. (Lin 1965)

Combining Eqs. (6.37), (6.58) and (6.59) gives

$$K = K_0 \exp[-(W+U/2)/RT] \ , \tag{6.60}$$

where $K_0 = N \cdot a \cdot b \cdot e$, taken to be constant.

From Eq. (6.60) it can be shown that

$$(W-U/2) = -R[d \ln(K)]/d(1/T) \ , \tag{6.61}$$

and therefore, based on the Arrhenius equation, Eq. (6.5), the apparent activation energy E is related to W and U by

$$E = (W+U/2) \ . \tag{6.62}$$

Lin (1965) measured the electrical resistivities of several hardwoods and softwoods as functions of moisture content from the dry condition to the vicinity of fiber saturation, over the temperature range from −50° to +150 °C. Figure 6.6 shows a typical family of curves given by Lin, each at a constant moisture content, of the logarithm of resistivity against the reciprocal of Kelvin temperature, in this case for Engelmann spruce. As was indicated in Section 6.2.2.2, the curves are essentially linear at the lower wood moisture contents but become less so as moisture increases.

The calculated activation energies decrease with increasing temperature above about 10% moisture content because of the curvilinearity in the relationship of log r and 1/T as shown in Fig. 6.6. Figure 6.25 shows how the apparent activation energies E for Engelmann spruce, as calculated by Lin using Eq. (6.5), decrease with increasing moisture content and increasing temperature (at moisture contents above 10%). Lin speculated that the latter effect is due primarily to reduction in the fiber-saturation point of wood with increasing temperature (Chap. 1.6) above the freezing point of water. He suggested that migration of water out of the cell wall into the cavity with rising temperature increases the resistance of the wood, thus offsetting some of the anticipated decrease in resistance with increasing temperature. This viewpoint is supported somewhat in that the curves of log r against 1/T are essentially straight lines below 0 °C, but curvilinear above this temperature (Fig. 6.6).

Lin (1965) used Eq. (6.49) to calculate U, which he equated to the ionization energy I at 21 °C, for each moisture content. He then calculated the value of W

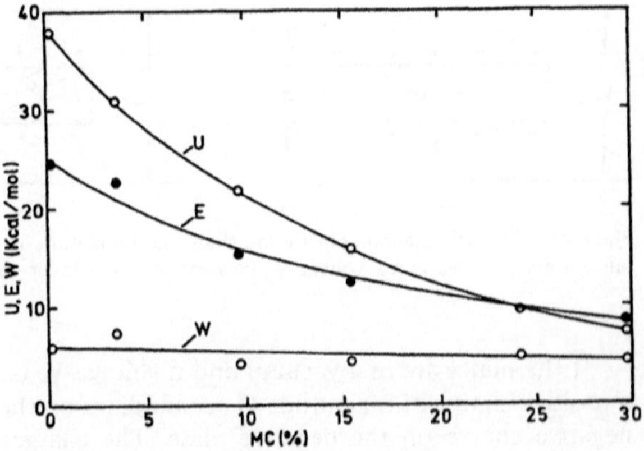

Fig. 6.26. Plotted points and calculated curves of dissociation energy *U*, activation energy *E*, and energy of mobility *W* for Engelmann spruce, taken from data of Lin (1965)

by use of Eq. (6.62), together with the value of E calculated from Eq. (6.51) for the same wood moisture content and temperature.

Figure 6.26 shows curves of E, U, and W, calculated as described above, for Engelmann spruce as functions of wood moisture content. The curves show that the apparent activation energy E and the ionization energy U both decrease strongly with increasing moisture content. The energy W, inversely related to the mobility [Eq. (6.59)], also appears to decrease somewhat, although not as consistently as do E and U. This slight decrease is therefore interpreted to mean that the mobility u increases slightly with increasing wood moisture content. Thus, according to Lin, both the mobility and degree of dissociation increase with increasing wood moisture content, although the latter increases more dramatically, in agreement with Hearle's theory.

6.3 Dielectric Properties

The dielectrical properties of wood are greatly affected by its moisture content, and are therefore important indicators of this parameter. They are normally measured using alternating current (AC) techniques. The dielectrical properties are also affected by other parameters such as wood density, temperature, grain orientation, and AC frequency. They also determine how much electrical energy can be absorbed by wood during high-frequency drying or gluing. The discussion which follows focuses on some of the more important factors which affect the dielectric properties of wood, particularly with respect to alternating currents.

6.3.1 Dielectric Principles

The *static* dielectric constant of an insulating material such as wood can be defined by considering Fig. 6.27a. This shows two thin parallel plates, each of area A and separated by a distance d, small compared with A.

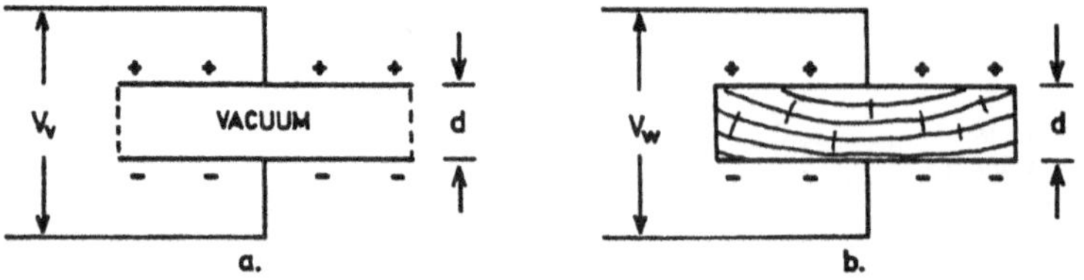

Fig. 6.27. Schematic diagram illustrating charge accumulation on two plates each of area A and distance d apart showing: **a** a voltage V_v in a vacuum, **b** a lower voltage V_w with a wood dielectric

If the plates are in a vacuum and a voltage V_v is impressed across the plates, a positive charge of magnitude Q accumulates on the positive plate and an equal negative charge on the negative plate. The charge density σ on each plate is therefore

$$\sigma = Q/A \ . \tag{6.63}$$

The electric field strength E_v between the plates is given by

$$E_v = V_v/d = D \ , \tag{6.64}$$

where D is called the electric displacement or flux density. It is proportional to σ and may be expressed as

$$D = K\sigma \ , \tag{6.65}$$

where K is a constant whose value depends on the units of D and σ.

The capacitance C_v of the plates with vacuum dielectric is given by

$$C_v = Q/V_v \ . \tag{6.66}$$

If a dielectric material such as wood is now interposed between the plates (Fig. 6.27b) after first disconnecting the voltage source, and assuming no leakage of charge through the wood, the voltage V_v decreases to a lower value V_w. The ratio V_v/V_w is defined as the static dielectric constant ε_w of the wood. Since the total charge Q on the plates is constant, the capacitance C_w is greater than C_v. Therefore, based on Eq. (6.66), the dielectric constant ε_w may then be written

$$\varepsilon_w = V_v/V_w = C_w/C_v \ . \tag{6.67}$$

The field strength is also reduced from E_v to E_w and from Eqs. (6.64) and (6.67)

$$\varepsilon_w = E_v/E_w \ , \tag{6.68}$$

since $E_w = V_v/d$. The dielectric displacement D, based on Eqs. (6.64) and (6.68), can be written

$$D = \varepsilon_w E_w \ . \tag{6.69}$$

The capacitance C_v, defined in Eq. (6.66), is related to the area A (cm^2) and spacing d (cm) of two parallel plates in a vacuum by

$$C_v = [A/(11.3\,d)] \times 10^{-12}\ \text{farads} \ . \tag{6.70}$$

Therefore, using Eq. (6.67), the capacitance C_w is given by

$$C_w = [A\varepsilon_w/(11.3\ d)] \times 10^{-12}\ \text{farads}\ . \tag{6.71}$$

The dielectric constant ε_w of a wood specimen can be calculated from Eq. (6.71) if C_w can be measured. It is difficult to measure C_w for wood using direct current techniques, and therefore *alternating current* measurements are generally used. These will be discussed next.

The dielectric constant is a measure of the polarization of atoms and molecules in the wood under the influence of an applied voltage V or voltage gradient E. At least four different kinds of polarization may occur in wood, each of which may contribute to the dielectric constant, depending on the frequency f of measurement. These are electronic, atomic, dipole and interfacial polarizations. The first two are of interest primarily from the viewpoint of optical and infrared spectroscopy, respectively. The last two are generally measured by electrical methods. The dipole mechanism involves permanent dipoles while the other three are caused by induced polarization.

Electronic polarization is caused by displacement of electrons with respect to the nucleus within the atom and occurs in the optical and ultraviolet region of frequencies. The dielectric constant ε in the optical range of frequencies is closely related to the optical index of refraction n, by the expression, $\varepsilon = n^2$.

Atomic polarization is caused by the stretching or bending of bonds between the atoms of a molecule. It occurs in the infrared range of frequencies, and is therefore studied by spectroscopic means.

Dipole polarization in wood is caused by *permanent dipoles* such as hydroxyl groups, or bound-water molecules in the cell wall, or free water in the cell cavities, etc. These dipoles rotate in the electric field and contribute to the dielectric constant at frequencies sufficiently low that they can follow the changes in the AC voltage. The *time constant*, that is the time required for the dipoles to rotate to about 2/3 of the total rotation they would have in a steady DC electric field, is in the order of microseconds (10^{-6} s) to picoseconds (10^{-12} s), depending on the moisture content and temperature of the wood.

Interfacial polarization is believed to be caused by the accumulation of charged ions at interfaces between different regions within the cell wall of wood. Just what these interfaces are in wood is not clear at present. They may be the interface between cells for example, or between the middle lamella and the secondary wall, or between crystalline and amorphous regions whose dielectric properties are markedly different from each other. This kind of polarization is believed to be responsible for the increases in the electrical resistance of wood with time, when DC methods are used, as in resistance moisture meters, for example. It is also possible that some interfacial polarization effects are caused by the accumulation of charges in the vicinity of the electrodes. Time constants may be in the order of minutes to milliseconds or less at room temperatures. It is often called the Maxwell-Wagner effect.

Figure 6.28 shows how the dielectric constant of an idealized material varies with the frequeny (log scale). Also shown are the contributions to the dielectric constant of each of the four polarization mechanisms cited above. The two regions of most importance for measuring the moisture content of wood, and also for heating wood electrically, are those in which the dipole and interfacial

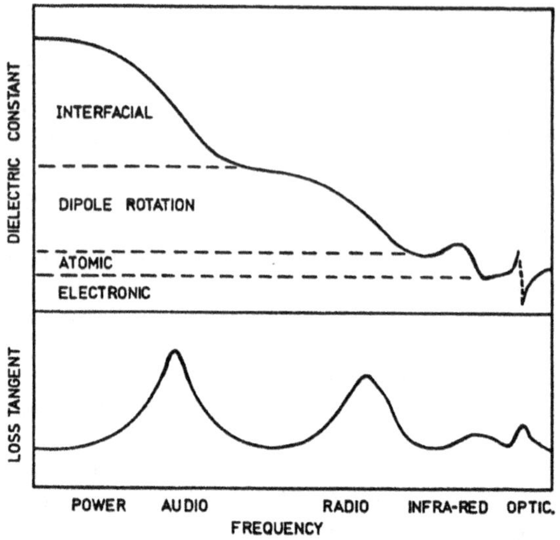

Fig. 6.28. Diagram showing the dielectric constant and loss tangent as functions of applied frequency (log scale) from power to optical frequencies, for a dielectric material. Four kinds of polarization and their effects on dielectric constant and loss tangent are illustrated

mechanisms are dominant. These are at frequencies in which electrical techniques are used, that is, below the infrared region. The dipole mechanism or Debye effect will be discussed first, followed by consideration of interfacial polarization or the Maxwell-Wagner effect.

6.3.1.1 Dipole Polarization (Debye Effect)

Permanent dipoles such as water molecules are of great importance in characterizing the dielectric properties of wood and their variation with wood parameters such as moisture content, density, temperature, and structural orientation. Therefore the dielectric behavior of wood in a sinusoidally alternating electric field will be described in terms of an idealized model which demonstrates the important characteristics of the variation of the dielectric properties of wood in the radio-frequency region.

Consider an idealized dielectric material containing permanent dipoles and subjected to a sinusoidally varying electric field of strength E and frequency f such that

$$E = E^* \sin(\omega t) \ , \tag{6.72}$$

where E^* is the peak amplitude, ω the angular frequency ($2\pi f$), and t is time. The electrical displacement D of the dipoles also varies sinusoidally with time, but may lag behind E by a phase angle δ such that

$$D = D^* \sin(\omega t - \delta) \ , \tag{6.73}$$

which is identical with

$$D = D^* \{(\cos\delta)[\sin(\omega t)] - (\sin\delta)[\cos(\omega t)]\} \ , \tag{6.73}$$

where D^* is the maximum amplitude of D.

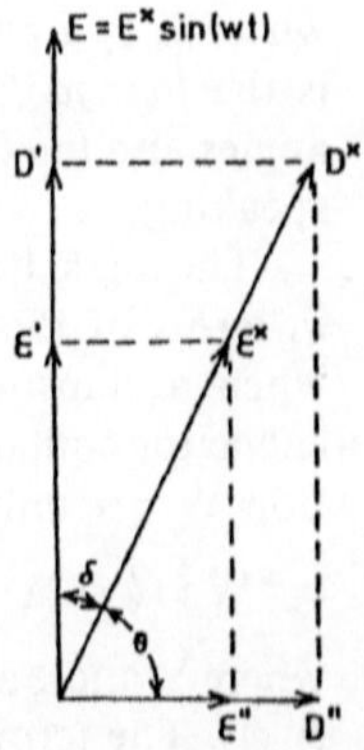

Fig. 6.29. Phase relationships among the dielectric displacement D* and its components D' and D'', the dielectric constant ε^* and its components ε' and ε'', and the voltage E, in relation to the phase angle θ and the loss angle δ

The displacement D can be separated into two parts, one D', in phase with ωt and the other D'', lagging ωt by $\pi/2$ radians. Thus in Eq. (6.73)

$$D' = D^* \cos(\delta) \tag{6.74}$$

$$D'' = D^* \sin(\delta) \ , \tag{6.75}$$

and Eq. (6.73) may be written

$$D = D' \sin(\omega t) - D'' \cos(\omega t) \ . \tag{6.76}$$

It is convenient to define two dielectric constant terms, ε' and ε'', based on Eqs. (6.69), (6.74) and (6.75). These are

$$\varepsilon' = D'/E^* = (D^*/E^*) \cos(\delta) \tag{6.77}$$

$$\varepsilon'' = D''/E^* = (D^*/E^*) \sin(\delta) \ . \tag{6.78}$$

The terms ε' and ε'' are the components of the *complex dielectric constant* ε^*, defined as

$$\varepsilon^* = D^*/E^* \ , \tag{6.79}$$

and in complex notation, using Eqs. (6.77) and (6.79)

$$\varepsilon^* = \varepsilon' - i\varepsilon'' \ . \tag{6.80}$$

Figure 6.29 illustrates the phase relationships among the components of D and E, in terms of the phase angle δ. Also shown are the dielectric constant components ε', ε'' and ε^*. The components D' and ε' are in phase with E and the components D'' and ε'' lag behind E by $\pi/2$ radians.

It is evident from Eqs. (6.77) and (6.78) that

$$\tan\delta = \varepsilon''/\varepsilon' \ , \tag{6.81}$$

sometimes called the *loss tangent*. It can be shown that the rate of energy dissipation in the dielectric is proportional to the dissipative component ε'' of the dielectric constant. The rate of energy dissipation W per unit area of plates in a capacitor is

$$W = K\omega(E^*)^2 \varepsilon'' = K\omega D^* E^* [\sin(\delta)] \ , \tag{6.82}$$

where K is a constant determined by the units used for D* and E*, and sin (δ) is the *loss factor.* The terms sin (δ) and tan (δ) are essentially equal for small angles and tan (δ) is often called the loss factor although this is incorrect, strictly speaking.

The angle $[(\pi/2)-\delta]$, labeled as θ in Fig. 6.29, is equivalent to the phase angle in terms of the relationship between the instantaneous voltage V and current I when a sinusoidally alternating voltage is impressed between the plates of a capacitor containing a dielectrical material. In this case the rate of energy dissipation W per unit area A, may be calculated from

$$W = (VI/A)\cos\theta = (VI/A)\sin\delta \ , \qquad (6.83)$$

where V and I are the root-mean-square values of the voltage and current, respectively. The term cos (θ) is called the power factor and is equal to sin (δ). The instantaneous current leads the displacement D by $\pi/2$ radians, and therefore leads the voltage V by the angle θ.

The dielectric coefficients ε^*, ε' and ε'' are all frequency-dependent for a dielectric material containing polar molecules. If all of these dipoles have the same time constant τ, the Debye equations, shown below, predict how ε' and ε'' vary with angular frequency ω. Thus

$$\varepsilon^* = \varepsilon_{ae} + (\varepsilon_0 - \varepsilon_{ae})/(1 + i\omega\tau) \qquad (6.84)$$

$$\varepsilon'(\omega) = \varepsilon_{ae} + (\varepsilon_0 - \varepsilon_{ae})/(1 + \omega^2\tau^2) \qquad (6.85)$$

$$\varepsilon''(\omega) = (\varepsilon_0 - \varepsilon_{ae})(\omega\tau)/(1 + \omega^2\tau^2) \ , \qquad (6.86)$$

where ε_0 is the static value of the dipole dielectric constant, that is the value of ε when all of the dipoles have sufficient time to completely rotate to their oriented positions, and ε_{ae} is the dielectric constant at such a high frequency ω that the dipoles cannot respond to the alternating field, and only the atomic and electronic polarizations contribute to ε.

Figure 6.30 shows curves of ε' and ε'' plotted against log ω for a dielectric material with a single relaxation time τ. The peak in ε'' occurs at the same value of ω at which ε' is decreasing most rapidly. Curves of tan δ [Eq. (6.81)] against log ω should also peak near those for ε'' vs log ω.

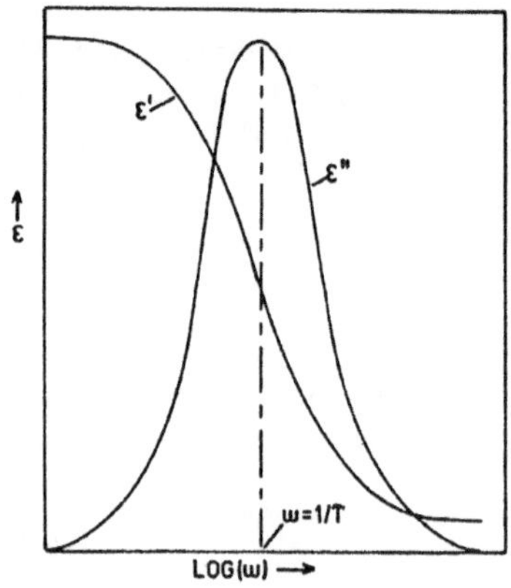

Fig. 6.30. Curves (not to scale) of dielectric constant components ε' and ε'' in relation to the logarithm of angular frequency ω for an idealized material with a single relaxation time τ

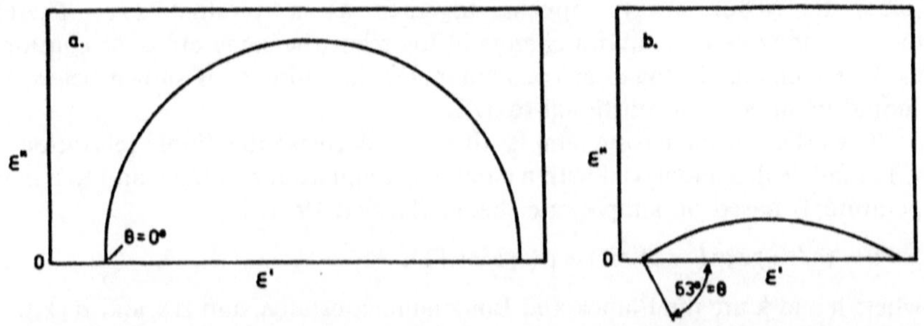

Fig. 6.31. Cole-Cole plots of ε'' against ε'. **a** For an idealized material with a single relaxation time, for which the angle θ is zero. **b** For a real material with a wide distribution of relaxation times, for which $\theta = 63°$

Cole and Cole (1941) have shown that when ε'', as defined by Eq. (6.86), is plottet against ε', defined by Eq. (6.85), over the entire frequency range in which anomalous dispersion occurs, a semicircle is obtained as shown in Fig. 6.31 a. However, for a material such as wood, which has a distribution of relaxation times, a curve more like that shown in Fig. 6.31 b is produced. These are referred to as Cole-Cole plots. These investigators proposed that the Debye equation for ε^*, Eq. (6.84) be modified to account for a distribution of relaxation times. Thus

$$\varepsilon^* = \varepsilon_{ae} + (\varepsilon_0 - \varepsilon_{ae})/(1 + (i\omega\tau)^{\beta}) \ , \tag{6.87}$$

where β is a parameter whose value may range from zero to unity. When $\beta = 1$, Eq. (6.87) reduces to the Debye equation, indicating a single relaxation time τ. A value for β of less than unity indicates a distribution of relaxation times, whose mean value is the principal relaxation time τ_0. As the distribution of relaxation times becomes broader, β becomes smaller.

The value of β can be obtained from the angle θ (Fig. 6.31 b) between the radius of the Cole-Cole circle where it joins the periphery, and the horizontal axis ($\varepsilon'' = 0$). Thus $\beta = (1 - \theta/90)$, when θ is measured in degrees. If θ is 63° as in Fig. 6.31 b, $\beta = 0.30$. Figure 6.31 a, based on the Debye equation shows that $\beta = 1$, consistent with the Debye model based on a single relaxation time τ. For wood and other dielectric solids β is usually less than one, indicating a distribution of relaxation times about τ_0.

Hill (1978) has analyzed several techniques that have been used to characterize the specific distributions of dielectric relaxation times in solids and liquids. He contended that none of them is generally applicable to all these materials, and proposed a more general expression for the distribution function, applicable to a greater range of materials. Nevertheless, the Cole-Cole plot has been an useful technique for analyzing the dielectric properties of wood.

Since increasing temperature reduces the time constant τ, plots of ε' and ε'' against frequency should shift toward higher frequencies as temperature increases. Furthermore, at constant frequency near f_0, it is anticipated that a plot of ε'' against temperature should peak at the temperature at which relaxation occurs at the frequency of measurement. This is based on the fact that ω and τ in Eqs.

(6.84) to (6.86) always appear together as a product $\omega\tau$. Therefore $\log \omega\tau = \log \omega + \log \tau$, and a change in $\log \tau$ has the same effect at constant ω as does a change in $\log \omega$ at constant τ. Furthermore, as is shown below, $\log \tau$ should be directly proportional to 1/T.

The effect of increasing temperature is to decrease the dipole relaxation time τ. For an idealized material with a single τ the equation relating τ and Kelvin temperature T, based on simple rate theory (Hasted 1973) is

$$\tau = (h/kT) \exp(-\Delta S/R) \exp(\Delta H/RT) , \tag{6.88}$$

where h and k are the Planck and Boltzmann constants, and ΔS and ΔH are the entropy and enthalpy changes, respectively, involved in the rotation of the dipole.

Taking the logarithm of both sides of Eq. (6.88), and the derivative of this with respect to (1/T), gives

$$d(\ln\tau)/d(1/T) = (\Delta H + RT)/R , \tag{6.89}$$

if ΔH and ΔS are assumed to be independent of temperature. The term $(\Delta H + RT)$ is equivalent to the apparent activation energy ΔE which can therefore be obtained from Eq. (6.89) as

$$\Delta E = \Delta H + RT = R[d(\ln\tau)/d(1/T)] . \tag{6.90}$$

The apparent activation energy ΔE is obtained from a plot of $(\ln\tau)$ against (1/T) using Eq. (6.90), the Arrhenius equation. Knowing ΔE, the enthalpy change ΔH can be obtained, since $\Delta H = \Delta E - RT$. Equation (6.88) can be rearranged to the following form to evaluate the entropy change ΔS

$$\Delta S = (\Delta H/T) + R[\ln(h/kT\tau)] . \tag{6.91}$$

The free energy change ΔG is obtained from

$$\Delta G = \Delta H - T\Delta S . \tag{6.92}$$

6.3.1.2 Interfacial Polarization (Maxwell-Wagner Effect)

It has been shown, based on the Debye model, that polar dipoles in materials cause anomalous dispersion, that is, a rapid change in the dielectric properties with frequency. The second kind of anomalous dispersion, caused by interfacial polarization, can be explained on the basis of the Maxwell-Wagner effect, which gives dispersion equations similar to those of Debye's model.

According to Hasted (1973) this effect can be visualized by considering a capacitor C, consisting of two parallel slabs of different dielectric materials (Fig. 6.32a), located between two condenser plates of area A (cm^2) and spacing d (cm). One of the materials is presumed to have a dielectric constant ε_1 and a conductivity K_1. The other is assumed to be a lossless dielectric with a dielectric constant ε_2 and conductivity K_2 of zero. The relative thicknesses of the two dielectrics are a_1 and a_2 respectively, such that $da_1 + da_2 = d$, or $a_1 + a_2 = 1$.

The capacitance C of the combined system (Fig. 6.32b) is

$$C = C_1C_2/(C_1 + C_2) . \tag{6.93}$$

Fig. 6.32. Schematic diagrams showing **a** composite dielectric material consisting of two parallel slabs of total thickness *d*, with dielectric properties ε_1, K_1 and ε_2, K_2, respectively, and **b** equivalent electrical network (see text)

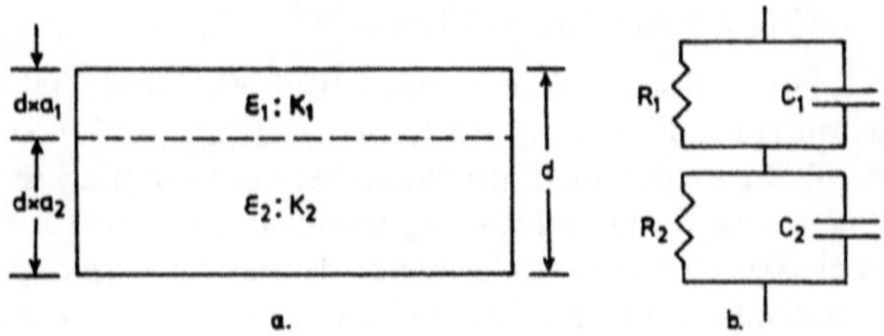

In this case C_1 and C_2 are given by

$$C_1 = B_0(A/d)(\varepsilon_1/a_1)[1-\exp(-t/\tau)] \tag{6.94}$$

$$C_2 = B_0(A/d)(\varepsilon_2/a_2) \tag{6.95}$$

$$B_0 = 10^{-12}/11.3 \ , \tag{6.96}$$

where C, C_1, and C_2 are all in farads, B_0 is the permittivity of free space (coulombs/V-cm), and τ is the time constant (s), given by

$$\tau = B_0(\varepsilon_1 a_2 + \varepsilon_2 a_1)/a_2 K_1 \ . \tag{6.97}$$

At time $t = 0$ after application of a DC voltage to the capacitor C, the capacitance C_1 is equivalent to $B_0(A/d)(\varepsilon_1/a_1)$. The instantaneous or short time value of C, C_i, therefore is

$$C_i = B_0[\varepsilon_1\varepsilon_2/(\varepsilon_1 a_2 + \varepsilon_2 a_1)] = B_0\varepsilon_i \ , \tag{6.98}$$

where ε_1 is the instantaneous dielectric constant, defined by

$$\varepsilon_i = \varepsilon_1\varepsilon_2/(\varepsilon_1 a_2 + \varepsilon_2 a_1) \ . \tag{6.99}$$

For long times, the capacitance C_1 approaches zero due to conduction of charges between its faces. Equation (6.93) then approaches the value of C_2 as given by Eq. (6.95) and

$$C_f = B_0(A/d)(\varepsilon_2/a_2) = B_0(A/d)\varepsilon_f \ , \tag{6.100}$$

where ε_f is the apparent final value of the dielectric constant measured after capacitor C_1 is shorted out by conduction of charges.

When a sinusoidally varying AC voltage is applied to the capacitor C, the apparent dielectric constant ε decreases with increasing angular frequency ω of the applied voltage. At low frequencies it approaches ε_f, equivalent to ε_2/a_2. At high frequencies it approaches ε_i, as given by Eq. (6.99). At intermediate frequencies it lies between these two limits.

Thus the system exhibits anomalous dispersion, similar to that shown by the dipole relaxation mechanism, but generally at much lower frequencies. The dielectric constant ε can also be separated into two components ε' and ε'', the real and imaginary components. The equations analogous to the Debye equations, Eqs. (6.84) to (6.86) are

$$\varepsilon = \varepsilon_i + (\varepsilon_f - \varepsilon_i)/(1 + i\omega\tau) \tag{6.101}$$

$$\varepsilon'(\omega) = \varepsilon_i + (\varepsilon_f - \varepsilon_i)/(1 + \omega^2\tau^2) \tag{6.102}$$

$$\varepsilon''(\omega) = (\varepsilon_f - \varepsilon_i)(\omega\tau)/(1+\omega^2\tau^2) \ . \tag{6.103}$$

For a system characterized by both interfacial and dipole mechanisms of anomalous dispersion, the terms ε_i in Eqs. (6.101) to (6.103) and ε_0 in Eqs. (6.84) to (6.86) are identical, provided there is no overlap in the frequency regions of the two kinds of anomalous dispersion. In other words the static value of the dipole dielectric constant ε_0 is equal to the high frequency value ε_i for interfacial polarization. In Fig. 6.28 this is the value of ε in the flat portion of the curve between the interfacial and dipole relaxation regions.

More refined models than the simple model described above have been used to explain anomalous dispersion due to interfacial polarization in heterogeneous substances such as wood. A number of these are described by Hasted (1973), but are not considered here.

Increasing temperature has been shown to reduce the time constant τ for the dipole relaxation mechanism. Temperature also reduces τ for the interfacial mechanism, and equations similar to Eqs. (6.88) to (6.92) may be applied to calculate ΔE, ΔH, ΔS and ΔG for this mechanism.

The dielectrical properties or real materials such as wood exhibit some of the general characteristics of the two kinds of anomalous dispersion depicted above, but they are more complex. Furthermore, water in wood has a pronounced effect on its dielectrical properties. We will consider first the dielectrical properties of water and of dry wood, followed by those of moist wood.

6.3.2 Dielectric Properties of Water

The dielectric properties of liquid water, including their relationships to its structure, have been discussed by Hasted (1973). Liquid water exhibits its principal anomalous dispersion in the microwave region. At 25 °C, its dielectric constant ε' decreases from near 80 below 10^9 Hz to about 5, above 10^{12} Hz. The peak frequency f_0 for ε'' is 2×10^{10} Hz, corresponding to a a wavelength of about 1.5 cm, and a relaxation time constant τ of 8×10^{-12} s. The Cole-Cole plot over this range (Fig. 6.33) almost coincides with that predicted by the Debye model given by Eq. (6.84), since β in the Cole-Cole equation (Eq. 6.87) is 0.987, close to the value of unity assumed in the Debye model.

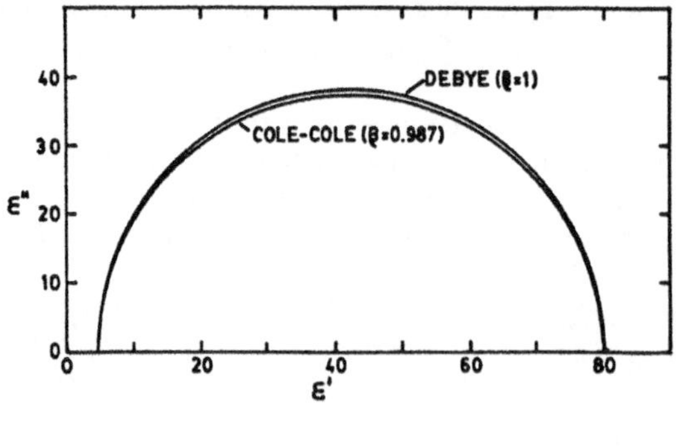

Fig. 6.33. Cole-Cole plots of ε'' against ε' for liquid water ($\beta = 0.987$) and for the ideal Debye case ($\beta = 1.0$) with optimized single relaxation time. (After Hasted 1973)

The peak frequency f_0 decreases with increasing temperature, and therefore τ decreases. An approximate equation describing this relationship is

$$\tau = 17.094 - 0.417\,C + 0.003183\,C^2 + 16.961\,C^3 \ , \tag{6.104}$$

where τ is in picoseconds and C in °C. The activation enthalpy ΔH^* is about 4.5 kcal/mol according to Hasted (1973), based on Eq. (6.88) written as

$$\tau = (A/T)\exp(\Delta H^*/RT) \ , \tag{6.105}$$

where A is a constant. Hasted points out that this is close to the enthalpy change ΔH_η calculated from the equation

$$\eta = B\exp(\Delta H_\eta/RT) \ , \tag{6.106}$$

which gives the temperature dependence of the viscosity η of water, assuming B is a constant. Hasted further indicates that, if $\Delta H^* = \Delta H_\eta$, τ should be proportional to η/T, which he notes has been observed. He suggests that this is as expected from the Debye relaxation theory which states that

$$\tau = 4\pi\eta a^3/kT \ , \tag{6.107}$$

where a is the molecular radius (of the water molecule in this case) and k is the Boltzmann constant.

Water in the cell cavities of wood may also occur as ice and it is instructive to note the changes in ε' and ε'' when water freezes. Ice may exist in several different phases but we will be concerned only with the form (ice I) which occurs at atmospheric pressure. Ice I is somewhat anisotropic and the dielectric constants differ by about 14% in the two major axes. Normally ice I is polycrystalline with a random distribution of the principal axes so as to appear isotropic at the macroscopic level (Hasted 1973).

The peak frequency corresponding to the principal relaxation of ice occurs in the kHz region at temperature slightly below the freezing point. For example, f_0 is about 3 kHz at $-10\,°C$ and decreases rapidly with decreasing temperature, giving an activation enthalpy ΔH^* of about 13 kcal/mol, almost three times that for liquid water. This is probably because of the higher energy barrier frozen water molecules must overcome to rotate in an electric field.

6.3.3 Dielectric Properties of Dry Wood

The dielectric components ε' and ε'' for dry wood vary with a number of intrinsic wood parameters, such as density, structural orientation, and chemical composition. They also vary, as anticipated from Eqs. (6.89) to (6.92), with frequency and wood temperature. Because of variations among woods with respect to density, structure, and chemical composition, there are substantial differences among species and also within species, even within a given tree. Lin (1967) has reviewed the literature on the dielectric properties of wood through 1966.

Several workers have reported on the dielectric properties of dry wood. Norimoto (1976) reviewed this work and extended it to include the effects of various wood parameters. Kröner (1944/1945) first pointed out that the simple

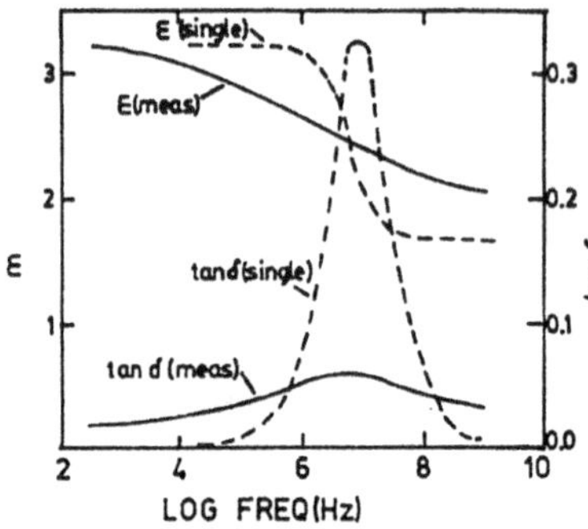

Fig. 6.34. Experimental dispersion curves of ε' and of tan δ for European beech as functions of frequency (log scale). Also shown are similar curves as calculated for a material with a single relaxation time. (Kröner 1944/45)

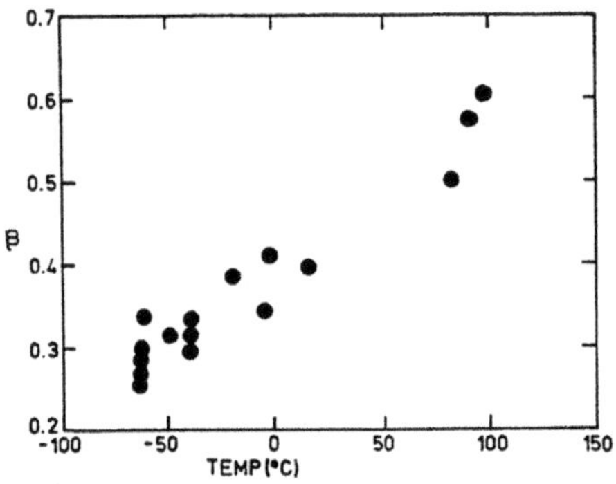

Fig. 6.35. Plotted points showing the increase in the parameter β with increasing temperature, as measured by Norimoto (1976)

Debye model given by Eqs. (6.84) to (6.86) does not give an adequate representation of anomalous dispersion in wood, because there is a distribution of relaxation times rather than a single value. Anomalous dispersion is observed but with a broader peak in ε'' than is predicted from the single relaxation time model of Debye. Equation (6.87), the Cole-Cole equation, accounts for the distribution of relaxation times.

Figure 6.34 shows the experimental dispersion curves of ε' and ε'', obtained by Kröner (1944/1945) on dry European beech cross-sections. Also shown are the corresponding Debye curves based on a single relaxation time τ corresponding to τ_0, the value calculated from the relationship $\tau_0 = 1/\omega_0$, where $\omega_0 (= 2\pi f_0)$ is the angular frequency, and f_0 is the frequency at which the measured value of ε'' is a maximum. In Fig. 6.34, f_0 is approximately 10 MHz, and τ_0 therefore is approximately 16 ns.

Norimoto (1976) reported on measurements of primary or dipole dispersion in oven-dry wood over the temperature range from $-60\,°C$ to $+100\,°C$. He noted that β, the parameter whose value is inversely related to the distribution of relaxation times, as given in Eq. (6.87), increased with increasing temperature (Fig. 6.35). He interpreted this to indicate that the distribution of relaxation times becomes narrower with increasing temperature. He also discussed other modifications of the Cole-Cole method for calculating the distribution of relaxation times.

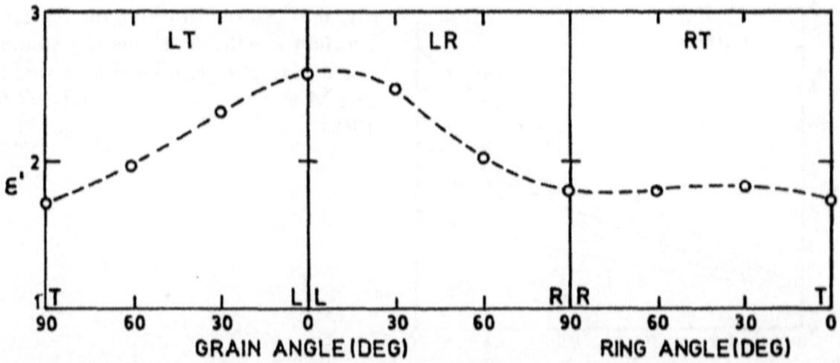

Fig. 6.36. Curves showing the variation of the dielectric constant ε' with grain angle in the *LT* and *LR* planes, and with ring angle in the *RT* plane. (After Norimoto and Yamada 1972)

Grain orientation affects the dielectric properties of wood. In general both ε' and ε'' are 20% to 50% higher when measured along than across the grain. Kröner and Pungs (1952) attributed this anisotropy to the gross cellular structure of wood. Others consider that the fibril orientation in the cell wall is the primary reason. Norimoto and Yamada (1971) have concluded that both the macroscopic (cell arrangement) and microscopic (cell wall) properties of wood contribute to the dielectric anisotropy. The difference in dielectric properties between the radial and tangential directions is not consistent among different woods. Figure 6.36, taken from Norimoto and Yamada (1972), shows how ε' varies with grain angle in the LT and LR planes, and also with respect to the growth ring orientation, in the RT plane for *Chamaecyparis obtusa* at 30 °C, 10 kHz and 0% moisture content. Nanassy (1964) reported that both ε and ε'' were highest when measured along the grain and lowest along the tangential direction, for dry spruce between 40 kHz to 50 MHz.

An interesting application of the variation of the dielectric constant ε' with grain orientation has been proposed by McLaughlan et al. (1973), who designed a noncontact slope-of-grain indicator based on the greater value of ε' along than across the grain. The apparatus consists of an oriented capacitance type of transducer located near the wood surface and rotating in a plane parallel to the surface. During each rotation, its axis is oriented successively along and across the grain direction of the wood. A sinusoidal signal is generated as a function of its orientation with respect to the longitudinal axis of the wood. Comparison of the phase relationship of this signal to another signal proportional to the orientation of the transducer axis gives an indication of the grain orientation at the surface. A commercial device using the principle has been developed (Samson 1984).

The dielectric constant ε', as anticipated, increases with wood density at constant moisture content. This is essentially because ε' for the cell-wall material is several to many times greater than the value (essentially unity) of the air in the cell cavities, for wood below fiber saturation. Figure 6.37 shows this relationship as given by Uyemura (1960), based on measurements across the grain on a number of woods of different densities at one MHz. Skaar (1948) found similar results at 2 and 15 MHz.

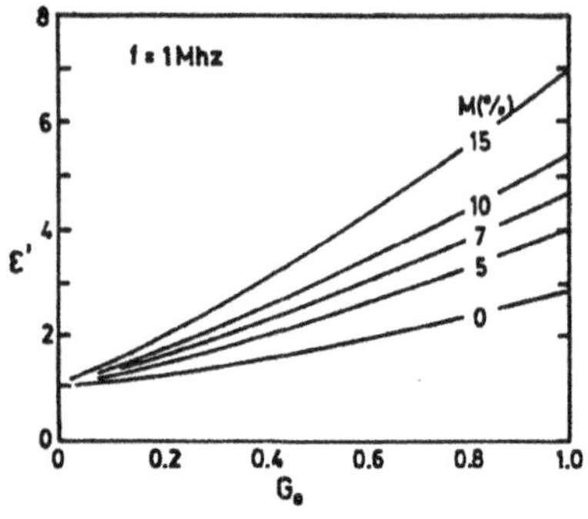

Fig. 6.37. Curves showing the increase in dielectric constant ε' with increasing dry volume specific gravity G_0, and with increasing wood moisture content M at a frequency of 1 MHz. (After Uyemura 1960)

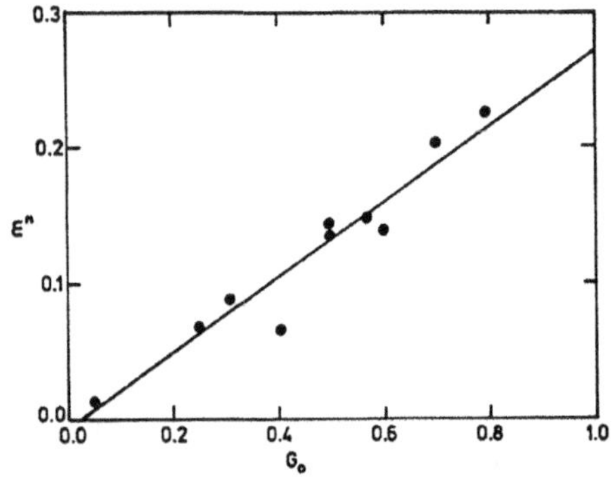

Fig. 6.38. Plotted points and regression line relating loss factor ε'' and dry wood specific gravity G_0, measured at 1 MHz frequency. (After Norimoto and Yamada 1970)

Venkateswaran (1974a) reported that the transverse dielectric constant ε' for oven-dry wood, measured at 10000 Hz and 20 °C, was highly correlated with transverse thermal conductivity. The latter increases with wood density and therefore ε' also increases with density. Venkateswaran (1974b) defined the specific permittivity or specific dielectric constant ε_0' for the cell wall of oven-dry wood as follows,

$$\varepsilon_0' - 1 = (\varepsilon' - 1)(\varrho_0/\varrho) \ , \tag{6.108}$$

where ε' is the measured dielectric constant for a wood of density ϱ, and ϱ_0, taken as 1.53 g/cc, is the density assumed for the dry cell wall.

Rafalski (1966) reported that ε' for oven-dry European beech, measured at 5 to 25 MHz, increased linearly with increasing density as the wood was compressed from its initial density of 0.70 to a maximum value of 1.40 g/cc.

The loss factor ε'' also appears to increase with increasing wood density. Rafalski (1966), in the study noted above, reported a linear increase in ε'' with increasing densification of the wood. Norimoto and Yamada (1970) observed a linear increase in ε'', measured parallel to the grain at 1 MHz and 31 °C, for ten wood species, covering a wide density range (Fig. 6.38). Similar increases have been reported for transverse measurements on wood (Lin 1967, Venkateswaran 1973).

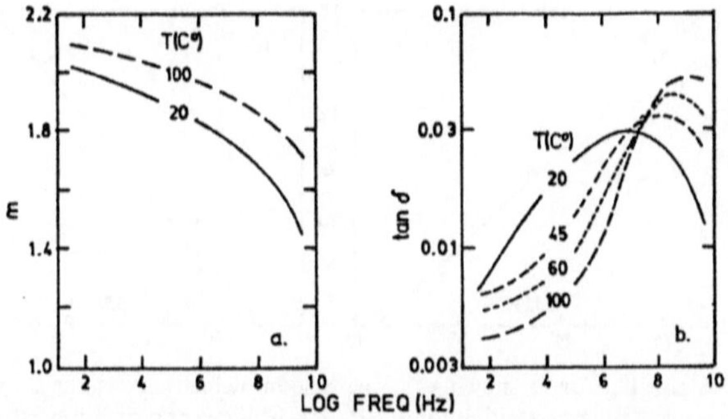

Fig. 6.39. Curves of dielectric constant ε' and of loss tangent as functions of log frequency and temperature for dry wood. (After Trapp and Pungs 1956)

According to Norimoto and Yamada (1970) dielectric dispersion in oven-dry wood occurs in at least two frequency ranges. The high frequency peak in ε'' is in the range of 10^7 to 10^8 Hz at room temperature, decreasing to about 10^4 Hz at $-50\,°C$, and to 30 Hz at $-100\,°C$ (Handa et al. 1982). They attribute this dispersion to the rotation of methylol (CH_2OH) groups in the disordered regions of the cell wall. They suggest that dispersion in the low frequency region may be due to interfacial polarization of conductive ions or to some other complex mechanism. It becomes more pronounced when the wood contains moisture. The dispersion associated with higher frequencies will be referred to as primary dispersion. That at lower frequencies will be called secondary dispersion.

It was indicated in the discussion of the Debye equations that curves of $\tan\delta$ against $\log\omega$ should shift toward higher frequencies with increasing temperature. Trapp and Pungs (1956) showed that this shift did occur for dry wood as expected (Fig. 6.39). Furthermore, they also demonstrated, as predicted by the Debye theory, that the shift in the primary peak frequency f_0 with temperature was such that a plot of $\log\tau_0$ against $1/T$ was linear. Based on this relationship and the points shown in their curve for oven-dry spruce wood, the slope $d(\log\tau_0)/d(1/T)$ is near 2500, giving an activation energy, using the Arrhenius equation, of approximately 11 400 calories per mole.

Trapp and Pungs (1956) found that the primary peak frequency f_0 decreased from 10^9 Hz at 100 °C to 10^7 Hz at 20 °C. Based on the Arrhenius equation and the value of activation energy of 11.4 kcal per mole, f_0 should shift to approximately 10^6 Hz at 0 °C assuming that the same dipole mechanism operates. Norimoto and Yamada (1969), working at lower frequencies and temperatures than Trapp and Pungs, found that ε'' peaked at 0 °C for a frequency of 1 MHz for *Chamaecyparis obtusa* in the oven-dry condition as predicted by Trapp and Pungs. Furhtermore, their values for activation energy (~ 10 kcal/mol) were similar to those reported by Trapp and Pungs. Handa et al. (1982) reported that this same peak occurred at $-103\,°C$ at 30 Hz for oven-dry beech wood.

The proportions of the basic chemical constituents of wood vary among as well as within woods. According to Norimoto (1976), based on his own work and a literature survey, both ε' and ε'' for cellulose, the most important constituent of wood, increase essentially linearly with decreasing crystallinity of the cellulose. Figure 6.40a, taken from Norimoto, shows how ε' (at a standard sample density

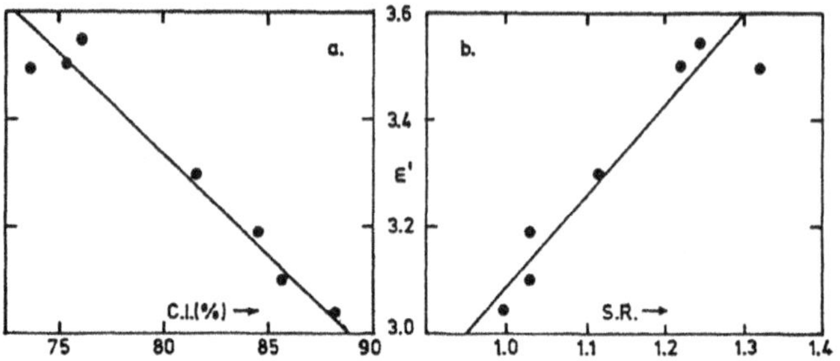

Fig. 6.40. Curves showing the increase in dielectric constant ε' measured at 1 kHz and $-32\,°C$ with **a** decreasing crystallinity index (*C. I.*), **b** increasing sorption ratio (*S. R.*). (After Norimoto 1976)

of 1.2 g/cc, measured at 1 kHz and $-32\,°C$) varies with the crystallinity index (CI) obtained from X-ray diffraction measurements. Norimoto also shows that ε' increases linearly with increase in the sorption ratio (SR), defined as the ratio of the amount of water sorbed by the cellulose to that sorbed by cotton at 20 °C and 63.5% relative humidity (Fig. 6.40b).

Measurements by Norimoto (1976) on the hemicelluloses, mannan and xylan, indicate that mannan behaves similarly to cellulose with respect to anomalous dispersion. However, xylan did not show an absorption peak in the one MHz and 0 °C range. Norimoto attributes this to the lack of a methylol group in xylan whereas it is present in both cellulose and mannan.

The lignin content of wood also affects its dielectrical properties. Venkateswaran (1974b) reported a linear decrease in the dielectric constant ε' of the vacuum-dry cell wall of wood from 6.3 to 4.3 when the acid-soluble lignin content increased from 15% to 35%. Measurements were made at 20 °C and 10000 Hz across the grain on 15 wood species. Norimoto (1976) reported dielectric constants of 5.4 and 4.0 respectively, for dry cotton and milled-wood lignin at 1 MHz and 20 °C. These relative values agree with the results reported by Venkateswaran, whose absolute values are higher, presumably because he used a much lower frequency.

6.3.4 Dielectric Properties of Moist Wood

It has been shown above that the dielectric properties of dry wood and of pure water are considerably different. It is not surprising, therefore, that moisture has a pronounced effect on the dielectric properties of wood. Moisture meters have been devised which use this effect to measure the moisture content of wood. Before considering such dielectric moisture meters, the effect of moisture in wood on its dielectrical properties will be discussed.

The intrinsic transverse dielectric constant ε' of the dry cell wall of wood is approximately 4 to 5 in the radio frequency range, whereas that of water is near 80. It is expected, therefore, that the dielectric constant of wood should increase dramatically with increasing moisture content. Furthermore, water also has a pronounced effect on anomalous dispersion in wood.

It has been indicated that there are at least two mechanisms for anomalous dispersion in wood. The primary or dipole rotation mechanism, observed in dry wood, is caused by the rotation of methylol groups, according to Norimoto (1976), with a characteristic peak frequency f_0 is the order of 10 MHz at room temperature. The secondary or interfacial polarization mechanism occurs at much lower frequencies at room temperature. It is inconspicuous in dry wood but becomes pronounced as the wood moisture content increases, according to Tsutsumi and Watanabe (1966). They found that the peak frequency f_0 at 15% moisture content was about 1000 Hz at room temperature for the secondary mechanism. This is about 6 orders of magnitude lower than the primary peak frequency of near 1 GHz for wood of 12.5% moisture content, as reported by Trapp and Pungs (1956). This secondary mechanism will be discussed in more detail after first considering the effect of moisture in wood on the primary mechanism associated with dipole rotation.

6.3.4.1 Primary Dispersion in Moist Wood

Trapp and Pungs (1956) measured ε' and $\tan\delta$ for European spruce over a wide range of temperatures (20° to 100 °C), moisture contents (0% to 95%), and frequency (40 Hz to 3.5×10^9 Hz), where the dipole mechanism dominates. Figure 6.41 shows that both ε' and $\tan\delta$ generally increase strongly with increasing moisture content at a given frequency at 20 °C. Furthermore, the frequency f_0 for which $\tan\delta$ peaks, shifts toward higher frequencies with increasing moisture content. This is as expected since the peak frequency for dipole rotation in liquid water at room temperatures is about 10^{10} Hz, three orders of magnitude higher than that for dry wood, which is near 10^7 Hz at 20 °C. Apparently the water in the cell wall reduces the mean time constant for dipole rotation and also increases the magnitudes of ε' and ε''.

Table 6.5 shows dipole relaxation times at several temperatures and moisture contents as reported by Trapp and Pungs (1956). The relaxation times at 20 °C are 16, 6.4, 0.80, 0.16 and 0.02 ns at corresponding wood moisture contents of 0%, 3%, 6.5%, 12.5%, and 20%, respectively. That of liquid water at 20 °C is approximately 0.01 ns, about half the value given above for wood at 20% moisture content. Also shown in the table are the corresponding peak frequencies.

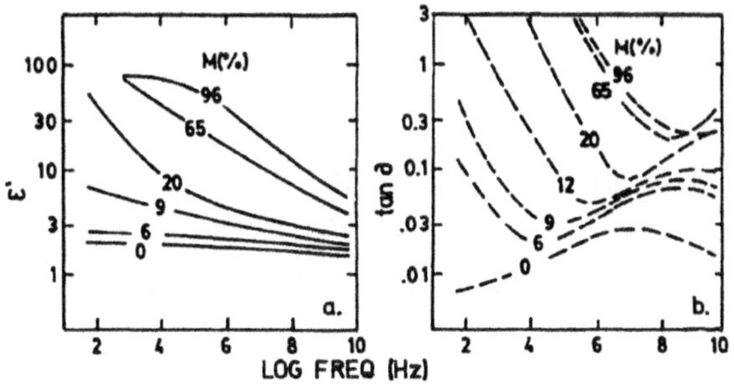

Fig. 6.41. Family of curves showing dielectric constant ε', and loss factor $\tan\delta$, of European spruce at 20 °C, as functions of frequency (log scale) at several wood moisture contents. (Trapp and Pungs 1956)

Table 6.5. Dipole relaxation times τ (ns) and peak frequencies f_0 (GHz) for European spruce at several moisture contents and temperatures. (Trapp and Pungs 1956)

Temp. (°C)		Wood moisture content (%)					Pure water
		0	3	6.5	12.5	12.5	20
20	τ	16.0	6.4	0.80	0.16	0.02	0.009
20	f_0	0.010	0.025	0.20	1.00	8.00	16.0
40	τ	3.2	1.6	0.23	0.032	–	0.0058
40	f_0	0.05	0.10	0.70	5.0	–	24.8
60	τ	1.3	0.40	0.11	–	–	0.0039
60	f_0	0.12	0.40	1.45	–	–	40.0
80	τ	0.4	0.11	0.032	–	–	0.0030
80	f_0	0.4	1.45	5.0	–	–	48.0
100	τ	0.16	0.053	–	–	–	–
100	f_0	1.0	3.0	–	–	–	–

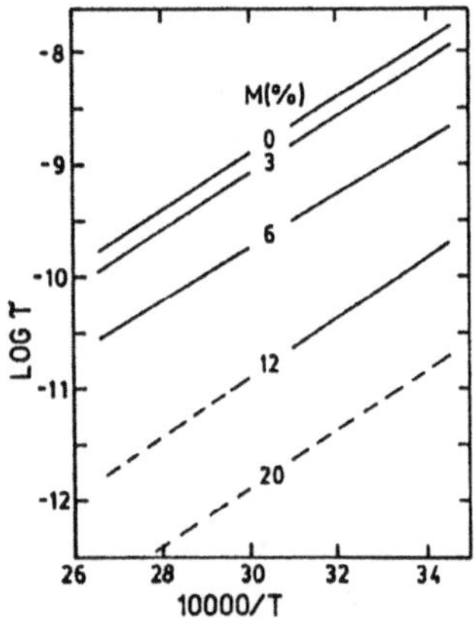

Fig. 6.42. Linear curves of the logarithm of relaxation time τ as functions of the reciprocal of Kelvin temperature for European spruce wood at several moisture contents. The portions of the curves shown as *broken lines* are estimated. (After Trapp and Pungs 1956)

Trapp and Pungs (1956) plotted curves of log (r) against 1/T (Fig. 6.42) from data given in Table 6.5. These appeared to be linear and, using the Arrhenius relation [Eq. (6.90)], the activation energy ΔE is in the order of 11 to 12 kcal/mol, apparently independent of wood moisture content. This is close to that of ice (13 kcal/mol) as noted earlier, but much higher than that for liquid water (5 kcal/mol).

It has been shown that the primary mechanism of anomalous dispersion, i.e., dipole rotation, occurs in the radio-frequency to microwave region at room temperatures and above. In this frequency region therefore, the dielectric constant ε' decreases with increasing frequency. It also increases with increasing wood moisture content as is shown in Fig. 6.43, taken from Skaar (1948), using a frequency of 2 MHz, and Uyemura (1960) at 1 MHz. The increase is nonlinear in the hygroscopic region, but appears to be linear above fiber saturation in both cases. The nonlinearity in the hygroscopic range is probably associated with the restricted freedom of rotation of the sorbed water at these moisture contents.

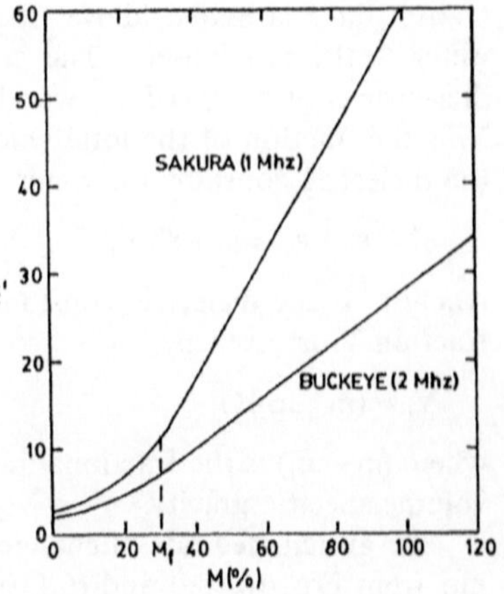

Fig. 6.43. Curves of dielectric constant ε' against wood moisture content for Japanese "sakura" at 1 MHz (Uyemura 1960) and for Ohio buckeye at 2 MHz (Skaar 1948)

Table 6.6. Time constants τ (ns) calculated for dipole polarization in Buna (*Fagus crenata*) wood at several temperatures and moisture contents. (Tsutsumi 1967)

Temp. (°C)	Wood moisture content (%)		
	3	7.5	13.0
40	16	–	–
0	320	80	–
−20	800	110	32
−40	1600	400	–
−70	110000	8000	1600

Tsutsumi (1967) also observed anomalous dispersion due to dipole orientation in Buna (*Fagus crenata*) over the moisture content range from 0.6% to 13%. He used frequencies from 30 Hz to 3 MHz and temperatures from −70° to +40 °C in order to observe the dispersion peaks in tan δ, the peaks shifting to lower temperatures at higher wood moisture contents. Table 6.6 tabulates the time constants at several different wood moisture contents and temperatures as reported by Tsutsumi. Although the actual values are somewhat different from those of Trapp and Pungs (1956) shown in Table 6.5, the orders of magnitude are similar, as are the trends with increasing temperature and moisture content. The activation energies ΔE obtained by use of Eq. (6.90) on data at temperatures of −20° and −70 °C are approximately 10, 9 and 8 kcal/mol at moisture contents of 3%, 7.5%, and 13%, respectively.

Bohno and Tsutsumi (1974) indicate that Cole-Cole plots of ε'' vs. ε' give values of θ in the order of 50° to 60° in the dipole dispersion region in the moisture range from 0% to 5% at 20 °C. The corresponding values of β range from 0.34 to 0.45, similar to those reported by Norimoto (1976) for dry wood at this temperature (Fig. 6.35).

As water is added above fiber saturation it condenses into essentially free water in the cell lumens. The contribution ε_l' of the lumen water to the total dielectric constant ε' of the wood is proportional to the product (80)(V_l), where V_l is the fraction of the total wood volume consisting of liquid water, and 80 is the dielectric constant for water. Thus, from the simple method of mixtures

$$\varepsilon' = \varepsilon_f' + \varepsilon_l' = \varepsilon_f' + 80\,V_l \ , \qquad (6.109)$$

where ε_f' is the dielectric constant of the wood at fiber saturation. The volume fraction V_l is given by

$$V_l = (m - m_f)\,G_f \ , \qquad (6.110)$$

where $(m-m_f)$ is the fractional moisture content of the lumen, and G_f is swollen volume specific gravity.

The anticipated rate of change of ε' with increasing moisture content m above m_f, from Eqs. (6.109) and (6.110), is given by

$$d\varepsilon'/dm = 80\,G_f \qquad (6.111)$$

since ε_f' and m_f are presumed constant. The actual slope $d\varepsilon'/dm$ above m_f in Fig. 6.43 for buckeye is approximately 32. Equation (6.111) anticipates a value of (80)(0.33) or 26, assuming a value for G of 0.33 for buckeye (*Aesculus* sp.). This is somewhat lower than the observed value of 32. The reason for the discrepancy is probably because the lumen water is in isolated cells and not in a continuous parallel path with the cell wall material as is assumed in Eq. (6.109).

Siau (1984) gives an equation for calculating the dielectric constant of wood, based on its porosity, which applies below fiber saturation. It can be modified for use at higher moisture content, giving a nonlinear relation between dielectric constant and wood moisture content above M_f.

Norimoto (1976) gives a more detailed model, based on wood structure, for calculating the effect of various wood parameters such as microfibril angle, latewood fraction, cell shape, ring curvature, cell arrangement, etc. From this model he has determined that all of these parameters affect the dielectric properties of wood.

The transverse dielectric constant of the cell wall of wood at 1 MHz has been calculated as a function of moisture content by Uyemura (1960) based on measurements on woods of varying densities at room temperature. He gives an exponential relationship of the form

$$\varepsilon' = A \exp(Bm) \ , \qquad (6.112)$$

where A and B are 4.80 and 9.19 for moisture contents below 0.076 (M = 7.6%), and 6.0 for moisture contents between 0.076 and fiber saturation. Based on Eq. (6.112), ε' increases almost sixfold from 4.8 for the dry cell wall to 27.9 at m = 0.25.

The dielectric constant of the gross wood also increases with moisture content, but not to the same extent as that of the cell wall. The increase is greater for more dense than for less dense woods, as Fig. 6.44, taken from Uyemura (1960) indicates. For example, from Fig. 6.44, ε' doubles between 0% and 25% moisture content for wood whose oven-dry specific gravity G_0 is 0.1, but quadruples over

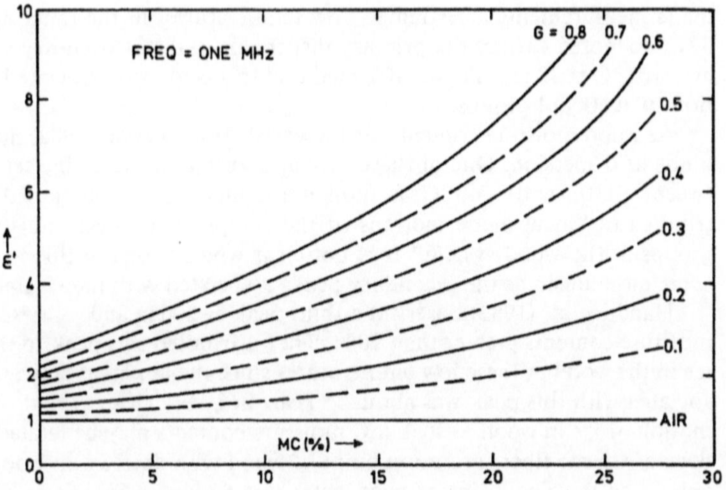

Fig. 6.44. Curves showing increase of dielectric constant ε′ with increasing wood moisture content and specific gravity. (After Uyemura 1960)

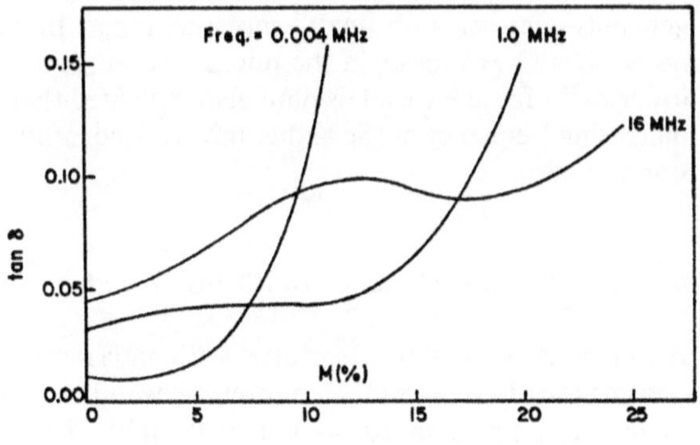

Fig. 6.45. Curves of loss tangent (tan δ) in relation to wood moisture content M for three different frequencies as given by Hearmon and Burcham (1954)

the same moisture range for wood of $G_0 = 0.6$, at 1 MHz. This is because the cell wall volume is greater for the more dense woods, and therefore the relative contribution of the cell wall is greater than for the low density woods.

It has been established that the loss tangent (tan δ) of wood also increases with increasing wood density (Venkateswaran 1973). Its variation with wood moisture content is more complex however, as Fig. 6.45 indicates. This shows that tan δ increases with wood moisture content in general. However, at 16 MHz it peaks near 12% moisture content before resuming the normal increasing pattern. This is probably due to the proximity of a region of anomalous dispersion at this combination of frequency, moisture content and temperature.

Handa et al. (1982) measured ε′ and ε″ of beech (*Fagus* sp.) wood over a wide range of moisture contents (0% – 40%) at frequencies from 30 Hz to 1 MHz. In order to observe the primary peaks in ε″ or in tan δ at these low frequencies they made measurements at extremely low temperatures, in the range from −196° to 0 °C. As noted earlier, the primary dispersion peak for oven-dry wood occurred at −103 °C at 30 Hz. They reaffirmed that this peak was associated with the rotation of methylol groups.

As wood moisture content was increased, Handa et al. (1982) noted two other kinds of dispersion. One of these, which occurred at increasing temperatures between −110° and −40 °C as moisture content increased (at 30 Hz), they attributed to "local mode motions of the complexes between moisture and polar groups in the wood system." It is not clear whether or not this is caused by the same mechanism as the secondary peaks associated with interfacial polarization.

Handa et al. (1982) observed a third peak at 30 Hz and −40 °C in wood with moisture contents greater than 10%. They attributed this peak to the presence of ice in the wood at these low temperatures since the apparent activation energy associated with this peak was about 13 kcal/mol, very close to that of ice. The formation of ice in wood at such low moisture contents may be related to the strong decrease in the fiber saturation point of wood with decrease in wood temperature below the freezing point of pure water (see Coldness Shrinkage in Chap. 2).

Measurements on Douglas-fir at 25 °C, using microwave frequencies from 10^9 to 10^{10} Hz (James and Hamill 1965) showed that ε' increased more than fivefold between wood moisture contents of 5% and 75%. However, $\tan\delta$ behaved in a complex manner, increasing with moisture content below fiber saturation but generally decreasing above this moisture range. In a later paper, James (1975a, b) noted that ε' was lower in the microwave range at all moisture contents than at lower radio frequencies. His data also indicated that $\tan\delta$ increased strongly with increasing frequency at the higher moisture contents, although the peak frequency was not seen.

6.3.4.2 Secondary Dispersion in Moist Wood

Anomalous dispersion associated with interfacial polarization is absent or inconspicuous in dry wood. However, as wood sorbs moisture the effect becomes increasingly pronounced, as Fig. 6.46, taken from James (1975a, b), indicates. This dispersion differs from that due to the dipole mechanism in wood in several respects. First, it occurs at much lower frequencies, in the audio range at room temperatures compared with the high radio frequencies for dipole polarization. Second, it yields much higher dielectric constants, approaching 10^6 according to Fig. 6.46a, compared with a maximum of near 80 for dipole polarization. Third, as has been mentioned, it is essentially nonexistent in dry wood, and is much more moisture-dependent than is dipole polarization.

Figure 6.46 was taken from James (1975a, b), who measured ε' and $\tan\delta$ for Douglas-fir (*Pseudotsuga menziesii*) and white oak (*Quercus alba*), as well as hardboard, for material conditioned to equilibrium at 80 °F at relative humidities of 30%, 65%, 80%, and 90%. Some wood samples were also water-soaked. Measurements were made in the three principal structural directions for the wood samples at temperatures between −20° and 90 °C over the frequency range from 20 Hz to 50 MHz.

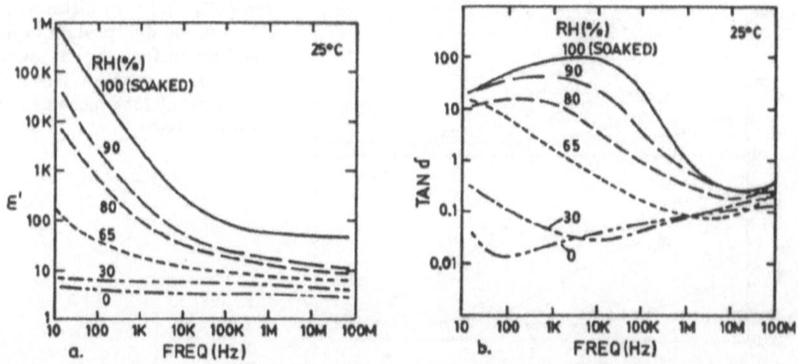

Fig. 6.46. Curves of dielectric constant ε' (**a**) and of tan δ (**b**) as functions of frequency (log scale) for Douglas-fir wood conditioned to various equilibrium relative humidities. (After James 1975a, b)

It is evident from Fig. 6.46 that anomalous dispersion occurs in the audio and sub-audio frequency ranges at room temperature. For example, the wood conditioned to 80% relative humidity (approximately 16% moisture content), peaks at about 200 Hz (Fig. 6.46b) with the peak frequency shifting to lower values as the wood moisture content decreases. This is about 7 orders of magnitude lower than the peak frequency of near 1 GHz (10^9 Hz) for the dipole mechanism as reported by Trapp and Pungs (1956) for wood in this moisture range (Fig. 6.41). It is therefore probably caused by interfacial polarization.

Data obtained earlier by James and Hamill (1965) on Douglas-fir at frequencies between 1 and 10 GHz showed no clear evidence of a peak in tan δ, although ε' does decrease with increasing frequency as anticipated.

James (1975a, b) also presents curves similar to those of Fig. 6.41 but at several different temperatures. The anticipated shifting of the peak (tan δ) frequency toward higher values with increasing temperature was observed. Although the peaks are broad, the peak frequency for wood conditioned to 80% humidity ($\simeq$16% moisture content) are estimated from the curves to be 50, 200, 500, and 1200 Hz at the respective temperatures of 5°, 25°, 45°, and 75 °C. Application of the Arrhenius equation to this data gave a value of 8700 cal/mol for the apparent activation energy ΔE, of the same order of magnitude as that for dipole polarization.

Tsutsumi (1967), in addition to measurements showing anomalous dispersion due to dipole orientation, also made measurements showing such dispersion in wood in the range of interfacial polarization. Figure 6.47 shows curves of ε' and tan δ against frequency for Tanagi (*Salix jessoensis*) wood at 15% moisture content and several temperatures. It is clear that the dispersion peaks shift to higher frequencies with increase in temperature. Estimated values of the peak frequencies are 300, 3000, 10000, and 20000 Hz at temperatures of −10°, 10°, 25°, and 45 °C. The Arrhenius equation yields a value for ΔE of 14 kcal/mol based on these data.

Tsutsumi also gives similar curves of ε' and tan δ for Sakura (*Prunus*) wood, also at 15% moisture content. The peak frequencies are approximately 20, 60, 450, and 2000 Hz at −20°, 0°, 20°, and 40 °C, respectively. In this case the calculated value of ΔE is approximately 12.5 kcal/mol.

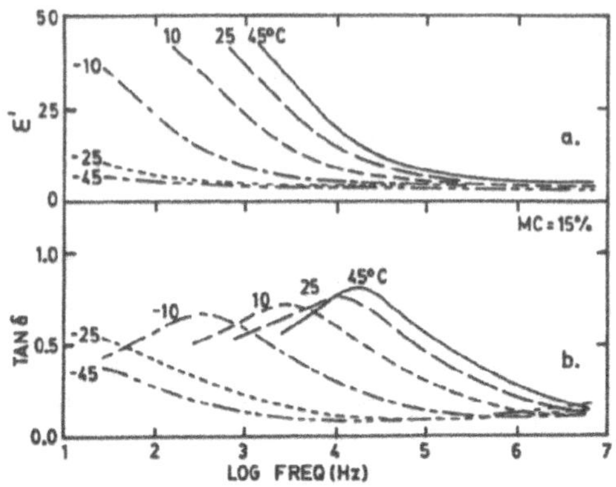

Fig. 6.47. Curves of dielectric constant ε' and of tan δ in relation to the logarithm of frequency at several temperatures, for wood of Tanagi (*Salix jessoensis*) at 15% moisture content. (Tsutsumi 1967)

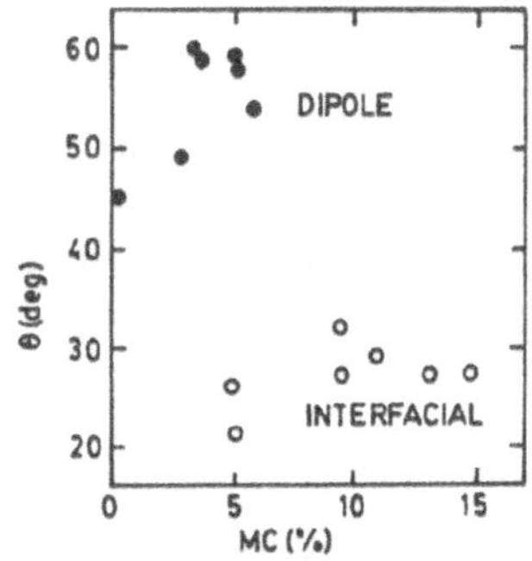

Fig. 6.48. Plotted points of the Cole-Cole parameter θ against wood moisture content for both dipole (high frequency) and interfacial (low frequency) polarization for Japanese shiogi (*Fraxinus commemoralis*) wood. (After Bohno and Tsutsumi 1974)

Cole-Cole plots of ε'' vs. ε' as given by Bohno and Tsutsumi (1974) for the interfacial polarization dispersion region give considerably lower values for the angle θ (Fig. 6.48) and therefore larger values of β than those obtained in the dipole dispersion region. Furthermore, θ increases and therefore β decreases as wood moisture content increases. This indicates a greater spread in the distribution of relaxation times for interfacial polarization as moisture content increases in this range. This would be the case if increasing wood moisture content released different "species" of ions with widely varying dispersion properties as wood moisture content increases.

6.4 Electrical Moisture Meters

The basic electrical properties of wood and the factors which affect them have been discussed in Sections 6.2 and 6.3. The importance of these factors in relation to electrical moisture meters has also been noted where appropriate. In this section we will discuss moisture meters based on the variation of these basic electrical parameters of wood with moisture content.

As noted earlier, there are two principal types of electric moisture meters, the *resistance* meter which generally measures the direct current resistance (DC) of wood between two electrodes, and the *dielectric* moisture meter which measures one or more of the dielectric properties of wood using alternating current (AC).

6.4.1 Resistance Moisture Meters

Resistance moisture meters are essentially DC megohm-meters, with scales calibrated in terms of wood moisture content based on one or more sets of calibration data such as are shown in Table 6.1. Adjustments for different kinds of wood are usually given in tables supplied by the manufacturer. In some cases they may be incorporated into the meter itself, accessible by means of a calibration or selection switch on the meter.

Historically, the oldest resistance moisture meter, known as the "blinker" moisture meter, produced about 1930, measured wood resistance in terms of the size capacitor, which could be charged to a given voltage in a period of 1 or 2 s, using a standard charging voltage in a series resistance-capacitance circuit. The charged capacitor was discharged through a neon bulb which blinked each time it was charged. The charging capacitor was calibrated in terms of wood moisture content. This was a bulky instrument, primarily because a number of large capacitors were incorporated into it.

Vacuum tube amplifiers were used subsequently, much reducing the size of the instruments. Modern instruments are made still smaller and more convenient, since they include solid state amplifiers and associated circuitry. Some instruments are equipped with species adjustments and with temperature correction circuits which automatically compensate for temperature variations in the wood.

As noted in Section 6.2.3.1 penetrating pin-type electrodes are normally used in resistance moisture meters. The time dependent resistance discussed in Sect. 6.2.3.4, which is particularly evident in more moist wood, may be minimized by use of low frequency AC or else by pulsed DC. At least one manufacturer has provision for switching to one of these modes when wood moisture content is 12% or higher. This is not practical at low moisture contents because the high resistance is difficult to measure using AC or pulsed DC.

The lower practical limit for most resistance moisture meters is in the order of 6% or 7%, although special instruments are available for measuring somewhat lower moisture contents. Moisture measurements in hot wood can be made at lower moisture contents because of the strong decrease in resistance with increasing temperature (Sect. 6.2.2.2). Conversely, the lower moisture limit for measurements in cold wood may be considerably higher.

Resistance moisture meters have been used for many years to monitor moisture conditions in lumber during drying. They sometimes are connected to electrodes embedded into the green lumber before drying, and measurement of resistance between electrodes, as well as wood temperatures, are made during drying. As noted earlier (Sect. 6.2.3), the contact resistance between the wood and electrodes generally increases with time in such cases. This results in moisture meter readings lower than the actual wood moisture content.

Forrer (1984) and Forrer and Vermaas (1987) have developed resistance measuring systems with improved electrodes which minimize the increase in apparent contact resistance with time. They also used low frequency AC excitation to reduce the effect of polarization at the electrodes. The system was automated by incorporation of a microprocessor and associated digital and analog electronic circuitry.

6.4.2 Dielectric Moisture Meters

It has been shown (Sect. 6.3.4) that both the dielectric constant and loss factor increase with increasing wood moisture content at a given AC frequency. Practical dielectric moisture meters usually operate in the audio or in the radio frequency region, although some also use microwave frequencies. There are two basic types of dielectric moisture meters, the *capacitance* type which essentially measures the dielectric constant and the *power-loss* type which measures the combined effect of dielectric constant and loss factor.

The capacitance type of moisture meter responds primarily to the capacitance between the electrodes, which is a function of the electrode configuration and the dielectric constant of the wood. Usually there is an air gap between the electrodes and the wood surface which reduces the sensitivity of the instrument to moisture content changes in the wood.

The power-loss type of dielectric moisture meter responds to increases in both the dielectric constant and loss factor with increasing wood moisture content. In other words, it measures the rate of electrical energy dissipation in the wood for a given input energy. This depends on the electrode configuration, which determines the electromagnetic coupling between the meter and the wood, as well as on the characteristics of the wood itself. The latter are complex functions of wood moisture content, temperature, density, structural orientation, frequency, etc., as is evident from Figs. 6.41 to 6.47.

Despite the many factors in addition to wood moisture content which affect dielectric moisture meter readings, they are useful instruments for estimating wood moisture content. They are often used in the wood industry to indicate wood moisture content during processing operations under reasonably standard conditions of species.

A publication by James (1975a, b) discusses the characteristics of electrical moisture meters for wood and the factors which affect them. Several other papers of interest are those of Mackay (1976), Dennis and Beall (1977), Beall (1979), Pfaff and Garrahan (1986), and James (1986).

References

Ahlgren PA, Wood JR, Goring DIA 1972 The fiber saturation point of various morphological subdivisions of Douglas fir and aspen wood. Wood Sci Technol 6:81–84

Anderson JH, Parks GA 1968 The electrical conductivity of silica gel in the presence of adsorbed water. J Phys Chem 72:3662–3668

Anderson NT, McCarthy JL 1963 Two-parameter isotherm equation for fiber-water systems. Ind Eng Chem Process Design Dev 2:103–105

Anderson RB 1946 Modifications of the Brunauer, Emmett and Teller equation I. J Am Chem Soc 68:686–691

Anderson RB, Hall KW 1948 Modifications of the Brunauer, Emmett and Teller equation II. J Am Chem Soc 70:1727–1734

Babbitt JD 1940 Observations on the permeability of hygroscopic materials to water vapour I. Observations at relative humidities less than 75%. Can J Res A-18 (6):105–121

Babbitt JD 1942 On the adsorption of water vapour by cellulose. Can J Res A-20:143–172

Babbitt JD 1950 On the differential equations of diffusion. Can J Res A-28:449–474

Babbitt JD 1951 On the diffusion of adsorbed gases through solids. Can J Phys 29:437–446

Babbitt JD 1958 Diffusion of adsorbed molecules. Nature 182:201–202

Barber NF 1968 A theoretical model of shrinking wood. Holzforschung 22:97–103

Barber NF, Meylan BA 1964 The anisotropic shrinkage of wood. A theoretical model. Holzforschung 18:146–156

Bariska M 1979 Die Methoden der Dimensionsstabilisierung bei Vollholzprodukten, ihr derzeitiger Stand und ihre Einsatzmöglichkeiten. Holzforsch Holzverwert 31:28–32

Barkas WW 1945 Swelling stresses in gels. Dep Sci Ind Res For Prod Res GB, Spec Rep 6, 60 pp

Barkas WW 1949 The swelling of wood under stress. Dep Sci Ind Res For Prod Res GB, 99 pp

Barker RE, Thomas CR 1964 Effects of moisture and high electric fields on conductivity in alkali-halide-doped cellulose acetate. J Appl Phys 35:3203–3215

Barrett JD, Schniewind AP, Taylor RL 1972 Theoretical shrinkage model for wood cell walls. Wood Sci 4:178–192

Barrie JA 1968 Water in polymers. In: Frank J, Park GS (eds) Diffusion in polymers. Academic Press, New York, London, 259–313

Beall FC 1979 Relative humidity and moisture content instrumentation. In: Proc symp on wood moisture content, temperature and humidity relationships. Virginia Polytech Inst State Univ, Blacksburg VA, US For Serv For Prod Lab, Madison WI, 47–54

Bello ED 1968 Effect of transverse compressive stress on equilibrium moisture content of wood. For Prod J 18 (2):69–76

Bodig J, Jayne BA 1982 Mechanics of wood and wood composites. Van Nostrand-Reinhold, New York, 736 pp

Bohno Y, Tsutsumi J 1974 (Effect of moisture content on dielectric behavior of wood.) Kyushi Univ For Rep 25:151–160

Boquet R, Chirife J, Iglesias HA 1980 On the equivalence of isotherm equations. J Food Technol 15:345–349

Bosshard HH 1956 Über die Anisotropie der Holzschwindung. Holz Roh- Werkst 14:285–295

Boutelje J 1973 On the relationship between structure and the shrinkage and swelling of wood in Swedish pine (Pinus silvestris) and spruce (Picea abies). Svensk Papperstidn 76:78–83

Boyd JD 1974 Anisotropic shrinkage of wood: Identification of the dominant determinants. Mokuzai Gakkaishi 20:473–482

Boyd JD 1977 Relationship between fibre morphology and shrinkage of wood. Wood Sci Technol 11:3–22

Bradley RS 1936 Polymer adsorbed films II. The general theory of condensation of vapours on finely divided solids. J Chem Soc 1936:1799–1804

Briggs GE 1967 Movement of water in plants. Bot Monogr 7. Davis, Philadelphia, 142 pp
Brown JH, Davidson RW, Skaar C 1963 Mechanism of electrical conduction in wood. For Prod J 13:455–459
Browne FL 1957 Swelling of springwood and summerwood in softwood. For Prod J 7:416–424
Brunauer S, Emmett PH, Teller E 1938 Adsorption of gases in multimolecular layers. J Am Chem Soc 60:309–319
Can For Prod Lab 1951 Canadian woods: Their properties and uses. Can For Serv, Ottawa Ont, 367 pp
Cave ID 1972 A theory of shrinkage of wood. Wood Sci Technol 6:284–292
Cave ID 1975 Wood substance as a water-reactive fibre reinforced composite. J Microsc 104:47–52
Cave ID 1976 Modelling the structure of the softwood cell wall for computation of mechanical properties. Wood Sci Technol 10:19–28
Cave ID 1978a Modelling moisture-related mechanical properties of wood I. Properties of the wood constituents. Wood Sci Technol 12:75–86
Cave ID 1978b Modelling moisture-related mechanical properties of wood II. Computation of properties of a model of wood and comparison with experimental data. Wood Sci Technol 12:127–139
Cech MY, Pfaff F 1977 Kiln operator's manual for eastern Canada. East For Prod Lab Rep OPX-192E, 189 pp
Chafe SC 1985 The distribution and interrelationship of collapse, volumetric shrinkage, moisture content and density in trees of Eucalyptus regnans F Muell. Wood Sci Technol 19:329–345
Chafe SC 1986 Collapse, volumetric shrinkage, specific gravity and extractives in Eucalyptus and other species I. The shrinkage/specific gravity ratio. Wood Sci Technol 20:293–307
Chalk L, Bigg JM 1956 The distribution of moisture in the living stem of Sitka spruce and Douglas fir. Forestry 29:6–21
Chen CL 1962 The physical properties of 101 Chinese woods. For Prod J 12:339–342
Chomcharn A, Skaar C 1983a Moisture and transverse dimensional changes during air-drying of small green hardwood wafers. Wood Sci Technol 17:227–240
Chomcharn A, Skaar C 1983b Dynamic sorption and hygroexpansion of wood wafers exposed to sinusoidally varying humidity. Wood Sci Technol 17:259–277
Choong ET 1963 Movement of moisture through a softwood in the hygroscopic range. For Prod J 13:489–498
Choong ET 1965 Diffusion coefficients of softwoods by steady-state and theoretical methods. For Prod J 15:21–27
Choong ET 1969 Effect of extractives on shrinkage and other hygroscopic properties of ten southern pine woods. Wood Fiber 1:124–133
Choong ET, Barnes HM 1968 Swelling behavior in corewood and maturewood of southern pine. Louisiana State Univ Sch For Utiliz Note 11, 5 pp
Choong ET, Manwiller FG 1976 Dimensional and relative hygroscopic properties of hardwoods from southern pine sites. Wood Sci 9:39–43
Christensen GN, Hergt HFA 1969 Effect of previous history on kinetics of sorption by wood cell walls. J Polym Sci A-1 7:2427–2430
Christensen GN, Kelsey KE 1959a Die Geschwindigkeit der Wasserdampfsorption durch Holz. Holz Roh- Werkst 17:178–188
Christensen GN, Kelsey KE 1959b Die Sorption von Wasserdampf durch die chemischen Bestandteile des Holzes. Holz Roh- Werkst 17:189–203
Clark J, Gibbs RD 1957 Studies in tree physiology IV. Further investigations of seasonal changes in moisture content of certain Canadian forest trees. Can J Bot 35:219–253
Clark JD, Williams JW 1933 The electric conductivity of commercial dielectrics and its variation with temperature. J Phys Chem 37:119–131
Cockrell RA 1946 Influence of fibril angle on longitudinal shrinkage of ponderosa pine wood. J For 44:876–878
Cockrell RA 1947 Explantation of longitudinal shrinkage of wood based on interconnected chain-molecule concept of cell-wall structure. Trans Am Soc Mech Eng 69:931–935
Cole KS, Cole RH 1941 Dispersion and adsorption in dielectrics I. Alternating-current characteristics. J Chem Phys 9:341–351
Cooper DNE, Ashpole DK 1959 The heats of sorption of cellulosic fibers deduced from calorimetric measurements. J Text Inst 50:T223–232
Cousins WJ 1976 Elastic modulus of lignin as related to moisture content. Wood Sci Technol 10:9–17

Crews DL 1965 Structural variables and the differential transverse shrinkage of wood. PhD Thesis, SUNY Coll For, Syracuse NY, 80 pp

Cudinov BS 1981 Zum Einfluß der Holzdichte auf den Fasersättigungspunkt. Holztechnologie 22:104–108

Davidson RW 1958 The effect of temperature on the electrical resistance of wood. For Prod J 8:160–164

DeGroot SR 1950 Thermodynamics of irreversible processes. Wiley Interscience, New York, 240 pp

Delgado GA 1970 The effect of hygroscopicity on the heat of wetting of wood. Unpubl Sen Res Problem, SUNY Coll For, Syracuse NY

Dennis JR, Beall FC 1977 Evaluation of a new portable radio-frequency moisture meter on lumber with drying gradients. For Prod J 27(8):24–29

Dent RW 1977 A multilayer theory for gas sorption I. Sorption of a single gas. Text Res J 47:145–152

De S, Wijesinghe LCA 1959 The shrinkage of rays and fibres in wood. Forestry 32:31–38

Dixon HH 1914 Transpiration and the ascent of sap in plants. MacMillan, London, 216 pp

Dole M 1948 Statistical thermodynamics of the sorption of vapors by solids. J Chem Phys 16:25–30

Doppert HL 1967 Adsorption of iodine from aqueous solutions by samples of tire yarn from regenerated cellulose. J Polym Sci A-2 5:263–270

Duff JE 1966 A probe for accurate determination of moisture content of wood products in use. US For Serv Res Note FPL-0142, 10 pp

Duff JE 1968 Moisture distribution in wood-frame walls in winter. For Prod J 18(1):60–64

Dunlap F 1912 The specific heat of wood. US For Serv Bull 110, 28 pp

Duran-Pulido JA 1977 Copper, chromium and arsenic ion migration and electrical conductivity relationships in wood. PhD Thesis, SUNY Coll Environ Sci For, Syracuse NY, 413 pp

Ecklund BA 1961 The fiber saturation point of wood as determined with radio isotopes. MS Thesis, Univ California, Berkeley CA

Eisenberg D, Kauzmann W 1969 The structure and properties of water. Oxford Univ Press, New York, 296 pp

Enderby JA 1955 Water absorption by polymers. Trans Faraday Soc 51:106–116

Erickson HD 1955 Tangential shrinkage of serial sections within annual rings of Douglas-fir and western red cedar. For Prod J 5:241–250

Espenas LD 1971 Shrinkage of Douglas-fir, western hemlock, and red alder as affected by drying conditions. For Prod J 21(6):44–46

Espenas LD 1974 Longitudinal shrinkage of western redcedar, western hemlock, and true fir. For Prod J 24(10):46–48

Evershed S 1913 The characteristics of insulation resistance. J Inst Electr Eng 52:51–73

Eyring HF 1936 Viscosity, plasticity and diffusion as examples of absolute reaction rates. J Chem Phys 4:282–291

Feist WC, Hon DN 1984 Chemistry of weathering and protection. In: Rowell R (ed) The chemistry of solid wood. Adv Chem Ser 207. Am Chem Soc, Washington DC, 401–451

Feist WC, Tarkow H 1967 A new procedure for measuring fiber saturation points. For Prod J 17(10):65–68

Forrer JB 1984 An electronic system for monitoring moisture gradients of drying wood. For Prod J 34 (7/8):34–38

Forrer JB, Vermaas HF 1987 Development of an improved moisture meter for wood. For Prod J 37(2):67–71

Forsaith CC 1926 The technology of New York State timbers. NY State Coll For, Syracuse NY, Tech Publ 18, 374 pp

Frey-Wyssling A 1940a Die Anisotropie des Schwindmaßes auf dem Holzquerschnitt. Holz Roh-Werkst 3:43–45

Frey-Wyssling A 1940b Die Ursachen der anisotropen Schwindung des Holzes. Holz Roh- Werkst 3:349–353

Gascoyne PRC, Pethig R 1977 Experimental and theoretical aspects of hydration isotherms for biomolecules. J Phys Chem Faraday Trans 1 73:171–180

Gerhards CC 1982 Effect of moisture content and temperature on the mechanical properties of wood: An analysis of immediate effects. Wood Fiber 14:4–36

Gibbs RD 1958 Patterns in the seasonal water content of trees. In: Thimann KV (ed) The physiology of forest trees. Ronald, New York, 43–69

Goldstein IS (ed) (1977 Wood technology: Chemical aspects. Am Chem Soc, Washington DC, 372 pp
Gorman TM 1985 Juvenile wood as a cause of seasonal arching in trusses. For Prod J 35(11/12):35–40
Goulet M, Fortin Y 1975 (Measuring the swelling of sugar maple wood during a cycle of moisture sorption at 21 °C.) Dép d'Exploit Utilis Bois, Univ Laval, Québec, Note Rech 12, 49 pp
Green DW, Link CL, DeBonis AL, McLain TE 1986 Predicting the effect of moisture content on the flexural properties of southern pine dimension lumber. Wood Fiber Sci 18 (1):134–156
Greenhill WL 1936 The shrinkage of Australian timbers I. A new method of determining shrinkages and shrinkage figures for a number of Australian species. CSIRO Div For Prod, Melbourne, Tech Pap 21, 54 pp
Greenhill WL 1944 The differential shrinkage of wood. Trans Am Soc Mech Eng 66:152–154
Grumach M 1951 The equilibrium moisture content of wood in superheated steam. CSIRO Div For Prod, Melbourne, Int Proj 17, Prog Rep 5, 5 pp
Hailwood AJ, Horrobin S 1946 Absorption of water by polymers. Analysis in terms of a simple model. Trans Faraday Soc 42B:84–92, 94–102
Halsey G 1948 Physical adsorption on nonuniform surfaces. J Chem Phys 16:931–937
Handa T, Fukuoka M, Yoshizawa S, Kamamoto T 1982 The effect of moisture on the dielectric relaxations in wood. J Appl Polym Sci 27:439–453
Hann RA 1969 Longitudinal shrinkage in seven species of wood. US For Serv Res Note FPL-0203, 12 pp
Harris JM 1977 Shrinkage and density of radiata pine compression wood in relation to its anatomy and mode of formation. NZ J For Sci 7:91–106
Harris JM, Meylan BA 1965 The influence of microfibril angle on longitudinal and tangential shrinkage in Pinus radiata. Holzforschung 19:144–153
Hart CA 1964 Theoretical effect of gross anatomy upon conductivity of wood. For Prod J 14:25–32
Hart CA 1984 Relative humidity, EMC, and collapse shrinkage in wood. For Prod J 34(11/12):45–54
Hasted JB 1973 Studies in chemical physics. Aqueous dielectrics. Chapman and Hall, London, 302 pp
Hawley LF 1931 Wood-liquid relations. USDA Tech Bull 248, 34 pp
Hearle JWS 1953 The electrical resistance of textile materials. J Text Inst 44:T177–198
Hearle JWS 1957 The relation between structure, dielectric constant and electrical resistance of fibres. J Text Inst 48:P40–51
Hearle JWS, Peters RH (eds) (1960 Moisture in textiles. Wiley Interscience, New York, 212 pp
Hearmon RFS, Burcham JN 1954 The dielectric properties of wood. For Prod Res, London, Spec Rep 8, 19 pp
Hearmon RFS, Burcham JN 1955 Specific heat and heat of wetting of wood. Nature 176:978
Henderson LG, Choong ET 1968 Variation in moisture content of standing sweetgum trees in Louisiana. Louisiana State Univ Sch For, For Note 81, 3 pp
Henderson SM 1952 A basic concept of moisture equilibrium. Agr Eng 33:29–33
Hergt HFA, Christensen GN 1965 Variable retention of water by dry wood. J Appl Polym Sci 9:2345–2361
Hermans PH 1946 Contributions to the physics of cellulose fibres. Elsevier, Amsterdam, 534 pp
Higgins NC 1957 The equilibrium moisture content-relative humidity relationships of selected native and foreign woods. For Prod J 7:371–377
Hildebrand R 1970 Kiln drying of sawn timber. Hildebrand Maschinenbau, Württ, Fed Rep Germany, 199 pp
Hill RM 1978 Characterization of dielectric loss in solids and liquids. Nature 275:96–99
Hiruma J 1915 (Experiment on the electric resistance of wood.) Bur For Dep Agr Commerce, Japan, Bull For Exp Sta Meguro, Tokyo, 59–65
Hittmeier ME 1967 Effect of structural direction and initial moisture content on swelling rate of wood. Wood Sci Technol 1:109–121
Hoadley RB 1967 Weather, water and wood. Univ Mass Coop Ext Serv, Amherst MA, Publ 15, 8 pp
Hoadley RB 1980 Understanding wood. A craftsman's guide to wood technology. Taunton Press, Newtown, 256 pp
Hsu NN, Tang RC 1975 Distortion and internal stresses in lumber due to anisotropic shrinkage. Wood Sci 7:298–307
Iglesias HA, Chirife J 1981 An equation for fitting uncommon water isotherms in foods. Lebensm Wiss Technol 14:105–106

Iglesias HA, Chirife J 1982 Handbook of food isotherms. Water sorption parameters for food and food components. Academic Press, New York, London, 347 pp

Imata J 1975 A study on the shrinkage and deformation of wood. Inst Vocat Training Tokyo Bull 4:21–75

James WL 1954 Electrical moisture meters for wood. US For Serv Rep FPL-1660, 13 pp

James WL 1965 Effects of wood preservatives on electrical moisture-meter readings. US For Serv Res Note FPL-0106, 20 pp

James WL 1968 Effect of temperature on readings of electric moisture meters. For Prod J 18(10):23–31

James WL 1975a Dielectric properties of wood and hardboard. Variation with temperature, frequency, moisture content, and grain orientation. US For Serv Res Pap FPL-245, 33 pp

James WL 1975b Electric moisture meters for wood. US For Serv Gen Tech Rep FPL-6, 28 pp

James WL 1981 Influence of electrode design on measurements of dielectric properties of wood. Wood Sci 13:185–198

James WL 1986 The interaction of electrode design and moisture gradients in dielectric measurements on wood. Wood Fiber Sci 18:264–275

James WL, Hamill DW 1965 Dielectric properties of Douglas-fir measured at microwave frequences. For Prod J 15:51–56

Johnson JA 1975 Hygro-thermoelastic stresses in the transverse section of wood treated as acylindrically orthotropic material. PhD Thesis, Univ Washington, Seattle WA, 193 pp

Johnson JA 1977 The removal of water from wood. In: Proc on practical applications of solar energy to wood processing. Virginia Polytech Inst State Univ, Blacksburg VA, For Prod Res Soc, Madison WI, 23–28

Jost W 1960 Diffusion of solids, liquids, and gases. Academic Press, New York, London, 558 pp

Kadita S 1960 Studies on the water sorption of wood. Wood Res 23:1–61

Kajita H 1975 (The bark-water relationship.) J Soc Mat Sci Jpn 24(264):862–866

Kajita H 1976 (The heat of wetting of wood in water I.) Kyoto Pref Univ For Bull 20:49–61

Kajita H 1977 (The heat of wetting of wood in water III.) Kyoto Pref Univ For Bull 29:85–94

Kajita H, Mukudai J 1971 (Studies on the wettability of wood VII.) Kyoto Pref Univ For Sci Rep 23:100–104

Kajita H, Yoneda M, Koshijima T, Mukudai J, Yata S 1977 (Heat of wetting of the constituents of wood in water. J Soc Mat Sci Jpn 26(284):417–423

Kanagawa Y, Hattori Y 1984 Studies on some Sulawesi woods: 2.1 Shrinkage properties. Nagoya Univ For Bul 8:54–78

Katchalsky A, Curran PF 1966 Nonequilibrium thermodynamics in biophysics. Harvard University Press, Cambridge, Oxford University Press, Oxford, 248 pp

Kato H, Nakato K 1968 (The transverse anisotropic shrinkage of wood and its relation to cell wall structure I. The lignin distribution in the radial and tangential walls of coniferous wood tracheids). Kyoto Univ For Bull 40:284–292

Katz JR 1917 Gesetze der Quellung. Kolloidchem Beih 9:1–182

Kauman WG 1956 Equilibrium moisture content relations and drying control in superheated steam drying. For Prod J 6:328–332

Kauman WG 1964 Zellkollaps im Holz I. Einflußgrößen bei der Entstehung des Zellkollaps und seine Rückbildung. Holz Roh- Werkst 22:183–196

Kawai S, Nakato K, Sadoh T 1980 Moisture movement in wood in the hygroscopic range. Coll Agr Kyoto Univ Mem 115:1–32

Kellogg RM, Sastry CBR, Wellwood RW 1975 Relationships between cell-wall composition and cell-wall density. Wood Fiber 7:170–177

Kellogg RM, Wangaard FF 1969 Variation in the cell-wall density of wood. Wood Fiber 1:180–204

Kelsey KE 1956 The shrinkage intersection point – its significance and the method of its determination. For Prod J 6:411–416

Kelsey KE 1957 The sorption of water vapour by wood. Aust J Appl Sci 8:42–54

Kelsey KE 1963 A critical review of the relationship between the shrinkage and structure of wood. CSIRO Div For Prod, Melbourne, Technol Pap 28, 35 pp

Kelsey KE, Clarke LN 1956 The heat of sorption of water by wood. Aust J Appl Sci 7:160–175

Keylwerth R 1949 Grundlagen der Hochtemperaturtrocknung des Holzes.ctr Holz-Zbl 75:953–954

Keylwerth R 1951 Formänderungen in Holzquerschnitten. Holz Roh- Werkst 9:253–260

Keylwerth R 1962 Untersuchungen über freie und behinderte Quellung von Holz 1. Freie Quellung. Holz Roh- Werkst 20:252–259

Keylwerth R 1964 Untersuchungen über freie und behinderte Quellung 4. Untersuchungen über den Quellungsverlauf und die Feuchtigkeitsabhängigkeit der Rohdichte von Hölzern. Holz Roh- Werkst 22:255–258

Keylwerth R, Noack D 1956 Über den Einfluß höherer Temperaturen auf die elektrische Holzfeuchtigkeitsmessung nach dem Widerstandsprinzip. Holz Roh- Werkst 14:162–172

King G 1960 Theories of multilayer adsorption. In: Hearle JW, Peters RH (eds) Moisture in textiles. Wiley Interscience, New York, 59–82

King J, Rollwitz WL 1983 Magnetic resonance measurement of flowing coal. Trans Instr Soc Am 22(4):69–76

Kirk TK, Cowling EB 1984 Biological degradation of solid wood. In: Rowell R (ed) The chemistry of solid wood. Adv Chem Ser 207, Am Chem Soc, Washington DC, 455–487

Koch P 1971 Process for straightening and drying southern pine 2×4's in 24 hours. For Prod J 21(5):17–24

Koehler A 1960 (1930) Longitudinal shrinkage of wood. US For Serv Rep FPL-1093, 23 pp

Kollmann FFP 1936 Technologie des Holzes. Springer, Berlin, 764 pp

Kollmann FFP 1940 Die mechanischen Eigenschaften verschieden feuchter Hölzer im Temperaturbereich −200 bis +200°C. VDI-Forschungsh 403(11):1–18

Kollmann FFP 1959 Über die Sorption von Holz und ihre exakte Bestimmung. Holz Roh- Werkst 17:165–171

Kollmann FFP 1962 Eine Gleichung der Sorptionsisotherme. Naturwissenschaften 49:206–207

Kollmann FFP, Côté WA, Jr 1968 Principles of wood science and technology I. Solid wood. Springer, Berlin, Heidelberg, New York, 592 pp

Kollmann FFP, Höckele G 1962 Kritischer Vergleich einiger Bestimmungsverfahren der Holzfeuchtigkeit. Holz Roh- Werkst 20:461–473

Kollmann FFP, Kuenzi EW, Stamm AJ 1975 Principles of wood science and technology II. Wood-based materials. Springer, Berlin, Heidelberg, New York, 703 pp

Kröner K 1944/45 Über dielektrische Untersuchungen an Naturhölzern und deren mechanischen und chemischen Abbaustoffen im grossen Frequenzgebiet. Holz Roh- Werkst 7:65–72

Kröner K, Pungs L 1952 Zur dielektrischen Anisotropie des Naturholzes im großen Frequenzbereich. Holzforschung 6:13–16

Kröner K, Pungs L 1953 Über das Verhalten des dielektrischen Verlustfaktors von Naturholz im großen Frequenzbereich. Holzforschung 7:12–18

Kübler H 1962 Schwinden und Quellung des Holzes durch Kälte. Holz Roh- Werkst 20:364–368

Kübler H 1970 A note on recovery of excessive shrinkage in wood. Wood Sci 3:62–64

Kufner M 1978 Elastizitätsmodul und Zugfestigkeit von Holz verschiedener Rohdichte in Abhängigkeit vom Feuchtigkeitsgehalt. Holz Roh- Werkst 36:435–439

Kühlmann G 1962 Untersuchung der thermischen Eigenschaften von Holz und Spanplatten in Abhängigkeit von Feuchtigkeit und Temperatur im hygroskopischen Bereich. Holz Roh- Werkst 20:259–270

Kuroda N, Tsutsumi J 1979 (Effect of applied voltage on electrical conductivity of wood.) Mokuzai Gakkaishi 25:757–762

Kuroda N, Tsutsumi J 1982 (Anisotropical behavior of electrical conduction in wood). Kyushu Univ For Rep 28:25–30

Labusa TP 1984 Moisture sorption: practical aspects of isotherm measurement and use. American Assoc Cereal Chemists, St Paul, 150 pp

Labusa TP, Mizrachi S, Karel M 1972 Mathematical models for optimization of flexible film packaging of foods for storage. Trans Am Soc Agr Eng 15:150–155

Langmuir I 1918 The adsorption of gases on plain surfaces of glass, mica and platinum. J Am Chem Soc 40:1361–1403

Langwig JE 1971 Trace element-electrical conductivity relationships in wood. PhD Thesis, SUNY Coll Environ Sci For, Syracuse NY, 181 pp

Langwig JE, Meyer JA 1973 Ion migration in wood verified by neutron activation analysis. Wood Sci 6:39–50

Langwig JE, Skaar C 1975 Electrical resistivities of 18 Venezuelan and 13 North American woods. Wood Sci 8:350–354

Leont'ev NL 1960 Pročnost' drevesiny pri različnoj vlažnosti. (The strength of wood at varying moisture contents.) Derevcobrab Promst 9(10):17–18
Limaye VD, Seaman LN 1933 The physical and mechanical properties of woods grown in India. Indian For Rec Silvic 18(10):1–70
Lin RT 1965 A study of the electrical conduction in wood. For Prod J 15:506–514
Lin RT 1967 Review of the dielectric properties of wood and cellulose. For Prod J 17(7):61–66
Lin RT 1973 Wood as an orthotropic dielectric material. Wood Fiber 5:226–236
Lindsay F, Chalk L 1954 The influence of rays on the shrinkage of wood. Forestry 27:16–24
Ma TP 1960 The mechanical and physical properties of important timbers of Taiwan. Taiwan For Res Inst, Ext Bull 19, 4 pp
Mackay JFG 1976 Effect of moisture gradients on the accuracy of power-loss moisture meters. For Prod J 26(3):49–52
Malmquist L 1958 Sorption as deformation in space. Kyltekn Tidskr 4:1–11
Malmquist L 1959 Die Wasserdampf-Sorption des Holzes vom Standpunkt einer neuen Sorptionstheorie. Holz Roh- Werkst 17:171–178
Malmquist L 1967 Untersuchungen zur empirisch-mathematischen Analyse der Sorption von Wasserdampf durch Holz. Holz Roh- Werkst 25:45–62
Manwiller FG 1975 Wood and bark moisture contents of small-diameter hardwoods growing on southern pine sites. Wood Sci 8:384–388
Matsumoto T 1950 (The anisotropic shrinkage of wood) Morioka Coll Agr For Iwate Univ Bull 26:81–88
McIntosh DC 1955 Shrinkage of red oak and beech. For Prod J 5:355–359
McLain TE, DeBonis AL, Green DW, Wilson FJ, Link CL 1984 The influence of moisture content on the flexural properties of southern pine dimension lumber. US For Serv Res Pap FPL-447, 40 pp
McLaughlan TA, Norton JA, Kusec DJ 1973 Slope-of-grain indicator. For Prod J 23(5):50–55
McMillen JM 1958 Stresses in wood during drying. US For Serv Rep FPL-1652, 33 pp
McMillin CW 1969 Specific heat of oven-dry loblolly pine wood. Wood Sci 2:107–111
McMillin CW 1970 Specific heat of some oven-dry chemical constituents of loblolly pine wood. Wood Sci 3:52–53
McMinn JW 1986 Transpirational drying of red oaks, sweetgum, and yellow poplar in the upper Piedmont of Georgia. For Prod J 36(3):25–27
Meierhofer U, Sell J 1979 Physikalische Vorgänge in wetterbeanspruchten Holzbauteilen 2. Tragende Holzbauteile im Freien unter Dach. Holz Roh- Werkst 37:227–234
Meredith R (ed) 1956 The mechanical properties of textile fibers. Wiley Interscience, New York, 345 pp
Meredith R 1957 Rigidity, moisture and fibre structure. J Text Inst 48:T163–174
Merten U (ed) 1966 Desalination by reverse osmosis. MIT Press, Cambridge, 289 pp
Meyer JA 1984 Wood-polymer materials. In: Rowell R (ed) The chemistry of solid wood. Adv Chem Ser 207. Am Chem Soc, Washington DC, 257–289
Meylan BA 1968 Cause of high longitudinal shrinkage in wood. For Prod J 18(4):75–78
Meylan BA 1972 The influence of microfibril angle on the longitudinal shrinkage-moisture content relationship. Wood Sci Technol 6:293–301
Mitchell PH 1981 Effect of heat treatment on selected properties of small clear specimens of Pinus taeda L and Pseudotsuga menziesii (Mirb) Franco. PhD Thesis, Virginia Polytech Inst State Univ, Blacksburg VA, 172 pp
Mörath E 1932 Studien über die hygroskopischen Eigenschaften und die Härte der Hölzer. Mitt Holzforsch Darmstadt (cited by Pentoney 1953)
Morschauser CR, Preston SB 1954 The effect of ray tissue on transverse shrinkage of red oak (Quercus borealis Michxf). Mich Wood Technol 1, 2 pp
Morton WE, Hearle JWS 1962 Physical properties of textile fibres. Butterworths, London, 608 pp
Morton WE, Hearle JWS 1975 Physical properties of textile fibres, 2nd edn. John Wiley, New York, 660 pp
Murphy EJ 1960 The temperature dependence of the conductivity of dry cellulose. Phys Chem Solid 15:66–71
Myer JE, Rees LW 1926 Electrical resistance of wood with special reference to the fiber saturation point. NY State Coll For, Syracuse NY, Tech Bull 19, 22 pp
Nagashima N, Suzuki E 1981 Pulsed NMR and state of water in foods. In: Rockland LB, Stewart GF (eds) Water activity: Influences on food quality. Academic Press, New York, 247–264

Nakato K 1958 (On the cause of the anisotropic shrinkage of wood X. On the relationships between the microscopic structure and the anisotropic shrinkage in transverse section.) Mokuzai Gakkaishi 4:183–186

Nanassy AJ 1964 Electric polarization measurements on yellow birch. Can J Phys 42:1270–1281

Nanassy AJ 1973 Use of wide-line NMR for measurement of moisture content in wood. Wood Sci 5:187–193

Nanassy AJ 1974 Water sorption in green and remoistened wood studied by the broad-line component of the wide-line NMR spectrum. Wood Sci 7:61–68

Nanassy AJ, Szabo T 1978 Thermal properties of waferboards as determined by a transient method. Wood Sci 11:17–22

Nelson RM, Jr 1983 A model for sorption of water vapor by cellulosic materials. Wood Fiber 15:8–22

Nelson RM, Jr 1984 A model for bound water diffusion in wood. Proc North Am Wood Drying Symp Mississippi For Prod Util Lab, 144–151

Nelson RM, Jr 1986a Diffusion of bound water in wood 1. The driving force. Wood Sci Technol 20:125–135

Nelson RM, Jr 1986b Diffusion of bound water in wood 2. A model for isothermal diffusion. Wood Sci Technol 20:235–251

Nelson RM, Jr 1986c Diffusion of bound water in wood 3. A model for nonisothermal diffusion. Wood Sci Technol 20:309–328

Newlin JA, Wilson TRC 1919 The relation of the shrinkage and strength properties of wood to its specific gravity. US For Serv Bull FPL-676, 35 pp

Noack D 1959 Beitrag zur Sorption des Holzes im Bereich höherer Temperaturen und Wasserdampfdrücke. Holz Roh- Werkst 17:205–211

Noack D 1964 Einfluß der Probenabmessungen auf die Bestimmung der Quellmasse von Holz. Holz Roh- Werkst 22:174–182

Noack D, Kleuters W 1960 Über die Bestimmung des Holzfeuchtigkeitsgehaltes mit Hilfe radioaktiver Isotopen (β-Strahlen). Holz Roh- Werkst 18:304–308

Noack D, Schwab E, Bartz A 1973 (Characteristics for a judgment of the sorption and swelling behavior of wood.) Wood Sci Technol 7:218–236

Norberg PH, Meier H 1966 Physical and chemical properties of the gelatinous layer in tension wood fibres of aspen (Populus tremula L). Holzforschung 20:174–178

Norimoto M 1976 Dielectric properties of wood. Wood Res 59/60:106–152

Norimoto M, Yamada T 1969 (The dielectric properties of wood II. Temperature dependence of dielectric properties of wood in absolutely dry condition.) Wood Res 46:1–9

Norimoto M, Yamada T 1970 The dielectric properties of wood IV. On dielectric dispersions of oven-dried wood. Wood Res 50:36–49

Norimoto M, Yamada T 1971 The dielectric properties of wood V. On the dielectric anisotropy in wood. Wood Res 51:12–32

Norimoto M, Yamada T 1972 The dielectric properties of wood VI. On the dielectric properties of the chemical constituents of wood and the dielectric anisotropy of wood. Wood Res 52:31–43

Okoh KIA 1976 Moisture sorption and electrical resistivity of some woods and barks. PhD Thesis, SUNY Coll Environ Sci For, Syracuse NY, 195 pp

Okoh KIA, Skaar C 1980 Moisture sorption isotherms of the wood and inner bark of ten southern US hardwoods. Wood Fiber 12:98–111

Onsager L 1931 Reciprocal relations in irreversible processes I, II. Phys Rev 37:405–426, 38:2265–2279

Oswin CR 1946 Kinetics of package life III. Isotherm. J Soc Chem Ind 65:419–421

Panshin AJ, de Zeeuw CH 1980 Textbook of wood technology, 4th edn. McGraw-Hill, New York, 722 pp

Paton JM, Hearmon RFS 1957 Effect of exposure to gamma-rays on the hygroscopicity of Sitka spruce wood. Nature 180:651

Peck EC 1950 Moisture content of wood in use. US For Serv Rep FPL-1655, 9 pp

Peck EC 1953 The sap or moisture in wood. US For Serv Rep FPL-768, 5 pp

Peirce FT 1929 A two-phase theory of the absorption of water vapour by cotton cellulose. J Text Inst 20:T133–150

Pentoney RE 1953 Mechanisms affecting tangential vs radial shrinkage. J For Prod Res Soc 3(2):27–32

Perem E 1954 Determination of the fiber saturation point of wood by centrifuging. J For Prod Res Soc 4:77–81

Perkitny T 1963 Swelling pressure of wood. Fifth FAO Conf Wood Technol, US For Prod Lab, Madison WI, 12 pp

Pfaff F, Garrahan P 1986 New temperature correction factors for the portable resistance-type moisture meter. For Prod J 36(3):28–30

Phillips EWJ, Adams EH, Hearmon RFS 1962 The measurement of density variation within the growth rings in thin sections of wood using beta particles. J Inst Wood Sci 10:11–28

Pittman RA, Tripp VW 1972 Wide-line nuclear magnetic resonance spectroscopy. In: O'Conner RT (ed) Instrumental analysis of cotton cellulose and modified cotton cellulose. Marcel Decker, New York, 379–399

Pratt GH 1974 Timber drying manual. Dep Environ Build Res, Princes Risborough GB, 152 pp

Preston JM (ed) (1953 Fibre science. Text Inst, Manchester GB, 421 pp

Preston RD 1942 Anisotropic contraction of wood and its constituent cells. Forestry 16:32–48

Prichananda C 1966 A study of some aspects of moisture sorption dynamics in wood. PhD Thesis, SUNY Coll For, Syracuse NY, 153 pp

Prigogine I 1961 Introduction to thermodynamics of irreversible processes 2nd edn. Wiley Interscience, New York, 119 pp

Purslow DF 1971 The use of a microwave moisture meter for studying moisture changes. J Inst Wood Sci (28)5(4):40–46

Quirk JT 1984 Shrinkage and related properties of Douglas-fir cell walls. Wood Fiber 16:115–133

Raczkowski J 1963 The swelling heat of wood. Fifth FAO Conf Wood Technol, US For Prod Lab, Madison WI, 4 pp

Rafalski H-J 1966 Über die dielektrischen Eigenschaften unterschiedlich verdichteten Rotbuchenholzes. Holztechnologie 7:118–122

Rasmussen EF 1961 Dry kiln. Operator's manual. USDA Agr Handb 188, 197 pp

Rees WH 1960 Heat of absorption. In: Hearle JWS, Peters RH (eds) Moisture in textiles. Wiley Interscience, New York, 33–58

Resch H, Ecklund BA 1963 Moisture content determination for wood with highly volatile constituents. For Prod J 13:481–482

Ritter GJ, Mitchell RL 1952 Fiber studies contributing to the differential shrinkage of cellulose. Pap Ind 33:1189–1193

Rogers CE 1967 Polymer films as coatings. In: Marchessault RH, Skaar C (eds) Surfaces and coatings related to paper and wood. Syracuse University Press, Syracuse, 463–490

Rosen HN 1974 Distribution of water in hardwoods: a mathematical model. Wood Sci Technol 8:283–299

Rosen HN 1979 Psychrometric relationships and equilibrium moisture content of wood above 212 °F. In: Proc symp wood moisture content, temperature and humidity relationships. Virginia Polytech Inst State Univ, Blacksburg VA, US For Serv FPL, Madison WI, 12–22

Sadoh T, Christensen GN 1967 Longitudinal shrinkage of wood I. Longitudinal shrinkage of thin sections. Wood Sci Technol 1:26–44

Salamon M 1966 Effects of drying severity on properties of western hemlock. For Prod J 16(1):39–46

Salamon M, Hejjas J, Hejja A, Pollok LB 1975 Sorption studies of softwoods. Can For Serv Publ 1346, 20 pp

Samson M 1984 Measuring general slope of grain with the slope-of-grain indicator. For Prod J 34(7/8):27–30

Schniewind AP, Kersavage PC 1962 Influence of rate of drying and rewetting on the dimensional changes of California black oak. For Prod J 12:29–33

Schroeder H 1972 Shrinking and swelling differences between hardwoods and softwoods. Wood Fiber 4:20–25

Sell J 1985 Long-term measurement of wood moisture content in structural members. For Prod J 35(11/12):27–29

Sharp AR, Riggin MT, Kaiser R, Schneider MH 1978 Determination of moisture content of wood by pulsed nuclear magnetic resonance. Wood Fiber 10:74–81

Siau JF 1971 Flow in wood. Syracuse University Press, Syracuse, 131 pp

Siau JF 1980 Nonisothermal moisture movement in wood. Wood Sci 13:11–13

Siau JF 1984 Transport processes in wood. Springer, Berlin, Heidelberg, New York, 245 pp

Siau JF, Babiak M 1983 Experiments on nonisothermal moisture movement in wood. Wood Fiber Sci 15:40–46

Siau JF, Bao F, Avramidis S 1986 Experiments in nonisothermal diffusion of moisture in wood. Wood Fiber Sci 18:84–89

Simpson WT 1966 The swelling pressure and reduced sorption of wood restrained from swelling. MS Thesis, SUNY Coll For, Syracuse NY, 79 pp

Simpson WT 1971a Equilibrium moisture content prediction for wood. For Prod J 21(5):48–49

Simpson WT 1971b Moisture changes induced in red oak by transverse stress. Wood Fiber 3:13–21

Simpson WT 1973 Predicting equilibrium moisture content of wood by mathematical models. Wood Fiber 5:41–49

Simpson WT 1979 Sorption theories for wood. In: Proc symp on wood moisture content and humidity relationships. Virginia Polytech Inst State Univ, Blacksburg VA, US For Prod Lab, Madison WI, 36–46

Simpson WT 1982 Importance of relative humidity and temperature control on wood moisture content. Wood Fiber 14:94–103

Simpson WT, Skaar C 1968 Effect of restrained swelling on wood moisture content. US For Serv Res Note FPL-0196, 5 pp

Skaar C 1948 The dielectric properties of wood at several radio frequencies. NY State Coll For, Syracuse NY, Tech Publ 69, 36 pp

Skaar C 1964 Some factors involved in the electrical determination of moisture gradients in wood. For Prod J 14:239–243

Skaar C 1972 Water in wood. Syracuse University Press, Syracuse, 218 pp

Skaar C, Babiak M 1982 A model for bound water transport in wood. Wood Sci Technol 16:123–138

Skaar C, Prichananda C, Davidson RW 1970 Some aspects of moisture sorption dynamics in wood. Wood Sci 2:179–185

Skaar C, Siau JF 1981 Thermal diffusion of bound water in wood. Wood Sci Technol 15:105–112

Smith SE 1947 The sorption of water vapor by high polymers. J Am Chem Soc 69:646–651

Spalt HA 1957 The sorption of water vapor by domestic and tropical woods. For Prod J 7:331–335

Spalt HA 1958 The fundamentals of water vapor sorption by wood. For Prod J 8:288–295

Stamm AJ 1927 The electrical resistance of wood as a measure of its moisture content. Ind Eng Chem 19:1021–1025

Stamm AJ 1929 The fiber-saturation point of wood as obtained from electrical conductivity measurements. Ind Eng Chem Anal Ed 1:94–97

Stamm AJ 1946 Passage of liquids, vapors and dissolved materials through softwoods. USDA Tech Bull 929, 80 pp

Stamm AJ 1959 Bound-water diffusion into wood in the fiber direction. For Prod J 9:27–32

Stamm AJ 1960 Bound-water diffusion into wood in across-the-fiber directions. For Prod J 10:524–528

Stamm AJ 1964 Wood and cellulose science. Ronald Press, New York, 549 pp

Stamm AJ 1971 Review of nine methods for determining the fiber saturation points of wood and wood products. Wood Sci 4:114–128

Stamm AJ 1977 Dimensional changes of wood and their control. In: Goldstein I (ed) Wood technology. Chemical aspects. ACS Symp Ser 43, Am Chem Soc, Washington DC, 115–149

Stamm AJ, Burr HK, Kline AA 1955 Heat stabilized wood (Staywood). US For Serv Rep FPL-R1621, 17 pp

Stamm AJ, Loughborough WK 1935 Thermodynamics of the swelling of wood. J Phys Chem 39:121–132

Stamm AJ, Loughborough WK 1942 Variation in shrinking and swelling of wood. Trans Am Soc Mech Eng 64:379–386

Stamm AJ, Seborg RM 1935 Adsorption compression of cellulose and wood I. Density measurements in benzene. J Phys Chem 39:133–142

Stanish MA 1986 The roles of bound water chemical potential and gas phase diffusion in moisture transport through wood. Wood Sci Technol 20(1):53–70

Stevens WC 1963 The transverse shrinkage of wood. For Prod J 13:386–389

Stone JE, Scallan AM 1967 The effect of component removal upon the porous structure of the cell wall in wood II. Swelling in water and the fiber saturation point. Tappi 50:496–501

Tarkow H, Turner HD 1958 The swelling pressure of wood. For Prod J 8:193–197

Tiemann HD 1906 Effect of moisture upon the strength and stiffness of wood. USDA For Serv Bull 70, 144 pp
Tiemann HD 1944 Wood technology. Constitution, properties and uses, 2nd edn. Pitman, New York, 328 pp
Trapp W, Pungs L 1956 Einfluß von Temperatur und Feuchte auf das dielektrische Verhalten von Naturholz im großen Frequenzbereich. Holzforschung 10:144–150
Troller JA, Christian JHB 1978 Water activity and food. Academic Press, New York, London, 240 pp
Troughton GE, Clarke MR 1987 Development of a new method to measure moisture content in unseasoned veneer and lumber. For Prod J 37(1):13–19
Tsoumis G 1960 Untersuchungen über die Schwankungen des Feuchtigkeitsgehaltes von lufttrockenem Holz. Holz Roh- Werkst 18:415–422
Tsutsumi J 1967 (Studies on dielectric properties of wood. Effects of frequency, moisture content and temperature on dielectric constant and dielectric loss constant.) Kyushu Univ For Bull 41:109–168
Tsutsumi J, Watanabe H 1966 Studies on the dielectric behaviour of wood II. Effect of moisture content on dielectric constant and dielectric loss factor. Mokuzai Gakkaishi 12:115–118
Turc CO, Cutter BE 1984 Sorption and shrinkage studies of six Argentine woods. Wood Fiber Sci 16:575–582
Urquhart AR 1960 Sorption isotherms. In: Hearle JWS, Peters RH (eds) Moisture in textiles. Wiley Interscience, New York, 14–32
US For Prod Lab 1955 Wood handbook. USDA FPL Agr Handb 72, 528 pp
US For Prod Lab 1958 Effects of wood preservatives on electrical moisture meter readings. US For Serv Rep FPL-R1682, 7 pp
US For Prod Lab 1974 Wood handbook. Wood as an engineering material. USDA FPL Agr Handb 72, 414 pp
Uyemura T 1960 (Dielectric properties of woods as the indicator of the moisture.) Gov For Exp Sta Tokyo Bull 119:95–172
Van den Berg C, Bruin S 1981 Water activity and estimation in food systems. Theoretical aspects. In: Rockland LB, Stewart GF (eds) Water activity. Influence on food quality. Academic Press, New York, London 1–61
Van Rysselberghe PV 1963 Thermodynamics of irreversible processes. Blaisdell, New York, 165 pp
Venkateswaran A 1970 Sorption of aqueous and nonaqueous media by wood and cellulose. Chem Rev 70:619–637
Venkateswaran A 1971 Application of dissociation hypothesis to electrical conduction in wood. Wood Sci 3:183–192
Venkateswaran A 1972 A comparison of the electrical properties of milled wood, milled wood cellulose, and milled wood lignin. Wood Sci 4:248–253
Venkateswaran A 1973 Effect of compression on electrical properties of wood and cellulose. Wood Sci 5:230–234
Venkateswaran A 1974a A note on densities and conductivities of wood. Wood Sci 5:60–62
Venkateswaran A 1974b The interdependence of the lignin content and electrical properties of wood. Wood Fiber 6:46–52
Vermaas HF 1974 A note on the mechanism of electrical conduction in wood. Holzforschung 28:145–148
Vermaas HF 1975 Experimental variables affecting the measurement of the DC resistance of wood. Holzforschung 29:140–144
Vermaas HF 1976 The dielectric constant of solid wood substance calculated with two different methods. Holzforschung 30:97–98
Vermaas HF 1983 DC resistance moisture meters for wood III. Temperature correction of moisture meter readings for a proposed new moisture meter for P pinaster and P radiata and calibration equations for a commercial moisture meter. S Afr For J 125:85–93
Vink AT 1965 Surinam timbers: a summary of available information with brief description of the main timber species of Surinam, 3rd edn. Surinam For Serv, Paramaribo, 253 pp
Vintila E 1939 Untersuchungen über Raumgewicht und Schwindmaß von Früh- und Spätholz bei Nadelhölzern. Holz Roh- Werkst 2:345–357
Voigt H, Krischer O, Schauss H 1940 Die Feuchtigkeitsbewegung bei der Verdunstungstrocknung von Holz. Holz Roh- Werkst 3:305–321

Volbehr B 1896 Untersuchungen über die Quellung der Holzfaser. Diss Univ Kiel, Kiel Germany (cited by Stamm 1964, p 200)

Vorreiter L 1963 Fasersättigungsfeuchte und höchste Wasseraufnahme der Hölzer. Holzforschung 17:139–146

Wagner JB 1917 Seasoning of wood. Van Nostrand, New York, 274 pp

Wangaard FF 1951 Tests and properties of tropical woods. For Prod Res Soc Proc 5:206–214

Wangaard FF 1957 A new approach to the determination of fiber saturation point from mechanical tests. For Prod J 7:410–416

Wangaard FF, Granados LA 1967 The effect of extractives on water-vapor sorption of wood. Wood Sci Technol 1:253–277

Ward RJ, Skaar C 1963 Specific heat and conductivity of particleboard as functions of temperature. For Prod J 13:31–38

Weichert L 1963 Untersuchungen über das Sorptions- und Quellungsverhalten von Holz verschiedener Rohdichte in Abhängigkeit vom Feuchtigkeitsgehalt. Holz Roh- Werkst 21:290–300

Welsh JK 1978 A model for time-dependent DC conduction in moist wood. PdD Thesis, SUNY Coll Environ Sci For, Syracuse NY, 248 pp

Wengert EM, Mitchell PH 1979 Psychrometric relationships and equilibrium moisture content of wood at temperatures below 212 °F. In: Proc symp on wood moisture content, temperature and humidity relationships. Virginia Polytech Inst State Univ, Blacksburg VA, US For Serv For Prod Lab, Madison WI, 4–11

Wexler A (ed) 1965 Humidity and moisture. Measurement and control in science and industry. Reinhold, New York, Vols 1–4, 687, 634, 562, 333 pp

White HJ, Eyring H 1947 Adsorption of water by swelling high polymeric materials. Text Res J 17:523–553

Wolf W, Spiess WEL, Jung G 1985 Sorption isotherms and water activity of food materials. Science and Technology Publ Ltd, Hornchurch Essex GB, 239 pp

Yao J 1969 Shrinkage properties of second-growth southern yellow pine. Wood Sci Technol 3:25–39

Ylinen A, Jumppanen P 1967 Theory of the shrinkage of wood. Wood Sci Technol 1:241–252

Yoshimura M 1961 A new technique for determining fiber saturation point from decay tests. Fac Agr Mie Univ Bull 24:297–309

Zimmermann MH 1983 Xylem structure and the ascent of sap. Springer, Berlin, Heidelberg, New York, 143 pp

Subject Index

List of Symbols

a	activity, relative vapor pressure, area/sorption site, constant
A	area, internal surface area, constant
AC	alternating current
ASE	anti-shrink efficiency
b	constant, sorption isotherm constant
B	constant, coefficient
BET	Brunauer-Emmet-Teller
c	specific heat, moisture concentration
C	constant, Celsius degrees, BET constant, capacitance
CHC	constant humidity condition
CI	crystallinity index
CMC	constant moisture condition
d	density index, thickness, degree of dissociation
D	constant, diffusion coefficient, diameter, electric displacement
DC	direct current
e	electrical charge per carrier
E	Young's modulus, activation energy, electric field intensity
EMC	equilibrium moisture content
ERH	equilibrium relative humidity
f	frequency, friction term in moisture transport
F	Fahrenheit degrees, force
g	grams
G	specific gravity, Gibbs free energy
h	relative vapor pressure, activity, hour, Planck's constant
H	percent relative humidity, enthalpy, energy level
HTD	high-temperature drying
I	current, ionization energy
J	flux of moisture or heat or energy
k	bulk modulus, conductance, Boltzmann's constant
K	Kelvin degrees, equilibrium constant, transport coefficient, constant
l	liter, longitudinal
L	capillary length, thickness, transport or coupling coefficient
m	fractional moisture content, meter
M	percent moisture content, mega, middle lamella
MC	moisture content
Mf	fiber-saturation point
MX	specific hygroexpansion coefficient
n	number of moles, number of layers of water
N	Avogadro's number, number of layers, number of charge carriers
NMR	nuclear magnetic resonance

p	partial water vapor pressure
P	atmospheric or hydrostatic or swelling or capillary pressure
PEG	polyethylene glycol
q	heat content, energy
Q	heat of sorption, heat of transfer
r	capillary radius, electrical resistance, radial
R	gas constant, electrical resistance, radial shrinkage
R2	regression coefficient of determination
RH	relative humidity
R/D	resorption-desorption ratio
s	percent shrinkage, seconds, Soret coefficient
S	percent swelling, entropy, strength, stiffness, tortuosity
SD	standard deviation
S1	S1 layer of secondary wall
S2	S2 layer of secondary wall
S3	S3 layer of secondary wall
t	time, tangential
T	Kelvin degrees, tangential shrinkage
T/R	tangential-radial shrinkage ratio
u	mobility of water molecule or of electrical charge carriers
U	moisture transport potential, energy of transfer, dissociation energy
v	volume, specific volume of water vapor or of water molecule
V	molar volume, specific volume, volume fraction, volts
w	mass or weight
W	heat of wetting, work of swelling, rate of energy dissipation
x	linear distance
X	molar fraction, moisture expansion coefficient, thermodynamic potential
y	depth of penetration of electrode
Y	humidity expansion coefficient
z	constant
Z	slope of sorption isotherm
β	Cole-Cole parameter
γ	surface tension
Δ	change in, difference
δ	loss angle
ε	swelling deformation, dielectric constant
η	viscosity
θ	fibril angle, phase angle, Celsius degrees, Cole-Cole parameter
μ	chemical potential, micro
π	osmotic pressure, pi (3.14159)
ρ	density
σ	stress, electric charge density
ϕ	spreading pressure
τ	time constant for dielectrical relaxation
ω	angular frequency

List of Subscripts

c	capillary
d	dew-point
e	earlywood
f	fiber saturation
g	green
h	hydrated, of hydration
l	latewood, longitudinal
m	moisture, at moisture content M
o	pure water, dry wood
p	constant pressure
q	heat
r	radial
s	dissolved, of solution
t	tangential
u	sublimation
v	water vapor
w	liquid water

GPSR Compliance

The European Union's (EU) General Product Safety Regulation (GPSR) is a set of rules that requires consumer products to be safe and our obligations to ensure this.

If you have any concerns about our products, you can contact us on ProductSafety@springernature.com

In case Publisher is established outside the EU, the EU authorized representative is:

Springer Nature Customer Service Center GmbH
Europaplatz 3
69115 Heidelberg, Germany

Batch number: 10761697

Printed by Printforce, the Netherlands